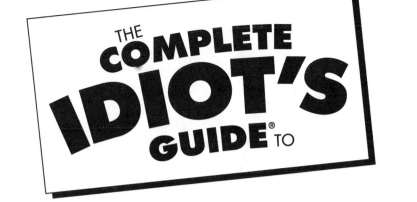

THE
COMPLETE
IDIOT'S
GUIDE® TO

Anatomy and Physiology

by Michael Lazaroff

ALPHA

A member of Penguin Group (USA) Inc.

To my Mom and Dad, for getting me started on the right road, and to my wife Joan and my daughter Emma, for keeping me there! Your aid and inspiration are, as ever, beyond value!

International Standard Book Number: 1-59-257203-0
Library of Congress Catalog Card Number: 2004100486

06   05   04   8   7   6   5   4   3   2

Interpretation of the printing code: The rightmost number of the first series of numbers is the year of the book's printing; the rightmost number of the second series of numbers is the number of the book's printing. For example, a printing code of 04-1 shows that the first printing occurred in 2004.

Printed in the United States of America

Note: This publication contains the opinions and ideas of its author. It is intended to provide helpful and informative material on the subject matter covered. It is sold with the understanding that the author and publisher are not engaged in rendering professional services in the book. If the reader requires personal assistance or advice, a competent professional should be consulted.

The author and publisher specifically disclaim any responsibility for any liability, loss, or risk, personal or otherwise, which is incurred as a consequence, directly or indirectly, of the use and application of any of the contents of this book.

Most Alpha books are available at special quantity discounts for bulk purchases for sales promotions, premiums, fund-raising, or educational use. Special books, or book excerpts, can also be created to fit specific needs.

For details, write: Special Markets, Alpha Books, 375 Hudson Street, New York, NY 10014.

Publisher: *Marie Butler-Knight*
Product Manager: *Phil Kitchel*
Senior Managing Editor: *Jennifer Chisholm*
Acquisitions Editor: *Mike Sanders*
Development Editor: *Michael Koch*
Production Editor: *Megan Douglass*
Copy Editor: *Michael Dietsch*
Illustrator: *Jody Schaeffer*
Cover/Book Designer: *Trina Wurst*
Indexer: *Angie Bess*
Layout/Proofreading: *Mary Hunt*

Contents at a Glance

Contents

Foreword

When we were first asked to be technical editors for Mr. Lazaroff's book, our first thought was that it was going to be fun. Our instincts were on the mark, for Mr. Lazaroff's combination of humor and anatomical knowledge made the book a pleasure to read. In our fields we have to draw upon a great deal of medical information, and we were both impressed by the depth of the author's knowledge of the subject.

Knowledge, alone, is not the only thing needed by a writer, especially in a subject that can be as dense and foreboding as anatomy and physiology. Luckily, Mr. Lazaroff has the ability to convey complex information in a manner that is not only easy to grasp, but is also lively and entertaining! He uses his playful humor to express ideas, but without distracting from the task at hand. Difficult areas such as chemistry and neuroscience are made fun in this book.

All too often writers approaching the subject of the reproductive system fall into one of two traps: being excessively clinical, or being overly bashful, and thus afraid to tackle the subject. Mr. Lazaroff's approach, on the other hand, is refreshing in that he covers the male and female systems in a relaxed and easy manner, without sacrificing content.

This book was written with the college anatomy and physiology student in mind. As such it deals with a great many areas important to the modern student looking for a career in the health and medicine fields. This book complements the typical college-level textbook. In fact, the reader would do well to approach this book first, before tackling the college text. Mr. Lazaroff has many useful tips that will help ensure that the concepts are under the student's belt before approaching the details that had to be left out here due to space considerations. Nonetheless, this book would also work quite well as a primary textbook in a high school—level course in anatomy and physiology.

Some anatomy and physiology students complain that the subject involves too much memorization. Mr. Lazaroff has a wonderful ability to tie concepts together so well that memorization is all but eliminated. So many ideas build upon one another that anatomical terms are extremely logical. Mr. Lazaroff always does his best to point out these logical connections, continually helping the reader to figure out terms easily. After trying this approach, the reader will quickly realize that memorization is the wrong way to tackle the subject!

One of the author's greatest strengths is his ability to connect a wide variety of ideas. Many anatomy and physiology texts are restrained by the "system-by-system" approach, which limits any understanding of the connections among the eleven systems. Mr. Lazaroff, however, elucidates the interconnections between these systems throughout his book, always projecting the "big picture" of the human body behind the biological, chemical, and medical details of its smaller parts.

If you have this book in your hands, whether you are nervous about an anatomy and physiology course, or just in the mood to learn more about the human body, you would do well to buy this book!

Tom K. Coffey, M.D.
Margaret Rudolf Coffey, M.D

Introduction

This has got to be one of the coolest subjects on the planet. No, really! It's all about you and how you work. You wouldn't have picked this book up if it hadn't been for your own fascination. If you are thinking of using this in conjunction with a college textbook, then look no further! I guarantee that there will be a bunch of "Ah ha!" moments as you read this.

From the curious child wondering about the rumblings in her or his stomach to the adult puzzled by aspects of an illness, everyone has had an unanswered question about the body. Over the years I learned how to tap into that desire, and to focus on areas that are usually left unanswered, bringing them into sharp relief. I have had many students come back from college to tell me how much they learned, and how easy their college courses were after the background I gave them!

The weird part about it all is how fun it can be! If you can't make the body gross, you just aren't trying! You'll find yourself having fun in the weirdest of places in this book, but never at the cost of the science behind it. The body is an intricate puzzle, filled with the weird and the wonderful, but beyond that, it is immensely fulfilling food for your mind. The way all the pieces fit together, any effort on your part (and I will always be there to help you along) will be rewarded.

As I was writing this book, I kept on finding myself thinking, as my fingers were flying over the keys to explain a particular topic, "This is so *cool!*" I love to learn, almost as much as I love to teach. I think that's *why* I love to teach, because I love sharing the joy of understanding new connections. Whether you are just curious about your body, or you are studying to make a career out of this, I know that in these pages you will find plenty of the excitement that made you turn to this subject in the first place.

No subject in this book is so hard that it can't be made easy and fun. I know that, in the end, you'll end up loving the subject as much as I do! So read on, and enjoy!

How This Book Is Organized

This book is divided into seven parts.

In **Part 1, "The Anatomy of Anatomy,"** I cover a number of areas that in turn are touched upon in every body system. The first chapter is a lesson in change, covering some of the enormous changes in medicine throughout history. The next chapter is designed to help you to work your way around the body, with body cavities, directional terms, and so on. Each of the next three chapters covers material that will weave in and out of all the other chapters: inorganic and organic chemistry (you might even laugh while reading this one!), cells, and tissues.

In **Part 2, "If the Foundation Is Good,"** with the basics out of the way, I get down to the largest of the systems, loved by athletes everywhere: skeletal and muscular. Far more than Halloween decorations, you will learn about the dynamic nature of bone tissue, followed by a chapter covering the myriad bones themselves. Since we are built to move, I then cover the joints and movements of the body, followed by chapters on how muscles move, before finishing with the muscles themselves. Next time you go dancing you'll be able to learn *what* you're shaking and *how* you shake it when you, uh, *shake it!*

In **Part 3, "Let's Make a Connection,"** the way the body gets materials from one end of its 100 trillion cells to the other is explored. Roads, roads, and more roads—in this case, blood vessels—circulate materials from one end to the other. Rather than thinking about a "circulatory system," you will see that there is actually an integration of the two systems, the cardiovascular and lymphatic systems, linked by the interstitial fluid around the body tissues. You'll finally learn what blood's got to do with it!

In **Part 4, "What Goes In Must Come Out,"** the how of the last section is expanded to include the what, in this case, what is actually picked up and delivered throughout the body. This is nothing more and nothing less than why and how we breathe, eat, drink, and, well, go to the bathroom! We would *die* if we didn't do these things, so it's about time you learned *how* you did it with your respiratory, digestive, and excretory systems! Don't you think?

In **Part 5, "Holding the Fort,"** the body's defenses are explored. We've got a lot going for us, but, then again, we've got a lot going for bacteria, viruses, and fungi, not to mention the occasional protist! With all the food, moisture, and warmth inside, it's a good idea to protect ourselves! From the first barrier, our skin (living layers topped by the dead), to the intricate methods of fighting anything foreign that breaches that boundary (the immune portion of our lymphatic system), we've found many ways to hold the fort!

In **Part 6, "Who's in Charge,"** you will see who's really in charge. How does our body talk to itself, so that the left hand knows what the right hand is doing (both literally and figuratively)? From the slow and easy chemical signals of our endocrine system, to the rapid commands and responses, the elegant complexity of the nervous system (neurons, the brain, peripheral nerves, and the senses), you will see the idea of connections writ large. Many conversations are going on all the time, and this section will bring many of those within earshot.

In **Part 7, "First Comes Love …"** regardless of one's sexual preference, you will learn the glorious differences between the sexes. You might even be surprised by the similarities! We really are cut from the same genetic cloth. Beyond that, the wonders of reproduction, from the first union, to the implantation and growth within the womb, to birth and breast feeding, this section will put it all together, showing the great link back to your parents, and forward to your children!

How to Use This Book

As you read this book, keep your mind not only focused on the ideas at hand, but think about what you have already learned. Every idea is connected to other ideas discussed earlier. You will notice throughout the book that I make reference to earlier chapters (see Chapter 10), by bringing earlier ideas into use in a more complex setting. Usually the point of reference is a particular term or concept.

At this point it might be helpful to refer to that chapter if you are unclear as to the reference. You will find, however, that a number of early concepts make so many appearances that you will easily find yourself smiling, thinking "been there, done that," at each repetition! This book is truly cumulative, and your understanding will really grow immensely as you move on.

Throughout the book are many sidebars, to which you will be introduced in the next section, which will be valuable signposts along the way. You might be surprised at the lack of a glossary, but that was purposeful. Glossaries, in and of themselves, can get in the way; my students who have weaned themselves off the glossary have always been more successful because of it. Don't worry, for I have an alternative that will be far more helpful to you.

There are so many terms in this book that it can seem daunting (just look at all the *italics*), but there are several tricks of the trade that make things much easier. If you find yourself thinking about all the memorization you'll have to do, you might be surprised! Memorization is the *wrong* way to study anatomy! I will teach you how to learn to derive meaning *from* the words, and you will soon see that after learning some basic background terms, that many of the new words in each chapter are merely extensions of earlier words. I want you to learn the ideas that connect everything, for then the words become window dressing to the underlying concepts. You'll find everything not only much easier to grasp that way, but even easier to remember!

For those of you who grew up on glossaries, I have an alternative: the index! If you can't remember something that you read in an earlier chapter, just go to the index and think of it as a big website that links terms to their explanations. Throughout the book I will be defining things like crazy. When you go back to those earlier pages, you will find it much easier to remember the words. Not only will the words look familiar, but they will also be in *context*. A glossary cannot do that, and, in fact, has the dangerous habit of focusing on definitions over their conceptual basis. Examples and connections, not to mention labeled diagrams, are at your fingertips, simply by using the index. I haven't used a glossary in over 20 years, and I've never looked back! Try using the index in every textbook you use, and you'll see how much better it is in every way!

Lastly, the key to remembering words is to know the meaning of the prefixes and suffixes. New terms will open themselves up to you that way! Take a word like *intercostal*. Now break it down into *inter*, which means *between*, and *costal*, which means *ribs*. At this point it should be no surprise that the *intercostal muscles* are between the ribs! Sometimes thinking this way might startle you, as when I discovered that the *adrenal glands* were named that way simply because they are on top of the kidney (*ad* = on top of, and *renal* = kidney)! This way we build our understanding by breaking down the words; after all, analysis is just that (*ana* = to build, and *lysis* = to break down)!

Extras

My students have told me that I have a bizarre, fun, cool, (*insert your adjective here*) way of looking at things. I love to go on tangents, and to make connections everywhere, because those tricks really help to get the ideas across. I can't really capture what it is like being in my classroom in the pages of this book, but these sidebars will help you not only get a feel for my classroom, but more importantly, a feel for anatomy and physiology. Unless you can link every new idea with something else you know, the ideas will lie fallow in your brain; by making connections you will guarantee easier access to the information stored in your mind.

The Big Picture

The Big Picture is all about making connections among body systems and between medicine and the world as a whole. It is a fallacy to think of body systems as independent entities, and so this sidebar will highlight those places where the body systems interact. It isn't possible to highlight all the connections in a book this size, and I encourage you to look for your own as well!

Medical Records

Case histories, sometimes of the weird and wonderful, can be found here. This sidebar brings many of the facts to life (or even death, as in the case of rigor mortis, for example). You will find many of the concepts in the chapters applied here to real-life medical situations. This is especially helpful to the future clinicians among you!

Flex Your Muscles

This sidebar is more than just interesting facts; it will help drive points home. The tips you find here will make your life easier by making ideas easier to remember.

Crash Cart

Every subject has pitfalls, quicksand, and blind curves. An occasional look at this sidebar will be as valuable as fastening your seatbelt. I've seen many students over the years fall into the same traps, so this sidebar allows me to help you by throwing you a rope!

Acknowledgments

To my wife Joan, I can't possibly say what I really feel here, for there isn't enough space. Just know that I value you, your help, and your insight more than you can possibly know. To my daughter Emma, who turned 11 while I was writing this book, I want to thank you for letting this book steal a summer, and a father, from you. You are always an inspiration to me; I am very proud of you. You're the best!

To both my parents, there are many more things than tadpoles on the kitchen counter and tortoises on the kitchen floor, for which I can thank you. Your intelligence, grace, and concern for others, not to mention the gruesome medical discussions over dinner (I was listening), trips to zoos and museums (reading every plaque), and just plain silliness at home, acted as the foundation that made me what I am today.

To two students, Neha Uppal and Zack Klomberg, who helped me with some details early on, thank you. And to all my students who have said, "Mr. Lazaroff, you should write your own book," thanks for your inspiration! Lastly I have many anatomy and physiology students to thank for their encouragement (more than I can name!), but special thanks goes to Shannon Loeck, who is studying to be a sports trainer. Her friendship, proofreading, and suggestions have been invaluable; I especially enjoyed her sarcastic comments at my unbelievable typos—word recognition software, my eye, it's better to *type!* Keep up the good work!

I have so many former students who have inspired me over the years, in part by their work in the classroom, in part by wisdom beyond their years, and in part by their general compassion (which many of our leaders would do well to emulate). I want to especially thank those of you who keep coming back to tell me what you have been doing (including, but by no means limited to, those of you who got into medical school); I am proud of all of you, and I wish I had the space to name more names.

Special Thanks to the Technical Reviewers

The Complete Idiot's Guide to Anatomy and Physiology was reviewed by two experts who double-checked the accuracy of what you'll learn here, to help us ensure that this book gives you everything you need to know about anatomy and physiology. Special thanks are extended to Thomas Coffey, M.D., and Margaret Rudolf Coffey, M.D.

Trademarks

All terms mentioned in this book that are known to be or are suspected of being trademarks or service marks have been appropriately capitalized. Alpha Books and Penguin Group (USA) Inc. cannot attest to the accuracy of this information. Use of a term in this book should not be regarded as affecting the validity of any trademark or service mark.

Part 1

The Anatomy of Anatomy

An area of study such as anatomy and physiology has a huge amount of material to cover. It might be fun to just launch right into the intricacies of the nervous system or hormonal feedback loops, but without some background info, too many students start looking like a deer caught in the headlights. *Help!*

If you start with the basics, everything will fall into place, and the complex can come across as amazingly simple! Don't think of this as the simple stuff, but as a real foundation, which will infuse everything else you'll learn in the other parts of the book. A beginning piano student might *hate* playing scales, only to realize later how often scale passages can be found in music. The difference here is that I always link the basics to other areas right away, and that, unlike practicing piano, you move on to the other stuff very quickly. First a little history, a body tour, then a quick chemistry tour, cells, and finally tissues will get you on the way to the body systems.

Any Way You Slice It ...

In This Chapter

- ◆ The relationship between anatomy and physiology
- ◆ The organs and functions of the 11 body systems
- ◆ Using directional terms to describe the body
- ◆ Planes, sections, and body cavities
- ◆ The classification of the abdomen into regions
- ◆ Feedback loops and their role in regulating the body

To understand both structure (anatomy) and function (physiology), it helps to think of a body as an enormously complex puzzle with thousands of pieces. Each piece alone is not enough for you to determine the nature of the puzzle, but when you see the connections between cells, the total structure gradually becomes clear, and the whole may be greater than the sum of its parts. If you think of cells as those puzzle pieces, you can carry the analogy farther. Apart from having to change the number of cells to about 100 trillion (100,000,000,000,000!), the analogy still fits rather nicely. For one thing, every puzzle piece, despite slight differences in shape and color, is still very similar in outward form (cells, with few exceptions, contain the same basic structures), not to mention materials (every cell, except gametes and red blood cells, contains a full set of DNA, and so on). The twentieth century has shown that the basis for all function is cellular, but even that is not enough. Those of you who can't wait to get to the guts of the subject mustn't forget that throughout this book we will be dancing with molecules. Despite this, most people, when they think about A & P, they think about organs. This is ultimately a difference between *reductionism*, which focuses on the inner molecular workings of the cell in order to learn about the whole, and the *holistic* approach, which looks at the big picture. When one studies anatomy and physiology, it is best to find a balance between the two. Still, puzzles don't make themselves, and houses need foundations, so let's get started, let's get small ….

For Starters: Anatomy Versus Physiology

Before you get the idea that this is going to be a professional wrestling match, I need to explain. There is a reason why so many textbooks have both words *anatomy* and *physiology* in their titles. Quite simply, they are inseparable. *Anatomy* refers to the structure of the body, anywhere from the basic skeletal shape to the individual shape of a bone cell (*osteocyte*). *Physiology* refers to the function of the body—for example how the bone cells are responsible for bone replacement (remodeling) or how individual muscle cells contract.

To illustrate this idea, have you ever picked up an old tool or machine with an odd shape and had absolutely no idea what it was for? When one *does* find out what a tool was used for, the odd shape suddenly makes perfect sense ("oh, cool")! Clearly the shape has something to do with the tool's job; the structure was related to the function. In the same sense, the anatomy will tell you a lot about the physiology, and vice versa.

Built from the Ground Up

Whether you consider human beings to be like onions or like cake, there is no denying that we have many layers (see Figure 1.1). Nature has a way of taking advantage of the smallest of parts in order to get the job done. For example, electrons are thrown about like hot potatoes during cellular respiration (the dreaded electron transport chain!) to the hydrogen ions (just a fancy way of saying a free proton). These hydrogen ions build up in our blood and spinal fluid every time we hold our breath. It is clear that we cannot escape the world of the *subatomic particle* (protons, neutrons, electrons).

Figure 1.1

The study of the body is best approached through an understanding of the impact of each of the levels of organization.

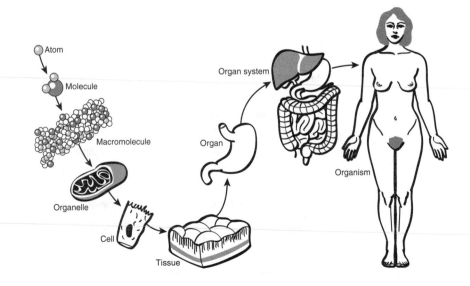

Atom

Molecule

Macromolecule

Organelle

Cell

Tissue

Organ

Organ system

Organism

From the positive protons, the negative electrons, and the neutral neutrons, we build atoms and small ions. (Those of you who are filled with terror at the mere mention of chemistry, fear not, that will all be explained in Chapter 2.) These atoms join to make molecules, which are the currency the cells use to do their business. All of the energy we use is in the form of molecules. We reproduce by copying molecules, and we grow by building molecules.

Molecules are grouped to form miniature organs called organelles that are the workhorses in every cell. These organelles build the molecules that need building and break down the ones that need breaking; they divide the labor of the cell. In turn these cells are found in groups of similar cells that work together to perform the same function; these groups of cells are called tissues.

There are four basic tissue types, and a group of different tissues that band together to carry out the same job is called an organ. Although there are some exceptions, almost every organ in the body is composed of all four tissue types. One organ alone is incapable of carrying out all the functions of that body system, so there are a number of organs in each body system (except for the integumentary, see Chapter 16). Even so, the body still requires 11 systems in order to get the job done! Is it any wonder that the whole shebang is called an *organ*ism?

Not Quite a Baker's Dozen

As I mentioned in Chapter 1, one of the best ways to remember the 11 body systems is through the acronym "SLIC MEN R RED" (see Figure 1.2). If you don't acknowledge the multiplicity of connections, it can be a bit hard to say where one organ system ends and another begins, but if you keep the connections in mind you will do fine! Let's start out with a look at the various systems, their major organs, and, well, what it is the system actually does!

The following table lists the major organs and systems of the body. Please note that the lists of organs and functions is not complete; more detail will be in subsequent chapters.

Major Organs and Their Functions

Organ System	Major Organs	Major Functions
Skeletal	Bones and ligaments	Support and protection
Lymphatic	Tonsils, lymph nodes	Drainage and immunity
Integumentary	Skin	Protection and defense
Cardiovascular	Heart and blood vessels	Transport materials
Muscular	Muscles	Movement, body heat
Endocrine	Pituitary, thyroid	Slow chemical control
Nervous	Brain, spinal cord	Fast nervous control
Respiratory	Lungs, trachea, nose	Gas exchange
Reproductive	Ovaries, testes	Producing offspring
Excretory	Kidneys, bladder	Filtering wastes
Digestive	Stomach, intestines	Break down and absorb food

Figure 1.2

The 11 body systems—SLIC MEN R RED!

(LifeART©1989–2001, Lippincott Williams & Wilkins)

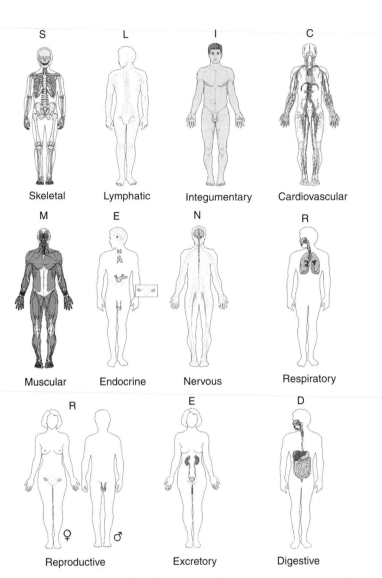

Despite this attempt to place organs and functions neatly into categories, a number of organs and functions refuse to cooperate. For example, the pancreas is both a digestive organ (producing digestive enzymes) and an endocrine organ (regulating blood sugar through the hormones insulin and glucagon).

Crash Cart

Even though I am starting off with a description of each of the 11 body systems, don't fall into the usual trap of most anatomy and physiology students. These systems are best understood in terms of how they interact with one another. I will make connections with other body systems in every chapter, but I encourage you to look for your own, for they are *everywhere*. Although you will have to learn details for each system, try to think outside the box, or at least connect the 11 compartments of the box!

Even functions can be troublesome. Maintaining body temperature falls within the muscular (heat production), cardiovascular (heat distribution), integumentary (heat dissipation and preservation), and nervous (control of all of the above) systems. Compartments are fine for storing boxes of food, but that line of thinking just doesn't work in anatomy and physiology. Living is a team sport, and our organs are the players.

In addition to SLIC MEN R RED, it helps to organize the systems according to six basic functions of living things: connection of systems, movement of matter in and out, control of the body, support and movement, protection, and continuity of the species. Even so, there is a little overlap, as shown in the following table.

Functional Grouping of the 11 Body Systems

General Body Function	Body System	Mode of Action
Connecting systems	Cardiovascular	Blood/vessels
	Lymphatic	Lymph/vessels
	Nervous	Nerve impulses
Movement of matter in and out	Digestive	Food, water
	Respiratory	O_2, CO_2
	Excretory	Wastes
Control of the body	Endocrine	Hormones
	Nervous	Nerve impulses
Support and movement	Skeletal	Bones
	Muscular	Muscles
Protection	Integumentary	Barrier
	Lymphatic	Immunity
Continuity of the species	Reproductive	Procreation

Which Way Is Up?

In order to make communication easier, it is helpful to have a universally accepted collection of terms. You've had conversations that have included the phrase "my left or your left." Could you imagine an emergency room with a doctor working on the wrong side of the patient? To avoid this problem there is an incredibly simple solution: Right and left are always the *patient's* right or left. This has the added benefit of clearing up any confusion of vantage point; in other words, the right side of the heart is the right side, regardless of your position! This is also clarified due to anatomical position (see Figure 1.3).

Flex Your Muscles

Don't stress out over the number of terms! Think in terms of opposites, like dorsal and ventral. Keep that in mind, for it will give you a frame of reference for most terms, as half of a pair! Also, remember the idea of the X, Y, and Z axes; directional terms are usually related to opposite ends of one of the three axes.

The terms that follow are crucial to navigate one's way around the body. These terms are not meant to be memorized and then forgotten! You will find these terms used throughout all of anatomy and physiology. Not only in terms of accurately locating structures, these terms are used repeatedly in naming parts of the body. For example, the bump on the outside (lateral) part of your ankle is the lateral malleolus, and the muscle in front of (anterior) your shin (tibia) is called the tibialis anterior. The key is not to hide these terms away, but to *use* them!

Figure 1.3

Anatomical position is used to highlight the layout of the body and minimize confusion of position. Note the directional terms.

(©2003 www.clipart.com)

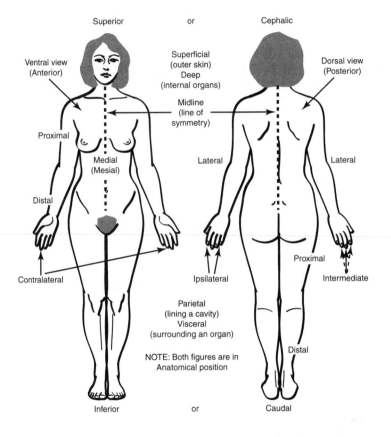

Directional Terms

Directional Term	Explanation
Terms Related to the Z Axis	
Ventral (anterior)	The front (belly side)
Posterior (dorsal)	The back of a biped
Prone	Ventral surface down (face down)
Supine	Dorsal surface down (face up)
Terms Related to the Y Axis	
Superior	Above
Inferior	Below

Directional Term	Explanation
Cephalic (cranial)	Head
Caudal	Tail
Terms Related to the X Axis	
Midline	The line of symmetry (bilateral)
Medial (mesial)	Toward the midline (middle)
Lateral	Away from the midline (to the side)
Ipsilateral	On the *same* side
Contralateral	On *opposite* sides
Left	The *patient's* left
Right	The *patient's* right
Terms Independent of Axis	
Proximal	Close to the trunk
Distal	Distant from the trunk
Superficial	Close to the surface
Deep	Further beneath the surface
Visceral	Lining the *outside* of an organ
Parietal	Lining the *inside wall* of a cavity
Intermediate	In between two structures

Note: In a biped, anterior and ventral are synonymous, but in a quadruped they are not. Think of where the dorsal fin is on a dolphin (dorsal); that's very different from the end of its tail (posterior).

If the list looks daunting, don't forget that with few exceptions, every term describes one half of an either/or situation, such as anterior/posterior or superior/inferior. In addition, most of the terms relate to one of the three axes—X, Y, and Z—with each term referring to the opposite sides of its axis. Stay tuned, for these words will come back again and again!

 **Flex Your Muscles**

As much as possible, use little clues in the words to help you remember: distal and distant, proximal and proximity, medial and middle, visceral and viscera (organs)! One of my favorite ways to remember prone (face down) is by thinking "he is *prone* to fall on his *face!*" In the same sense, supine can be remembered by thinking "soup's up," and supinated (see Chapter 7) hands look like a soup bowl!

Planes, Trains, and Sections

Don't forget those axes, because they're coming into play again. The advent of sophisticated scanning technology, such as the CT scan (computerized tomography, sometimes called a CAT scan) and the MRI (magnetic resonance imaging), has allowed a level of noninvasive diagnosis unprecedented in medical history. They allow us to see what the patient's body would look like if it were cut into sections, thus exposing any potential problems. The only other way to see the same thing would be quite a lot more invasive … to actually slice the patient up like so much lunch meat!

Sectional views are extremely important in anatomy. A remarkable breakthrough in anatomical study came with the Visible Human Project (see Appendix B), in which a man on death row willed his body to science, and upon his death his body was frozen, encased in wax, and sliced into more than 1,800 one-millimeter-thick *transverse* slices. Those slices were digitally photographed, and the data was then used to construct animations, three-dimensional views of organs, and blood vessels without the organs, not to mention every conceivable type of section.

It turns out that these sections are based upon planes that are parallel to the three axes, and which divide them into sections (see Figure 1.4). The various sections appear in a table that also includes the relevant axis for each.

Planes and Sections

Plane/Section	Orientation	Divides the Body Into
Plane/Section Related to the Z Axis		
Frontal	Superior to inferior left to right	Anterior and posterior
Plane/Section Related to the Y Axis		
Sagittal	Superior to inferior anterior to posterior	Left and right
Midsagittal	Down the midline	Exactly in half
Parasagittal	To one side of the midline	Unequal sections
Plane/Section Related to the X Axis		
Transverse	Left to right anterior to posterior	Superior and inferior
Plane/Section Independent of Any Axis		
Oblique	Diagonal	Dependent on how oblique the angle is

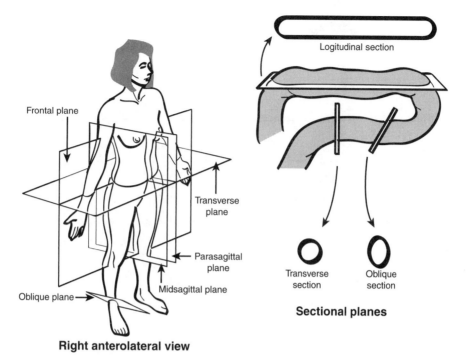

Frontal plane

Logitudinal section

Transverse plane

Parasagittal plane

Midsagittal plane

Oblique plane

Transverse section

Oblique section

Sectional planes

Right anterolateral view

Figure 1.4

The planes indicated on the body divided it into different sections, enabling you to see the internal organs.

Body Cavities

Remember the terms *visceral* and *parietal*; they are important because they refer to body cavities. We are essentially hollow critters with a whole lot of organs inside. These organs can be found in several hollow cavities. These cavities are divided into two groups, yet another pair of opposites: dorsal and ventral (see Figure 1.5).

Dorsal Body Cavities

The dorsal cavities are almost exclusively dedicated to the *central nervous system* (CNS). Our cranial cavity contains, as you would imagine, the brain. When you get to the skeleton (see Chapter 6) you will see that about a third of the bones of the skull are called cranial bones because they surround and protect the brain.

Each vertebra has a large *foramen* (hole in the bone) toward the dorsal end of the bone, known as the *vertebral foramen*, which is for the spinal cord. By stacking up the bones in the spinal column you create a long, hollow tube for the spinal cord; this is the vertebral cavity.

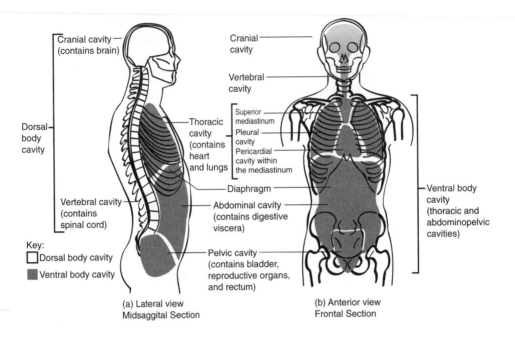

Figure 1.5

This view of the body, from a midsagittal section, shows all of the major dorsal and ventral body cavities.

Medical Records

We have all probably grown up learning that creatures with a backbone are called *vertebrates*, because of the vertebrae that protect our spinal cord. What most people never learned is that a long-standing evolutionary trend led to animals with their brains in their heads. This trend, to use our directional terms, is called *cephalization*, and it necessitated the evolution of a bony plate to protect the brain, long before the vertebra evolved. As such, it makes a lot of sense that some biologists would rather call us *craniates*!

The Superior Ventral Body Cavity

The ventral body cavity is the same hollow cavity that is found in the turkey we eat every Thanksgiving; this cavity is called the *coelom*, and it is the perfect place to put the stuffing in the turkey! In most vertebrates, including birds, there is no division to this cavity. Mammals, however, have a muscular division known as the *diaphragm*. The thoracic cavity, in turn, is divided into three separate cavities. Right down the middle (remember *medial*?) is a sheet of connective tissue known as the *mediastinum*, which divides the thoracic cavity in half. By dividing the cavity in half, two smaller cavities are formed, called the pleural cavities, which hold, you guessed it, the lungs. This cavity is lined, along the inside of the ribcage, with a serous membrane (see Chapter 4) to lubricate the lungs; this pleura, to get directional again, is called the parietal pleura. Can you guess the name of the pleura that surrounds the lungs themselves? Yup, the visceral pleura.

The Big Picture

The *mediastinum* is not content sitting alone in the thoracic cavity. There is quite a lot of activity going on in and around it! The major blood vessels to and from the upper and lower body (don't forget the heart) must travel through the mediastinum. Hitching a ride are the *thoracic duct* and the *thymus* of the lymphatic system. Don't forget the *vagus nerve* (parasympathetic nervous system), the *esophagus* (digestive), and the *trachea* (respiratory).

The only other portion of the mediastinum to talk about is the (not so) curious widening at the base. The mediastinum widens, creating a sack just above the diaphragm and between the lungs. This sack is also lined with a lubricating serous membrane, and it gets its name, *pericardium*, from its location (*peri* = around and *cardium* = heart). Don't forget its location—at the bottom of the thoracic cavity, and almost exactly in the middle—for most people when they pledge allegiance put their right hand over the superior lobe of their left lung!

The Inferior Ventral Body Cavity

I don't think the inferior ventral body cavity suffers from any sort of complex, simply because so much is going on down here! Every body system, except the respiratory system, which hogs so much of the thoracic cavity, can be found down here, some exclusively so (excretory and reproductive—with the exception of the breast, penis, and testes, which are outside both cavities)! This large cavity is called the *abdominopelvic cavity*.

Some of you may have expected this to be called the *abdominal cavity*, so it would be helpful to know the difference between the two. For one thing, there's only one true cavity; this cavity is probably what you considered to be the abdominal cavity. The abdominopelvic cavity's parietal surface runs from the bottom of the diaphragm, all along the inside of the abdominal wall, down to the inside walls of the pelvis.

At this point you have to use your imagination, because the division of the abdominopelvic cavity into the abdominal and pelvic is an imaginary one! The division between the abdominal cavity and the *pelvic cavity* is an oblique plane that follows the downward (from dorsal to ventral) slope of the pelvis. The pelvic cavity contains all the deep reproductive organs (as well as the bladder and the rectum); the only other reproductive organs (breast, penis, testes) are all superficial. Superior to that is the abdominal cavity, which is primarily filled with digestive organs; just about every other organ system is represented. One small correction, from above: The kidneys are not actually in the abdominal cavity, but rather behind the abdominal wall (also known as the peritoneum).

Abdominal Regions and Quadrants

In emergency rooms across the country there are patients who either come in, or are brought in, complaining of unknown abdominal pains. As you work your way through this book you will learn more and more about the locations of the many organs of the abdominopelvic cavity. Given all the organs therein, it would be very helpful if a doctor or nurse had an idea what organs were where!

Beyond that, patients have a habit of saying somewhat unhelpful things like "my stomach hurts," while pointing to their belly buttons (which has my favorite anatomical name of all—the *umbilicus*), which is nowhere near the actual stomach, but is instead superficial to the ileum of the small intestine. To clear things up, the abdomen is usually divided into different areas (see Figure 1.6), each of which has its own particular organs.

Figure 1.6

Emergency situations require a quick understanding of the regions and quadrants of the abdomen, given the locations of the various organs within.

(LifeART©1989–2001, Lippincott Williams & Wilkins)

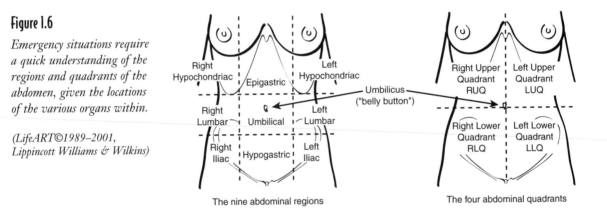

The nine abdominal regions

The four abdominal quadrants

The most precise division is into nine separate regions; this makes the patient look a bit like a big tic-tac-toe game. Using some of the terms and prefixes to which you will soon be accustomed, I start off with the most central region, called the *umbilical* region, for obvious reasons! Above that is the *epigastric* region (*epi* = above and *gastric* = stomach), and below is the *hypogastric* (*hypo* = below). Being bilateral, the other six regions need to be divided into left and right, but don't forget it's the *patient's* left and right!

These six regions are all named according to the bones that are nearby. The regions on either side of the umbilicus are only near the lumbar vertebrae, so is it any wonder they are called the left and right *lumbar regions*? Below those regions is the iliac portion of the pelvic bone, making the regions the left and right iliac regions. The last two regions, the left and right *hypochondriac regions*, which are so named because they are underneath the costal cartilage portion of the ribcage (*hypo* = under and *chrondriac* = cartilage).

Flex Your Muscles

Whatever you do, *don't memorize the regions!* You need to *learn* them! Use the names, because the names have directional meaning that will also help you to remember not only each region's name but also some of the organs within! Also, don't forget that it's always the *patient's* left and right!

The other way of dividing the abdomen is the far less detailed, but still useful method of dividing into four quadrants. The naming is also far simpler, for we divide the y axis into upper and lower, and the x axis into left and right. These quadrants are often abbreviated for simplicity's sake (for example, LLQ = left lower quadrant, RUQ = right upper quadrant, and so on). These are more useful, for example, when the patient complains of pain in a larger area that would involve multiple regions. (That is, the LUQ contains all of the left hypochondriac, as well as part of the epigastric, umbilical, and left lumbar regions!)

Feedback Loops

Quite a bit of what I discussed in this chapter has been anatomical in nature, so I thought this would be a good point to throw in some physiology. Have you ever wondered how a critter such as ourselves is able to survive in so many different environments? The key to our survival is our ability to maintain a constant internal environment. In order for this to happen, it is essential that we are able to alter our condition, in one direction or the other, in order to maintain a balance. Living systems maintain a balance that is called a state of *equilibrium*; given that the equilibrium in living systems is constantly changing, in other words dynamic, it is given a new name—*homeostasis* (*homeo* = same and *stasis* = to stay).

In order to maintain homeostasis, the body must monitor its internal environment, decide upon a course of action to alter that environment, and then actually alter it. The body does this through a series of *feedback loops*, both positive and negative, which are used to control the body's environment. Think about how your parents or teachers have reacted to your behavior over the years. Whenever you did something good, parents and teachers had a habit of praising your good behavior ("Good Job!"). On the other hand, their reactions were quite different when you did something wrong ("Stop that!"). Praise is *positive feedback*, because it is used to increase the incidence of one type of behavior. Punishment is *negative feedback*, because it is used to stop the frequency of a certain behavior.

The Cast

In order to maintain homeostasis, there are several players involved. The first character is that of the *receptor*, which is responsible for receiving information (a stimulus) about the internal environment. Chemical receptors monitor changes in the concentration of certain molecules in our bodies, and neural receptors are nerve endings that send information to other parts to the body.

When information about the internal environment is received, the next character, the *control center*, decides upon the course of action. Given the responses to our body's environment can be either hormonal (slow chemical control) or nervous (rapid neural control), the control center can take different forms. If the response is hormonal, then the control center is usually an operon (see Chapter 3) in the nucleus of endocrine organ cells, which will allow the production of the hormone on demand, while neural responses use some portion of the nervous system.

The last player in this little drama does all the grunt work. The *effector* receives instructions from the control center, and actually carries out the response that causes a change in the internal environment. Effectors could be anything from the part of a cell that makes protein (ribosomes), to the contracting cells of the heart. The combination of all three parts of a feedback is another glorious example of multiple body systems working together in harmony. Figure 1.7 provides a quick glance at the three parts of the feedback loop.

Figure 1.7

There are the three basic parts of a feedback loop: receptor, control center, effector.

(©*Michael J. Vieira Lazaroff*)

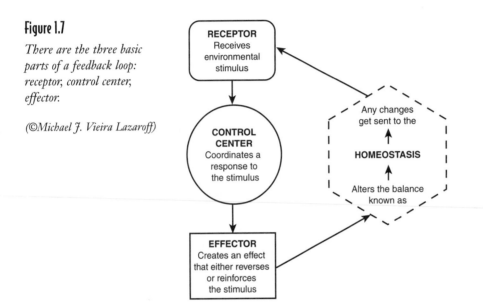

Accentuate the Positive

Positive feedback loops are rare in the body. When the receptors receive some stimulus about the body's internal environment, a positive feedback loop will direct a response to increase the stimulus. If our body temperature regulation functioned through a positive feedback loop, the body temperature would get higher, and higher, and higher … until we died! Positive feedback loops are useless in terms of maintenance. Given such a weird feedback system, when would we ever use such a thing?

A perfect example of the use of a positive feedback loop is during labor contractions (see Figure 1.8). When a woman goes into labor, the muscles in her uterine walls start to contract. These contractions cause a distention of the uterine walls; stretch receptors in the uterus walls send a message to the hypothalamus (a hormonal portion of the brain [see Chapter 18]). The hypothalamus had previously produced a hormone called *oxytocin* (OT), which is then stored in the neurohypophysis, or posterior pituitary gland. The hypothalamus sends a signal to the neurohypophysis to release the oxytocin into the bloodstream.

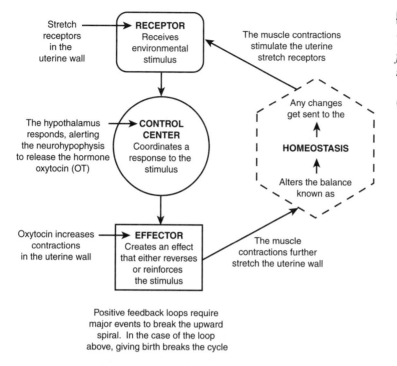

Figure 1.8

This is a sample of a positive feedback loop, which is rare in the body.

(©Michael J. Vieira Lazaroff)

The oxytocin, in turn, stimulates the contraction of the uterine walls. Further contraction stimulates the stretch receptors, which trigger the release of more oxytocin, which causes more contractions, which stimulate the stretch receptors, and so on, and so on .... As a result of this feedback loop, labor contractions get progressively stronger over time; some people have described labor contractions as being similar to climbing a mountain, starting with the foothills and ultimately reaching the steepest portion right before the summit.

Clearly something like labor is a rather unusual situation, so positive feedback loops must therefore be rather rare in the body. In fact, given the nature, feedback loops must be broken by an extraordinary event, or they run the risk of killing us. In the case of labor, that extraordinary event is the birth of the baby.

Eliminate the Negative

The job of maintaining homeostasis falls upon negative feedback loops. Negative feedback loops are everywhere, controlling everything from body temperature to the menstrual cycle. They are so ubiquitous in the endocrine system that I'll pay special attention to them there. What is crucial about the nature of negative feedback loops is the fact that their response to a stimulus is to reverse that stimulus. Think about the thermostat in your house. When the temperature in the house drops too low, the heat goes on; but when the temperature rises above a certain level, the heat goes off. A thermostat is able to keep the temperature of a house constant.

In the same way, negative feedback loops are responsible for controlling blood glucose levels (see Figure 1.9). At this point you may find me getting rather jealous, because, as a diabetic, just one part of my feedback loop (the beta cells in my pancreas) is missing, and so I have to take insulin injections to stay alive. In any case, what follows is a description of what happens in a normal person.

Figure 1.9

Far more common are negative feedback loops, which are essential to the maintenance of homeostasis.

(©Michael J. Vieira Lazaroff)

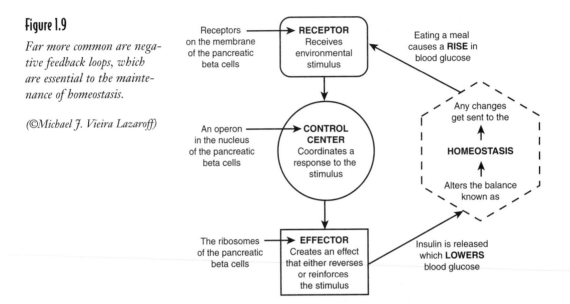

When eaten, complex carbohydrates are broken down into simple sugars called *monosaccharides* (see Chapter 3). Those monosaccharides are absorbed in the bloodstream in the villi of the small intestine (see Chapter 14). At this point, blood glucose levels rise. This may not seem serious, but if those levels stay high, they can cause severe damage to the kidneys, eyes, and nerves (the areas where diabetics can develop serious complications over time). It is therefore important to not let the levels get too high. In addition, the liver stores the excess glucose from the blood for times when it is needed; after all, we're not in the habit of eating *every time* we need energy!

Receptors on the beta cells of the pancreas send a message to the nucleus of the cell, telling it to make insulin. Inside the nucleus there's a gene, or DNA recipe, for making insulin; the insulin, once it is made and released, will lower the blood glucose level. But first the insulin needs to be produced. As you will learn in more detail in Chapter 3, the body preserves its resources by making proteins, such as insulin, only when they're needed. This production is regulated by the use of operons on the DNA, which literally turn the genes on and off. When the beta cell receptors turn the insulin operon on, the cell directs its ribosomes to make insulin.

As a result of releasing insulin, the blood glucose levels drop, and this drop will ultimately break the cycle of releasing insulin. Eventually, the body's blood glucose levels rise again, causing the release of insulin, which lowers the blood glucose levels, and so on, and so on, and so on .... Pretty cool, huh? This is a classic example of a negative feedback loop.

As a diabetic, I have to take my insulin first, injecting it under the skin (using a hypodermic needle—*hypo* = under, *dermic* = skin), and give it time to be picked up by the lymphatic system and delivered to the bloodstream, where it will do its work. Then I need to monitor the time I eat and the amount of food, particularly carbohydrates, that I eat so that I don't eat more than my injected insulin can handle. Rather than my body releasing the right amount of insulin after the food, I have to do everything backward!

How do I know how much insulin to take, and whether I've eaten too much? Simple, I test my blood glucose by sticking my finger, up to eight times a day! Folks, you don't know how lucky you've got it!

The Least You Need to Know

- The human body is best understood by looking at all of the levels of organization, from the atomic to the systemic.

- The 11 body systems—"SLIC MEN R RED"—work together, and it is misleading to think about the body systems as being separate.

- Directional terms—organized into opposites, and based on the X, Y, and Z axes— are crucial in that they allow us to accurately locate any part of the body, regardless of our perspective.

- The various planes allow us to divide the body into various section, and sectional images produced by CAT and MRI scans are an essential modern diagnostic tool.

- The abdomen is divided into regions and quadrants to ease diagnosis, especially in emergency situations.

- The body must maintain a constant state of balance known as homeostasis, and it does this through a number of negative feedback loops.

Hit the Bricks I: Molecules

In This Chapter

- ◆ Atomic structure
- ◆ Covalent and ionic bonds
- ◆ Polarity of water
- ◆ Acids, bases, and neutralization reactions
- ◆ Monomers, polymers, and organic molecules

For those of you who faced chemistry in school only out of necessity, your memories, which may include some panic, are probably not fond ones. If you ask the average college students in the sciences, they usually split easily into two groups: those who love biology and hate chemistry, and those who hate biology and love chemistry. True, some people love both, but a certain amount of the biology students' hatred of chemistry was probably due to a lack of connections with biology.

In this chapter, I start exploring the body from the smallest levels—the atomic level and the molecular level. Unlike so many other approaches, however, everything you learn here will be exploited for its connections to the human body. You'll see that organic molecules are nothing more than a typical meal away, and you will see that you truly *are* what you eat.

These glorious, and often delicious, organic molecules will be explored for their roles in your cells. For that you will need to learn more about the basics of cell structure, and how cells hold themselves together in your organs. Lastly, I will explore the processes of energy capture in our cells, DNA replication, and protein synthesis.

Atomic Structure

First of all, let's get the basics out of the way. Long before we had any way to actually measure, and even visualize, things on the atomic level, some forward-thinking people believed
that we were made of smaller particles. We now know them to be atoms, but further evidence
led us to conclude that such small particles are indeed made of smaller particles (and even
those are made of smaller particles … *aaargh*, but we will leave those for the physicists).

There are three basic subatomic particles: the proton, the neutron, and the electron. There
are two ways to think of them: in terms of location, and in terms of charge. Which is
more important? Both! *Protons* are found in the nucleus and have a positive charge. Also
found in the nucleus are *neutrons* which, as their name suggests, are neutral. Last are the
negative *electrons*, which are found in areas surrounding the
nucleus called *orbitals*.

Flex Your Muscles

In terms of remembering
the charges of the sub-
atomic particles, use the
names themselves as clues.
A *proton* is positive, just as
the *protagonist* in a story is the
hero. The *neutron* is *neutral*,
which is probably the easiest to
remember. Lastly, the *electron* is
negative, as is the outcome of so
many *elections* … sorry, I
couldn't resist that one!

An orbital is the three-dimensional area where the electrons
can be found orbiting (see Figure 2.1). Despite the name, they
are not like planetary orbits, but rather areas. The combination of these shapes has led to a second name for orbitals: electron clouds or electron shell. Finally, the electrons in different
orbitals have a different amount of energy, which is the origin
of the third name for orbitals: energy levels. Energy levels
help explain how glow-in-the-dark toys work: Absorbed light
energy excites the electron to a higher energy level (the first
and lowest level is the closest to the nucleus), and when the
electrons drop to their original energy level, they release the
energy in the form of light, and thus the toy glows! Cool, huh!

Figure 2.1

*2-D and 3-D models
of orbitals. Atoms fill
in electrons in the
order shown at left,
which explains bond-
ing patterns.*

(©*Michael J. Vieira
Lazaroff*)

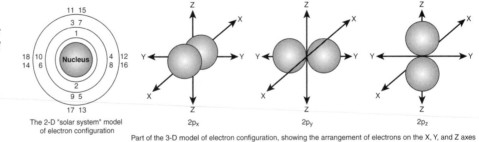

The 2-D "solar system" model
of electron configuration

Part of the 3-D model of electron configuration, showing the arrangement of electrons on the X, Y, and Z axes

One other note about these subatomic particles before I continue: Protons are important in
that they determine what element an atom is. Electrons determine how an atom reacts, or
bonds, with other atoms. Neutrons, true to their name, are neutral, even in temperament!
The number of neutrons can vary in an atom without changing either the element or the way
it bonds. Atoms of an element with different numbers of neutrons are called *isotopes* (remember, *iso-* means same, in this case the same element) of that element.

Not to be outdone, however, the number of neutrons in an atom's nucleus can make it unstable. An unstable atom is radioactive, and radioactive atoms transform themselves into other

atoms by changing the number of particles in the nucleus. This process is called *radioactive decay*. An understanding of radioactive decay has made radiation very useful in both medical diagnosis and treatment.

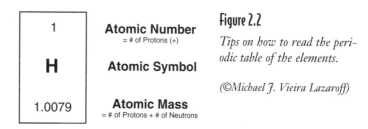

Figure 2.2

Tips on how to read the periodic table of the elements.

(©Michael J. Vieira Lazaroff)

All the elements are arranged on a map of sorts, which is arranged according to the subatomic particles of each atom, and this map is called the periodic table of the elements (see Figure 2.3). The number of protons in an atomic nucleus determines which element it is. That number is equal to the atomic number of an element. The combination of the number of protons and neutrons in an atomic nucleus makes up the atomic mass number. The number of electrons is equal to the number of protons. In this sense a typical atom would be neutral, having an equal number of positive and negative charges.

Periodic Table

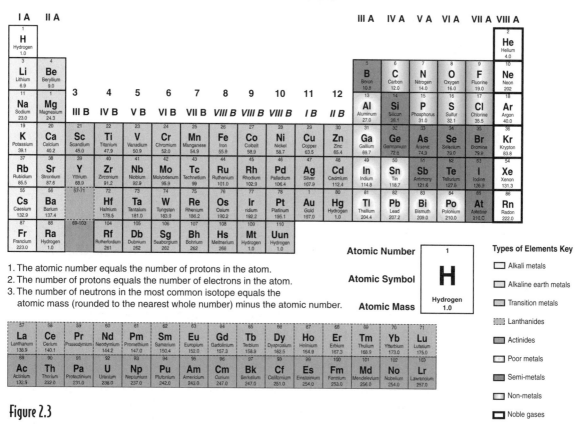

Figure 2.3

The periodic table of the elements.

Flex Your Muscles

To figure out the number of neutrons in an element, or more precisely, one of the isotopes of that element, subtract the atomic number from the atomic mass number of that isotope. The atomic mass numbers on the periodic table usually contain decimals because more than one isotope, each with its own atomic mass, can be found for each element. Thus the typical carbon, C^{12} with a mass of 12, is also found with its close cousins, C^{11}, C^{13} and C^{14}.

Covalent and Ionic Bonds

Because electrons are on the outside edge of the atom, it makes sense that they affect how the atom bonds. Don't forget that the molecules that atoms make are not just the building blocks of our bodies, but also the moving parts! To better understand these molecules, it is important to know the maximum number of electrons that can exist on any one energy level:

Energy Level	Number of Electrons
First	2
Second	8
Third	8
Fourth	18
Fifth	18
Sixth	32
Seventh	32

Knowing that, it is important to consider that atoms behave as if they want a full outer energy level. Consider the lonely hydrogen atom, with one proton and one neutron. Its single electron cannot fill its outer energy level, which is the first level. By sharing electrons with another atom, however, the outer levels of each atom can be filled (that is, the electron of the second atom becomes the second electron of the first atom, and vice versa). This bond, in which electrons are shared, is called a *covalent bond*, and the molecule produced (H_2) is hydrogen gas (see Figure 2.4).

Another way that atoms can fill their outer level is by donating and accepting electrons and forming ionic bonds. A classic example is that of table salt (NaCl = sodium chloride). A quick look at the periodic table will tell you that the sodium atom (Na) has 11 protons. Being in the first column of the third row, Na has only one electron in its outer energy level. To get a full outer shell, it needs to either gain seven electrons or lose one. Following the path of least resistance, it will lose one. Now that it has 10 electrons, but 11 protons, it will have a positive charge and become a positive ion (Na^+).

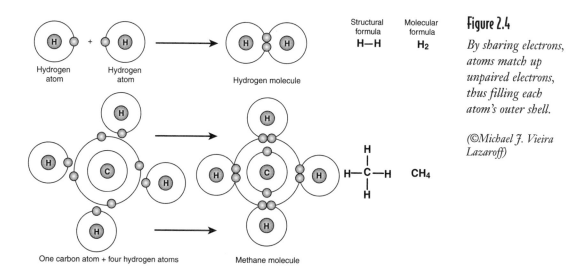

Structural
formula
H—H

Molecular
formula
H₂

Figure 2.4

By sharing electrons, atoms match up unpaired electrons, thus filling each atom's outer shell.

(©Michael J. Vieira Lazaroff)

$$H-\underset{\underset{H}{|}}{\overset{\overset{H}{|}}{C}}-H \quad CH_4$$

Taking another peek at the periodic table, chlorine has 17 protons, but it needs another electron to make its outer shell full. Once again taking the path of least resistance, it gains one electron. Now that it has 18 electrons, one more than the number of protons, it becomes a negative chlorine ion (Cl⁻). All of this switching of electrons reminds me of that old joke:

Two atoms walked into a bar.
One of them said "I think I lost an electron."
The other one asked, "Are you sure?"
"I'm positive."

The Big Picture

These ions will become very important later on. In order for nerves to transmit and receive messages, they must move ions through the nerve cell membranes (see Chapter 19). Na⁺ ions are also used to trigger the contraction of muscle cells. Calcium ions (Ca²⁺) are used in muscle cells as the key that unlocks the sarcomere, which is the basic unit of muscle contraction (see Chapter 8). Hydrogen ions (H⁺) are what make an acid an acid. Acids and bases are ionic compounds, and they are crucial in numerous chemical pathways, including the monitoring of breathing in the brainstem (see Chapter 13).

Now that both atoms have a charge, they are attracted to each other, because opposite charges attract (see Figure 2.5). This is not as strong a bond as a covalent bond; it's more like flirting than an actual marriage. As a matter of fact, the bond is so weak that ionic compounds easily dissolve in water.

Figure 2.5

By donating and accepting electrons with one another, each atom achieves a full outer shell but gains a charge in the process.

(©Michael J. Vieira Lazaroff)

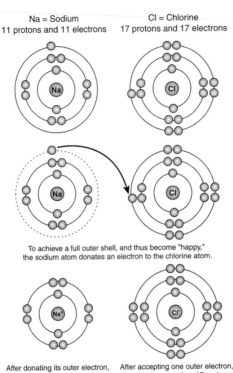

Na = Sodium
11 protons and 11 electrons

Cl = Chlorine
17 protons and 17 electrons

To achieve a full outer shell, and thus become "happy," the sodium atom donates an electron to the chlorine atom.

After donating its outer electron, the sodium atom has 11 protons and 10 electrons, thus leaving the ion with a net charge of 1+.

After accepting one outer electron, the chlorine atom has 17 protons and 18 electrons, thus leaving the ion with a net charge of 1-.

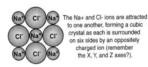

The Na+ and Cl- ions are attracted to one another, forming a cubic crystal as each is surrounded on six sides by an oppositely charged ion (remember the X, Y, and Z axes?).

Water, Water, Everywhere ...

When the atoms form covalent bonds to become a water molecule (H_2O), something interesting happens. The hydrogen atom only has a single electron, and it must put its electron near the other atom in order to form the bond. With the electron on one side the nucleus of the hydrogen atom is exposed, as if it's sticking its butt out.

Since the nucleus of a typical hydrogen atom just has a single proton (the two other isotopes also have neutrons), any covalently bonded hydrogen atom, with its exposed proton, has a positive charge. The oxygen, on the other hand, has only electrons on its outside; a covalently bonded oxygen atom has a negative charge. (Nitrogen atoms behave the same way.) As a result, water molecules have both positive and negative ends; this is the definition of a polar molecule.

Before you shrug off the polarity of water, you should know that that simple fact explains an extraordinary number of things crucial to your body. First of all, the polarity enables the formation of hydrogen bonds between pairs of water molecules (see Figure 2.6). A hydrogen bond is the result of the weak attraction between a positively charged, covalently bonded hydrogen and an oppositely charged, covalently bonded oxygen or nitrogen. These hydrogen bonds are very important in the structure of proteins and nucleic acids.

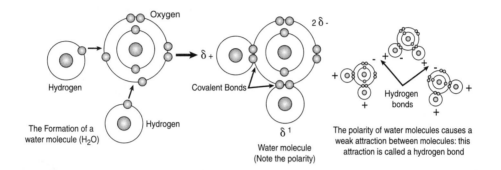

Figure 2.6

Opposite charges make water polar, and the weak bonds between polar water molecules are called hydrogen bonds.

(©Michael J. Vieira Lazaroff)

As a polar molecule, water molecules form hydrogen bonds with each other. This enables water to carry a large amount of heat. The body takes advantage of this in terms of heat regulation; bringing heat to your skin involves little more than dilating your capillaries near the skin—the plasma in the blood, being mostly water, carries the heat. If you have ever tried to see how many drops of water fit on a penny you know intuitively that the hydrogen bonds enable water to form surface tension; you will see later that the lungs need a special protein to counteract this tendency of water (see Chapter 13).

Finally, water's polarity enables it to act as a wonderful solvent; it has been called the universal solvent, because an enormous number of substances dissolve in water. When substances are dissolved in another substance, the mixture is called a solution. The solute is the solid, liquid, or gas that is dissolved by the solvent. The polarity of water enables the positive end of the water molecules to surround the negative portion of the solute, and water's negative end surrounds the positive portion of the solvent. Ions, with their inherent charge, dissolve beautifully in water. Blood takes beautiful advantage of this.

Acids and Bases

When you consider the effect of water's polarity on the dissolving of ions, that almost 92 percent of blood plasma is water, and that approximately two thirds of your body is made of water, it becomes apparent that ions are very important in the body. Two types of ions, called *acids* and *bases*, deserve special mention here. To understand acids and bases, you must first think about the structure of the hydrogen atom, for it is the concentration of hydrogen ions in solutions that determine whether a substance is an acid or a base, a concentration that is measured on the pH scale (see Figure 2.7).

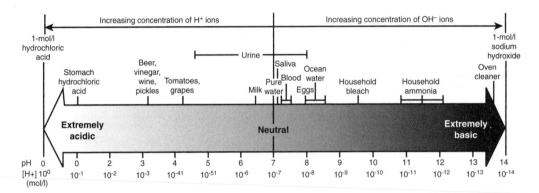

Figure 2.7

The pH scale.

Flex Your Muscles

An increase of 1 in the numbers on the pH scale, such as pH 3 to pH 4, is a tenfold decrease in the amount of hydrogen ions in the solution, because pH 3 = 10^{-3} H^+ ions, and pH 4 = 10^{-4} H^+ ions. Acids have a pH less than 7, with bases having a pH more than 7; this explains why acids with a lower pH number are actually stronger!

Because hydrogen atoms are made of a nucleus with only one proton, around which orbits one lonely proton, this pair of subatomic particles is prone to separate. If this happens to a covalently bonded hydrogen atom, the electron tends to stay with the other atoms in the molecule, while the lonely proton will go off on its own. Remember the charge of an ion that *donates* an electron? That ion will be positive, so a hydrogen ion is H^+, and the ion that kept the electron from the hydrogen will be negative.

Solutions that donate hydrogen ions, and thus have a higher concentration of those ions, are called acids, and they have a lower pH number (less than 7); acidic foods will taste sour, as you can tell from the citric acid in lemons. Solutions, on the other hand, that accept hydrogen ions, and thus have a higher pH number (more than 7), are bases; basic (also known as alkaline) substances will taste bitter, as anyone who has tasted soaps will know.

Bases sometimes contain OH^- (hydroxide) ions, such as sodium hydroxide (NaOH). The combination of H^+ and OH^- ions makes water (H_2O). Given that water has an equal concentration of both acidic ions and basic ions, water is considered neutral, and it has a pH of 7, or right in the middle!

In the same sense, reactions that combine equal concentrations of acids and bases are called neutralization reactions:

$$HCl + NaOH \rightarrow H_2O + NaCl$$

Thus the reaction of hydrochloric acid and sodium hydroxide produces water and a salt, in this case sodium chloride. Given that the body functions for the most part within a narrow pH range, hovering around neutral, any fluctuations need to be monitored and adjusted. These neutralization reactions occur all the time in your body.

You Are What You Eat

Have you ever heard that old expression, "You are what you eat"? You are made of four types of organic molecules: carbohydrates, lipids, proteins, and nucleic acids. Those should sound familiar, because three of them can be found on every food label (you might only have recognized two because lipids appear on the label as fats). Most people might take that idea and stop right there ... but I'm not most people ...

The Big Picture

Fluctuation in pH is essential in controlling your breathing. When you exercise you initially don't get rid of enough carbon dioxide (CO_2), which mixes with your blood, producing carbonic acid, which is carried by the spinal fluid, and which in turn is monitored by the medulla oblongata, thus triggering deeper breathing. Also, the low pH of the stomach acid (pH 2) requires the release of a great deal of mucus by the small intestine, which in turn raises the pH via a neutralization reaction.

So? Let's run with it! When you eat meat, the proteins in them are chopped up and rearranged into your own proteins. If anyone ever called you "chicken" in grade school, think about what you ate! Even your DNA is made up of the recycled DNA that is in what you eat! Very little of you is unique; why build from scratch what you can recycle from the garden patch?

Monomers and Polymers

This recycling will make more sense when you understand the basic structure of organic molecules. Think about an organic molecule as being like a jigsaw puzzle. All the pieces are fairly similar, but when they are connected they make something new (another emergent property, as explained in Chapter 1) that is greater than the sum of the parts. These smaller, repeating molecules are called *monomers* (*mono* = one, and *mer* = part), and the larger molecules that they make up are called *polymers* (*poly* = many).

When we talk about food we often use terms that reflect the concept of monomers and polymers. Complex carbohydrates (polymers) are better for you than simple sugars (monomers). Digestion is the act of converting those polymers into monomers.

The act of building polymers from monomers is basically the act of making covalent bonds between two adjacent monomers, or between a monomer and a polymer. More specifically, it involves the connection between two hydroxides (-OH and HO-) to make a bridge. Using oxygen's ability to covalently bond twice, an –O- bond forms. This means that one molecule drops a hydrogen (-H) and the other drops a hydroxide (-OH); these two fragments combine to make HOH, also known as H_2O. To build such a bond involves the removal of a single water molecule. Since the removal of water is called dehydration, and another word to describe building is called synthesis, we call this chemical reaction dehydration synthesis.

Digestion, or the breakdown of polymers into monomers, requires the opposite reaction. This involves breaking bonds by adding a water molecule, or rather the parts of a water molecule (H- + -OH) to the polymer to restore the hydroxides you saw before (thus from –O- to –OH and HO-). This reaction is called *hydrolysis* (*hydro* = water and *lysis* = to break). Enzymes that do this are called *hydrolytic enzymes*, and hydrolytic enzymes are made of proteins.

Carbohydrates

A human being is, in the most basic sense, a machine. In order for a machine to work, it needs fuel. Carbohydrates make up the main fuel (see Figure 2.8). They come in the form of single sugars called *monosaccharides* (*mono* = single and *saccharide* = sugar); these are the monomers, the most famous of which is glucose, which is important in cellular respiration (see Chapter 3). A double sugar is a *disaccharide* (table sugar, or sucrose, is a disaccharide), and larger sugars, often referred to as complex carbohydrates, are called *polysaccharides*; starch is an example of a polysaccharide.

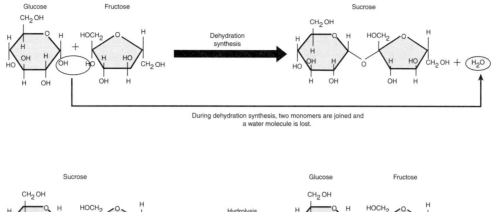

During dehydration synthesis, two monomers are joined and
a water molecule is lost.

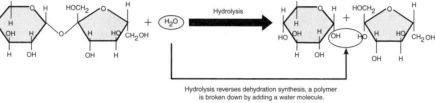

Hydrolysis reverses dehydration synthesis, a polymer
is broken down by adding a water molecule.

Figure 2.8

On the top: Two monosaccharides (monomers) are joined via dehydration synthesis to make a disaccharide (polymer).

On the bottom. Hydrolysis is used to separate a polymer into monomers (disaccharide to two monosaccharides).

Lipids

Fat, *schmat*; lipids are your energy-storing molecules. You cannot function like a solar calculator, able to function only when food is available. The development of photosynthesis is a classic solution to this problem, for without storing energy (in this case in the form of sugar), every plant would die at night! A more efficient storage, however, is in the form of lipids (see Figure 2.9). Carbohydrates carry 4 kilocalories of energy per gram (kal/gram), whereas lipids hold 9 kcal/gram.

In addition to energy storage, lipids perform another valuable function simply because they don't like to mix with water! If you have ever taken some of Mom's homemade chicken soup out of the fridge, you have seen the proteins forming a Jell-O–like substance, covered with a yellowish-white layer of fat on top. Not only are lipids less dense than water, and thus they float, but they are hydrophobic, which means they don't mix with water. By being hydrophobic, they are very useful on your skin, for the oil provides a crucial barrier to prevent you from losing too much moisture!

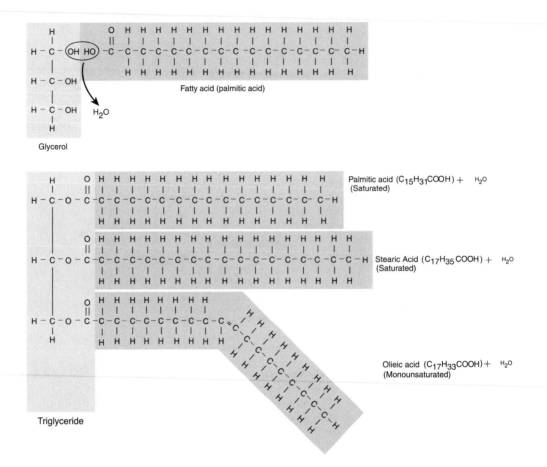

Figure 2.9

Note the polymerization of fat. Also shown are both saturated fatty acids (with only single bonds) and unsaturated (with a double bond).

The Big Picture

Nature requires that animals have quick access to sugar when they need it. A rabbit doesn't stop to eat grass before it runs away from a fox. Our vertebrate ancestors evolved ways of storing energy. Beyond simply making and storing fat from sugars, you build the monomers into a large polymer called *glycogen*, made of many glucose molecules. This glycogen is made and stored in the liver. This makes sense given that the food absorbed in the capillaries of the small intestine passes through the liver before it returns to the heart. In addition, the pancreas regulates both the building (dehydration synthesis) and breakdown (hydrolysis) of glycogen, which is stored in individual muscle cells, making it easier to escape from the jaws of death!

Proteins

The Human Genome Project's task, completed in 2003, was to determine the number and function of *all* the genes (the human genome) in human DNA. The project determined that there are more than 30,000 genes in the human genome. This relates to protein in that each and every gene is a *recipe* for a specific protein.

Proteins really are the wonder molecules. Proteins are responsible for much of our basic structure; they comprise most of the hormones in our body; they make our muscles contract; and they are what make it possible to distinguish you from invaders (bacteria and viruses), not to mention the means by which you destroy those invaders (via antibodies). This variety of function is possible because, of all the organic molecules, proteins are the most complex in their structure.

Proteins are large polymers made up of varying combinations of 20 basic *amino acids*, which act as the monomer. The covalent bond that forms between two amino acids are called *peptide bonds*, which is why so many smaller proteins are often referred to as peptides or polypeptides. Of those 20, there are some that you make in your body and some that it is essential you get from your diet. As such, those that you must eat are called *essential amino acids*, and those that you manufacture are called *nonessential amino acids*.

 **Flex Your Muscles**

The most efficient source of protein per Calorie consumed in your diet is meat. Meat is the muscle of the animal, and it is filled with a tremendous amount of the proteins actin and myosin (and others, see Chapter 8), which allow the muscle to contract. If, however, you do not want to eat meat, not all is lost. Vegetarians will be happy to hear that plants also make protein. Be careful, however, because some plants are richer in protein and especially essential amino acids. With care, it is possible (through certain combinations, such as beans and rice) to get a full complement of all the amino acids without eating any meat at all.

Part of a protein's complexity lies in the levels of its structure (see Figure 2.10). The four levels of a protein's structure are as follows:

- ◆ **Primary structure**   The basic chain of amino acids covalently bonded, also known as its amino acid sequence.

- ◆ **Secondary structure**   The shape that forms when amino acids form hydrogen bonds with one another, and form shapes known as an alpha helix or a pleated sheet.

- ◆ **Tertiary structure**   The shape that forms when a *secondary* structure folds in on itself through both hydrogen and sulfur bonds. This gives a protein its globular shape.

- ◆ **Quaternary structure**   This is simply the result of two or more *tertiary* structures bonded together (for example, hemoglobin has four *tertiary* chains in its *quaternary* structure).

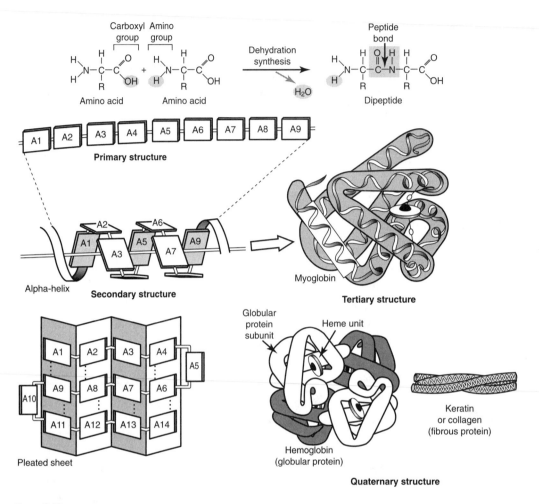

Figure 2.10

There are four levels to a protein's structure, from primary to quaternary.

One result of this globular shape is the formation of an active site on molecules called *enzymes* (see Figure 2.11). Molecules called *substrates* fit into the active site like a hand into a glove. As a result of opposite charges between the two molecules repelling, and like charges attracting, a protein changes its shape. Depending on the type of enzyme, either a polymer substrate is split into monomers (hydrolysis) or monomers are joined into a polymer (dehydration); the new molecule produced is called the product. Despite all this building (anabolism) and breaking (catabolism), the enzyme remains unchanged, free to be used over and over again!

The globular shape of a protein also allows for an incredible variety of shapes, such as the zipper-like proteins that hold neighboring cells together (without them you would literally fly apart from the force of a single sneeze!). This also makes them the ideal molecule to mark a cell (by placing it on the outside of the cell membrane, see Chapter 3) as belonging to one person. Proteins really are globs of fun!

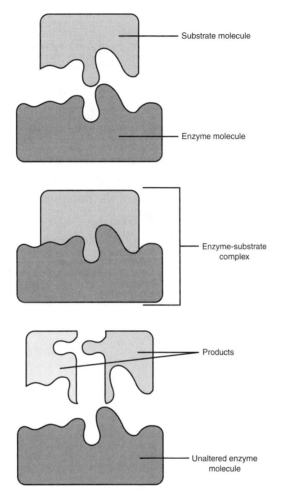

Figure 2.11

When a substrate fits into the active site of an enzyme, the substrate changes, but the enzyme remains unchanged.

Substrate molecule

Enzyme molecule

Enzyme-substrate complex

Products

Unaltered enzyme molecule

The Big Picture

Most hormones are protein based, and the manufacture of lipid-based hormones requires proteins. Brain chemistry is based on small chains of amino acids called *neuropeptides*. The proteins on cell membranes are essential to the function of the immune system, for they enable the white blood cells to determine whether a cell is ours. Viruses can attach to such proteins and infect a cell; this is why most viruses cannot be spread from species to species. The similarity of proteins, however, does occasionally allow certain viruses to jump species (as HIV and SARS—severe acute respiratory syndrome—did) and thus affect humans. The experts call such viruses *emergent viruses*, and they help explain why humans will never fully conquer disease.

Nucleic Acids

I always propose a scenario to my students, but they never take me up on it. I select two students at random and then ask them if they would mind whether I just took out one kidney from each of them and then replaced it with the other person's kidney. They never are happy about that idea, and rightly so! Knowing about organ transplants, they realize that exchanging organs can lead to tissue rejection. But what is it about the tissue that causes the rejection by the immune system? Simply put, it is the proteins on the outside of every cell, the combination of which is uniquely your own.

Your DNA is just as unique as your protein, simply because *your DNA is the recipe for making protein* (for more on that process, see Chapter 3). The monomer for nucleic acids, so called due to their ties to the nucleus, is called a nucleotide, which is composed of other units: phosphate, sugar, and a base. Given that there are four bases, there are four possible nucleotides for either DNA or RNA. Some of the differences (including the names of the bases) can be found in the following table.

A Quick Comparison of DNA and RNA

Characteristics	DNA	RNA
Sugar	Deoxyribose	Ribose
Nitrogenous bases	Adenine	Adenine
	Guanine	Guanine
	Thymine	Uracil
	Cytosine	Cytosine
Base pairing	A-T, G-C	A-U, G-C
Shape	Double helix	Single helix
Location	Nucleus only	Nucleus and cytoplasm

Due to the base pairing of DNA, A-T and G-C, one long chain of nucleotides (a single helix), made from a phosphate-sugar backbone with the bases facing on the other side, forms hydrogen bonds with the bases of a complementary chain of nucleotides (see Figure 2.12). These two single helices form one double helix. RNA forms a single helix because it is used to make a complementary copy of only one side of the DNA molecule. That copy is then used to make multiple copies of the protein for which the DNA coded.

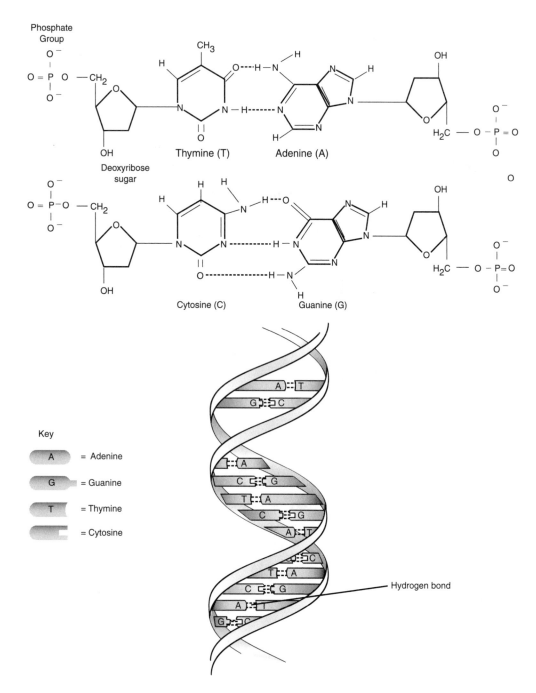

Figure 2.12

The phosphate-sugar backbone of each strand enables the complementary base pairs to form, thus creating the double helix structure of DNA.

The Least You Need to Know

♦ Protons determine the element an atom is, neutrons determine which isotope an atom is, and electrons determine how the atom bonds.

♦ When atoms share electrons, they form covalent bonds. When atoms donate, or accept electrons, they become ions.

♦ Acids and bases are ions whose strength is measured on the pH scale (acids = less than 7, neutral = 7, bases = greater than 7).

♦ Organic molecules—carbohydrates, lipids, proteins, and nucleic acids—can be found as monomers or long chains known as polymers.

♦ Carbohydrates and lipids are the major energy-carrying molecules, while lipids also help prevent water loss.

♦ Proteins perform the widest variety of tasks in the body. And nucleic acids, DNA and RNA, are used to make proteins.

Hit the Bricks II: Cells

In This Chapter

◆ Basic cell structure and organelle functions

◆ Passive and active transport

◆ DNA replication and protein synthesis

◆ Cellular respiration

◆ The cellular basis of disease

We are all glorified amoebas! Whaaaat? You might be thinking, "them's fightin' words," but it's true! Regardless of what we do, we are not, in reality, all that different from the lowly amoeba. Every living thing, in order to exist within its environment, needs to be able to exchange materials with that environment. Since an amoeba is so simple (see Figure 3.1), with its cell membrane as its outer edge, it makes perfect sense that all exchange happens through its membrane.

Gases, oxygen and carbon dioxide, pass through the amoeba's moist cell membrane, as they do through the membranes of the alveoli and alveolar capillaries in our lungs, which are kept moist by the blood in the capillaries and the mucus in the respiratory bronchioles (see Chapter 13).

Just as food is absorbed through the amoeba's membrane, or at least through the membrane of the vacuole, food is also absorbed through the membranes of the villi of the small intestine (see Chapter 14). For those die-hards who like to throw about terms like *intracellular* and *extracellular* digestion, don't forget that our roaming macrophages (see Chapter 17) indeed consume and digest exactly like the amoeba!

Figure 3.1

Our cousin, the not-so-lowly amoeba!

(©2003 www.clipart.com)

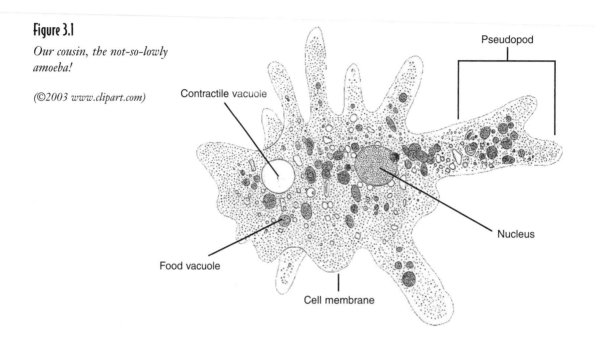

Contractile vacuole

Pseudopod

Nucleus

Food vacuole

Cell membrane

Just as an amoeba excretes its waste (its own form of urine) through its membrane, the capillaries of our nephrons (in our kidneys) filter out the blood's waste. In so many ways, what we do shows all the signs of our evolutionary history. To explore the big, you need to learn of the small.

Talkin' on My Cell

In the modern world people are so used to the ubiquitous nature of cell phones that I wonder if they even remember what a cell really is? People probably don't realize that a cell phone is called that because it is connected to other phones through a network, but that each phone exists on its own. Terrorist "cells" function on the same concept, each one existing on its own, yet cooperating, without knowledge of each other, to accomplish a task.

Even within such small terrorist groups there is one person in charge, that person acts as a nucleus does, controlling the cell. Beneath the nucleus are other parts of the cell who have specific tasks, as organs do, and so they are called *organelles* (see Figure 3.2). There is a world of wonder within, so let's begin.

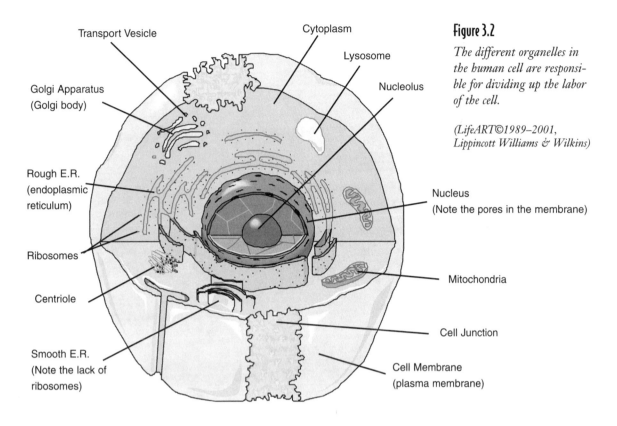

Transport Vesicle

Golgi Apparatus
(Golgi body)

Rough E.R.
(endoplasmic
reticulum)

Ribosomes

Centriole

Smooth E.R.
(Note the lack of
ribosomes)

Cytoplasm

Lysosome

Nucleolus

Nucleus
(Note the pores in the membrane)

Mitochondria

Cell Junction

Cell Membrane
(plasma membrane)

Figure 3.2

The different organelles in the human cell are responsible for dividing up the labor of the cell.

(LifeART©1989–2001, Lippincott Williams & Wilkins)

Membranes

On the outside of a cell is a barrier called the cell membrane. Think of a prison, with its outside wall. But as a barrier, a prison wall would only be so useful if no one could get in or out, so every wall has gates, and those gates have guards. The role of the guard is to decide who gets in and who gets out. This is the role of the cell membrane, and to accomplish this it has two basic routes: passive transport (which requires no energy) and active transport (which does require energy). Since some materials move easily, while others require effort, and some don't pass through at all, we call such a membrane *semipermeable*. Cell membranes are made of phospholipids, and you may remember from Chapter 2 that lipids do not mix well with water. Molecules that repel water are called *hydrophobic*, and molecules that are attracted to water are called *hydrophilic*. Phospholipids are unusual because they have both a hydrophobic end (known as the tail) and hydrophilic end (known as the head). What is unusual about these molecules is the way they behave in water. Rather like a wagon train under attack, phospholipids line themselves up with the hydrophilic ends facing the water, and the hydrophobic ends hiding in the middle facing each other. This double layer, called a *phospholipid bilayer* (see Figure 3.3), forms naturally whenever phospholipids mix with water.

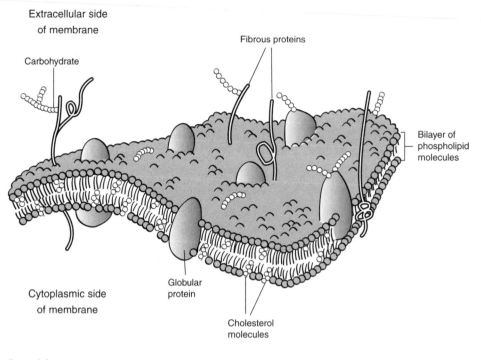

Figure 3.3

A cell membrane is a phospholipid bilayer embedded with proteins and glycoproteins.

However, cell membranes are not just a continuous layer of phospholipids; embedded within the membrane are an incredible variety of proteins and glycoproteins that are crucial to a cell's ability to function. Rather than being fixed, these molecules float around the membrane, adding distinctiveness to it, just as brightly colored tiles in a mosaic. For this reason, the structure of a cell membrane has been described as a fluid mosaic.

The Big Picture

Cell recognition by the immune system is a glorious example of the big picture depending on small details. DNA is the recipe for, and controls the production of, protein. Since all human beings (except for identical twins) have a unique combination of genes in their DNA, it appears logical that the immune system would look at the DNA to make the decision of "self or foe?" However, that would not work, because to "scan" the DNA, each cell would need to be invaded. On the other hand, because DNA is the recipe for proteins, and unique DNA means unique proteins, and cells have proteins on their membranes, the immune cells need only to check out the surface proteins!

These proteins have multiple roles in the life of a cell. Since most of the membrane is lipid based, molecules that are lipid soluble pass through with ease, but other molecules need help. Proteins act as several kinds of channels that help move materials in and out, such as the acetylcholine (ACh) on muscle cells (see Chapter 8). These proteins also help to identify the cell as belonging to the body, and not being foreign (such as a bacterium or virus). Foreign antigens (or proteins) on a cell will cause it to be singled out and destroyed by the immune system.

Passive Transport

Cells aren't stupid. Why use energy when there is a perfectly acceptable way of doing something free; cells often take the easy way out. There is a natural habit of molecules to move, either in air or in a gas, from an area of high concentration to an area of low concentration, just as it's easier to roll a ball downhill than up. This movement of molecules from an area of high concentration to an area of low concentration is called *diffusion*. The direction in which the molecules move is called the concentration gradient (the angle of a hill is its grade), and it is always easier to go downhill! Of all the molecules that diffuse through the membrane, one deserves special mention, for every living being needs it: water. The diffusion of water through a semipermeable membrane is called *osmosis*.

Every cell lives in a solution; in our bodies this solution is one of the following: blood plasma, interstitial fluid, or lymph. As I mentioned in Chapter 2, a solution has two parts: the solvent, which is the fluid doing the dissolving (in the case of cells, the solvent is water), and the solute, which is the substance that dissolves into the solvent. Salt water is a typical example of a solution.

Solutions are described in terms of the concentration of solute, and that description always involves a comparison to the concentration in the cell's cytoplasm. The three types of solution are called *hypotonic*, *isotonic*, and *hypertonic*. Although the name depends on the concentration of solute, the real concern here is the amount of water, and the direction the water moves. For example, if a cell's cytoplasm is 30 percent solutes, it is 70 percent water. A solution of the same concentration would be *isotonic* (*iso* = same and *tonic* = solute). A *hypotonic* (*hypo* = less) solution would be less solute than the inside of the cell, but if the solute concentration goes down, the water concentration goes up (that is, 25 percent solute means 75 percent water or solvent). In such a case the water would diffuse through the membrane from high to low, which in this case would mean that osmosis moves the water inside the cell.

A hypertonic solution would do the opposite (*hyper* = more, so more solute outside would mean less water outside). A hypertonic solution would create an osmotic pressure that would move the water outside the cell. We will see how the body uses this when we study the kidney.

The only thing I haven't mentioned yet is the idea of facilitated diffusion. As a kid, did you ever have a balloon from a carnival? Do you remember leaving it floating up to the ceiling at night, only to see it on the floor the next morning? Why did it do that? The answer, of course, is diffusion of the helium out of the balloon, given the low concentration in the atmosphere.

Have you ever popped a balloon? That loud sound, and sudden rush of helium out of the hole is still diffusion, but the hole helped it along; this is what is known as facilitated diffusion. You can also see this when you blow up a balloon and let go, only to see the balloon jet-propelled around the room, making farting noises as it goes. The cell accomplishes this same process through open, tubular, protein channels in the cell membrane.

Active Transport

Diffusion is a wonderful concept, and it is elegant the way cells take advantage of it, but it alone is not enough. In some cases the cell needs more of a substance than it could get by pure equilibrium. In such a situation, the cell needs to move materials in the opposite direction from diffusion, against the concentration gradient. This is not just movement uphill, but upstream, since diffusion will be trying to go in the opposite direction at the same time. As you can imagine, this type of transport takes energy. It is the opposite of passive transport, for it doesn't just happen, but rather the cell takes an active part in the transport—hence the name active transport.

The Big Picture

One thing that most textbooks seem to leave out, for some bizarre reason, is that active transport and facilitated diffusion often go side by side. The reason for this is simple: the higher the difference in concentration (in other words, the steeper the concentration gradient), the faster molecules diffuse. With active transport maintaining a high concentration on one side of the membrane (called *polarization*), and facilitated diffusion allowing the diffusion (which in this case is called *depolarization*), when it is triggered by opening the channel, to happen *very* fast, it is now possible to have cells trigger a very fast response. You will see examples of this not only in the nervous system (Chapter 19), but also in the muscular system (Chapter 8).

On a small scale this happens in the movement of certain ions, such as sodium (Na^+), potassium (K^+), and calcium (Ca^+) ions, across the cell membrane. The calcium ions are not only important in bone formation (see Chapter 5), but also in terms of muscle contraction (see Chapter 8). The movement of Ca^+ ions is controlled by the twin actions of active transport and facilitated diffusion, which greatly increases the speed of the muscle contraction (see the preceding "The Big Picture" sidebar). The movement of Na^+ and K^+ ions is so often seen together that the active transport mechanism is called the sodium/potassium (Na^+/K^+) pump.

Surface Area to Volume Ratio

Clearly membranes, as the gateway for all materials in and out of the cell, are *very* important to the cell. So much so, that the amount of membrane on an individual cell is integral to its survival. If a cell membrane's surface area is not large enough to move all of the materials in and out efficiently, the cell will die. So it makes sense that a larger cell would have a bigger surface area, and thus be better. Right?

Wrong! It turns out that when a cell grows larger, and its volume increases, although the surface area does grow larger, the ratio between the surface area and the volume actually grows smaller! At first this doesn't seem to make sense, but a quick look to the numbers in the following table shows that it has to be true. This illustrates a favorite concept of mine—if something doesn't make sense at first, you don't have enough information. The body does make sense, as long as you have enough information about it. This is an excellent argument for what is called pure scientific research, for any information gathered is likely to be useful someday, even if researchers can't say what it would be useful for right now.

Surface Area to Volume Ratio

Cube	Cube Surface Area	Cube Volume Cube	SA/V
X cm	L×W×6 = SA	L×W×H = V	SA/V
1 cm	1cm×1cm×6 = 6 cm^2	1cm×1cm×1cm = 1 cm^3	6:1
2 cm	2cm×2cm×6 = 24 cm^2	2cm×2cm×2cm = 8 cm^3	3:1
3 cm	3cm×3cm×6 = 54 cm^2	3cm×3cm×3cm = 27 cm^3	2:1

Although you might think that "a cube is a cube is a cube," the math shows us that this is *not* so! There are two conclusions you can draw from this: As the volume increases the surface area increases, but more importantly, as the volume increases the *surface area to volume ratio decreases.* This might not sound very important, but I assure you the implications are enormous. If a cell gets too big, it cannot absorb enough of what it needs, and also cannot rid itself of enough of its waste. In the body you will find examples everywhere, in every organ system, where the body will make every attempt to maximize the surface area to volume ratio. With the cells of our body being so small, is it any wonder that we have a hundred trillion of them?

Endocytosis and Exocytosis

What if what you want to eat is just too dang big? Think of the roaming macrophages (*macro* = big and *phage* = eater) in the body; they search out and destroy dead cells, not to mention eat bacterial and viral invaders. They can't simply wait for the cells to break down into small pieces, so they do the next best thing; they engulf the cells whole, only to break them down inside the cell! This process is called *phagocytosis.*

Flex Your Muscles

Endocytosis and exocytosis are easy to tell apart because of the prefixes. Both involve the movement of materials either into or out of the cell. The "endo" in endocytosis sounds like "enter," and that is exactly what it means: materials *enter* the cell. The "exo" in exocytosis sounds like "exit," and that is exactly what it means: materials *exit* the cell.

Phagocytosis is part of a larger group of processes known as endocytosis, or *entering* the cell (see Figure 3.4). Endocytosis is quite simply the act of bringing materials too large to be absorbed through the membrane into the cell by surrounding the materials with the cell membrane, thus forming a vacuole around it. Phagocytosis is a form of active transport because energy is needed to perform it.

A smaller form of endocytosis is known as *pinocytosis* (it should be easy to remember that because pins are so small). Pinocytosis is sometimes called cell drinking, because only a small portion of the cell membrane is used, and only a small amount of the extracellular (outside the cell) fluid is engulfed, along with any dissolved materials. Examples of this can be found in capillaries and renal tubules.

The Big Picture

Pinocytosis is rarely explained as an important function in most textbooks. Think for a moment, however, about the bloodstream. As the major means of transport, it is very important that materials be able to move in and out. Pinocytosis comes to the rescue here, as blood plasma enters the thin capillary walls and then the vehicle carrying the plasma travels from the lumen side of the wall to the interstitial side, where it is released by exocytosis. Clearly, pinocytosis is a major force in transport in our bodies!

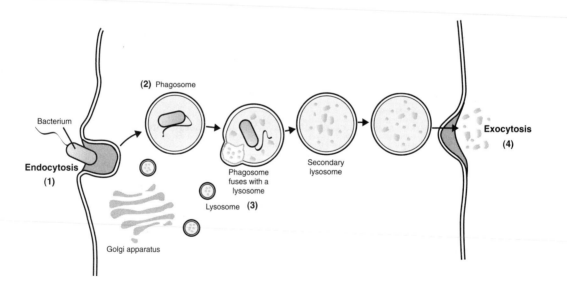

Figure 3.4

This figure shows the continuum of endocytosis and exocytosis: (1) phagocytosis (a form of endocytosis) of food, (2) formation of a vacuole, (3) digestion of the contents through fusion with a lysosome, and (4) exocytosis of the waste.

The other process to be reckoned with is exocytosis. As its name suggests, this process involves materials exiting the cell. The process is the opposite of endocytosis, in that a vacuole or vesicle fuses with the inside of the cell membrane, and the contents are then dumped into the extracellular fluid. These contents may be wastes from intracellular digestion, hormones, proteins for export, digestive enzymes, and even mucus!

Organelles

Just as the body has multiple organs that perform specialized functions, each cell has its own miniature organs, aptly named *organelles*, which are found throughout the jelly-like fluid called the cytoplasm. In addition to having different functions, these organelles vary in size, shape, and *origin*. The cell manufactures most organelles, but some actually reproduce by themselves within the cell! The basic organelles, their function, and their origin is in the table below.

Medical Records

A brilliant explanation for the evolution of eukaryotic cells (such as ours) from prokaryotic cells (those without a nucleus or membrane-bound organelles, such as bacteria) is called the *endosymbiont hypothesis*. This brilliant hypothesis suggests that certain organelles (chloroplasts in plants, and mitochondria in all eukaryotes) were originally prokaryotic cells that once invaded larger prokaryotic cells. This theory arose because chloroplasts (in plants and algae) and mitochondria (in all eukaryotic cells) have their own DNA and they grow and reproduce independent of the life of their "host cell."

The Function and Origin of Organelles

Organelle	Function	Origin
The Builders		
Nucleus	Regulates protein synthesis	Replication
Nucleolus	Produces ribosomes	DNA
Ribosomes	Protein synthesis	Nucleolus
Rough E.R.*	Protein synthesis, and making transport vesicles	Build themselves
Smooth E.R.	Detoxification, lipid synthesis, Ca^{2+} storage	Build themselves
Golgi body	Modification and storage making transport vesicles	Rough E.R.

continues

The Function and Origin of Organelles (continued)

Organelle	Function	Origin
The Breakers		
Lysosomes	Intracellular digestion; breakdown of old organelles	Golgi apparatus
Microbody	Breakdown of H_2O_2	Golgi apparatus
Mitochondria	Cellular respiration	They divide!
Various		
Vacuole	Storage, transport	Rough and smooth E.R., endocytosis
Cilia	Movement of materials over cell surface	Rough E.R.
Flagellum	Locomotion (sperm)	Rough E.R.
Centriole	Anchoring for mitosis	Ribosomes
Cytoskeleton		
Microfilaments	Movement, anchoring	Ribosomes
Microtubules	Movement, anchoring	Ribosomes
Intermediate	Anchoring	Ribosomes filaments

E.R. = endoplasmic reticulum; ribosomes on the rough E.R. make it "rough."

Unit of Structure and Function

Division of labor is an amazing thing. During the nineteenth century people saw the rise of mass production. Through the simple use of standard parts, and workers with very specialized tasks, people were able to greatly increase production. In the unicellular organism is an incredibly large variety of tasks that need to be accomplished. If every cell in a multicellular organism had to carry out every task, the resulting inefficiency would severely limit the organism's ability to survive.

As organisms grow larger, new problems develop; evolution crafted many sophisticated solutions to these problems. For one thing, it was no longer always possible to have every cell have simple access to exchange gases with its environment. Diffusion is an amazing thing, but it has practical limitations. A large organism in which every cell is able to exchange gases with its environment must have an enormous surface area. By allowing cells to become specialized, in a wonderful process called differentiation, cells emphasize specific characteristics by turning various genes on and off during development (see the section in this chapter on the operon). This enables cells to have widely different shapes and functions. In this way, cells can be the basic unit of structure and function.

Don't Be Cellfish!

What, exactly, are the functions of a cell? It is impossible to list them all in a book so small, but I will illustrate a number of basic functions here that are common to all cells. Much of the difference between cells is due to the amount of the cell's life that is devoted to specific functions. For example, all cells make protein, but goblet cells make a tremendous amount. All cells have actin molecules as part of their cytoskeleton, but muscle cells have an enormous amount of highly organized actin, thus allowing them to contract. Despite the differences, what follows are the basic functions that all cells have in common.

DNA and Protein Synthesis

DNA is an incredibly powerful molecule, especially given its inherent simplicity. As a double helix, with the simple base pairing rule (see Chapter 1) of A-T (adenine-thymine) and C-G (cytosine-guanine), it controls the production of all protein in the body, and ultimately the development of our body itself! But how is it possible that every cell in the body (except red blood cells) has a complete set of DNA?

When DNA unzips, each half combines with free nucleotides, courtesy of the digestive and cardiovascular systems, to form new base pairs (see Figure 3.5). The result of an A-T pair splitting, then the A side matching with a T, and then the T side matching with an A, is two identical A-T pairs, each an identical copy of the original. When Watson and Crick discovered the structure of DNA, they also cracked the code for its making an identical replica of itself, a process known as replication (yup, another name that makes sense!).

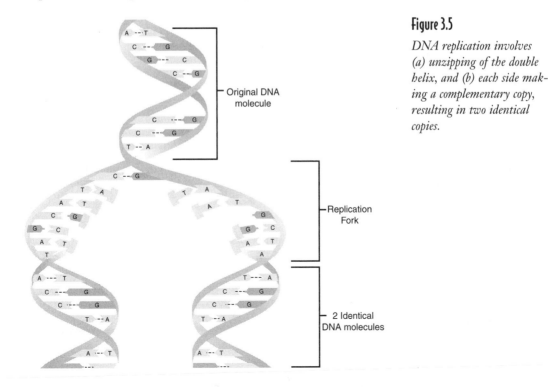

Figure 3.5

DNA replication involves (a) unzipping of the double helix, and (b) each side making a complementary copy, resulting in two identical copies.

Original DNA molecule

Replication Fork

2 Identical DNA molecules

The sequence of bases in DNA act as a recipe for protein. But that recipe is so precious that the cell cannot risk taking it out of the nucleus. It's like an architect, let's call him Bob, who brings the only copy of the blueprint to a construction site, only to lose it when the concrete foundation is poured (over the blueprint!). If only Bob had made a copy … Luckily cells aren't that stupid. In a process called transcription (see Figure 3.6), one side of the DNA double helix is copied by a single strand of RNA (called *messenger RNA* or *mRNA*); in the same sense, one may transcribe spoken English to written English, but it's still the same language! This transcription makes a copy of the recipe, but it still stays in the language of nucleic acids.

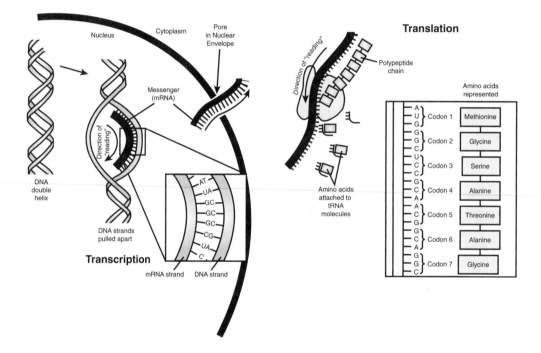

Figure 3.6

mRNA transcription and protein synthesis.

The next step occurs outside the nucleus at the ribosomes. (Free ribosomes make proteins for use in the cell, while those on the rough E.R.—and it is the ribosomes that make it "rough"—make proteins for export via transport vesicles). The mRNA molecule makes its way out of the nucleus—through some convenient pores in the nuclear envelope—to the ribosome. Some other RNA molecules called tRNA (transfer RNA) carry amino acids, once again courtesy of the digestive and cardiovascular systems, to the ribosomes.

Every group of three bases in DNA, conveniently called a triplet, codes for an amino acid. The three RNA bases that transcribed the DNA triplet are called an *mRNA codon*. Each mRNA codon matches with a specific *tRNA anticodon*, and that same tRNA carries a specific amino acid. You might remember, from Chapter 3, that amino acids are the monomers that make up protein polymers. In that way, DNA to mRNA to tRNA to amino acid, DNA acts as a recipe for the production of protein.

Do we make every protein all the time? Apart from using far too much energy, and not having enough amino acids to around, there are some very dangerous things that can happen if you make too much of some proteins. One example is that of insulin.

With too much insulin in the body, which sometimes happens when a diabetic waits too long after an injection to eat, your blood glucose drops so low that your body gets shaky and nervous, and ultimately your brain cells start to die.

Clearly it would be useful to have a way of turning a gene, such as the one for insulin, on and off. Every cell has the gene for insulin, but during development most cells turn that gene off permanently—all except the beta cells in the pancreas. Imagine being a diabetic and knowing that every cell in your body with DNA has the recipe to make insulin, but you *still* have to inject it!

How does the cell turn its DNA on and off? By a wonderful mechanism called an *operon* (see Figure 3.7). In order to make protein, one must first unzip the DNA using an enzyme called RNA polymerase. The RNA polymerase attaches to a place on the DNA molecule called the promoter which is before the start of the gene that needs to be transcribed. After the DNA is unzipped, the mRNA is made. But what if something were blocking the attachment?

After the promoter, and before the gene, is a place called the operator, which can be blocked by another protein called the active repressor. That active repressor is, of course, coded for by a gene on another part of the DNA. When that repressor is deactivated, the RNA polymerase does its job, and ultimately the protein is made.

The repressor always has an active site where another molecule can attach. In some cases, a repressor is deactivated when the molecule is in the active site; in other cases, the molecule in the active site activates the repressor. Think about lactose. To digest lactose you need to have the enzyme lactase. The only time you need lactase is when we need lactose. Do you see the connection? The lactose deactivates the repressor on the lactase gene. Once the lactase is made, the lactose is broken down, including the lactose on the repressor, thus reactivating the repressor and turning the gene off again. You will see more of operons when you see the feedback loops of the endocrine system.

Multiplying by Dividing

Growth is a lot more than just getting bigger, because as you learned in the discussion of surface area to volume ratio, being small is better. So in getting bigger you have an equal responsibility to be small. In order to grow you must therefore make more cells, each of which is small. A cell's simple solution to getting bigger is to divide itself into two equal halves; with each half being smaller, the surface area to volume ratio will be higher. That's a neat trick, and it's known as mitosis.

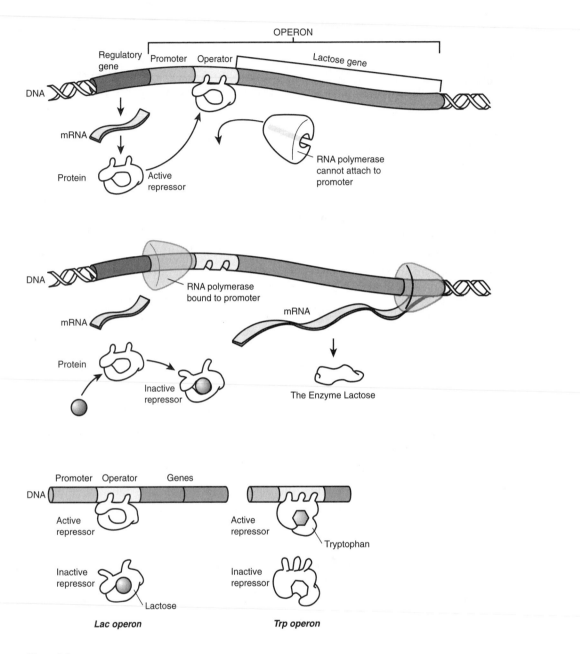

Figure 3.7

An operon is the means by which a cell turns its genes on and off. Note the two different types of repressors.

The only thing needed for each cell to survive, and then be able to divide in turn when it gets too big, is a complete copy of the DNA. Given DNA's wonderful ability to copy itself, the nucleus will be gradually copying all three billion base pairs of DNA. This period of growth and replication is the longest part of the cell cycle, and it is known as *interphase* (meaning the phase between separate episodes of mitosis).

DNA is so important, and so long, and so fragile, that it is important to keep the two copies from being damaged when the cell actually physically divides. The steps the cell goes through to ensure that are the phases of mitosis: prophase, metaphase, anaphase, and telophase (see Figure 3.8). The first phase is called *prophase*; in this phase uncoiled DNA molecules, called *chromatin*, is coiled up to form chromosomes, and the nuclear envelope, or nuclear membrane, disintegrates. This coiling enables the chromosomes to separate cleanly, without damage to any of the DNA.

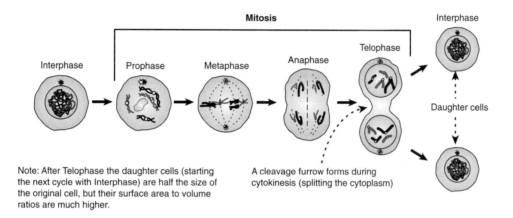

Note: After Telophase the daughter cells (starting the next cycle with Interphase) are half the size of the original cell, but their surface area to volume ratios are much higher.

A cleavage furrow forms during cytokinesis (splitting the cytoplasm)

Figure 3.8

The five phases of the cell cycle: interphase, prophase, metaphase, anaphase, telophase.

(LifeART©1989–2001, Lippincott Williams & Wilkins)

The next phase, *metaphase*, is when the chromosomes line up along the equator of the cell, or the middle of the cell. Each chromosome has a central portion called the *centromere*, which is a narrowing of the chromosome. The centromere is also the point of attachment for the spindle fibers. Spindle fibers are long chains of microtubules that connect to centrioles at each pole of the cell.

Think of these spindle fibers as a fishing line, with the centriole at the pole being the fishing pole, and the chromosome being the fish. When the spindle fibers shorten, the chromosomes will move toward opposite poles. This phase, where the chromosomes are pulled apart, is called *anaphase*. The final phase, *telophase*, involves the actual splitting of the cell into two distinct cells. Telophase has two basic parts: *cytokinesis*, or splitting the cytoplasm, and finally the uncoiling of the chromosomes and reformation of the nuclear membrane. At that point, that's all folks!

There is one other process that deserves mention here. Meiosis is similar in many ways to mitosis, except that the chromosome number is divided in half, from 46 (or 23 homologous pairs, meaning that they carry the same types of genes) to 23 (the pairs are split up). This is the process of gamete formation, or gametogenesis, and it only happens in the mature female's ovaries, and in the mature male's testes. I will discuss this in more detail in Chapter 23.

Respiration

One of the most important jobs of all cells is the harnessing of energy. Our primary energy source is glucose, which we derive from our food. There are two steps in the harnessing of energy from glucose: one, *glycolysis*, happens in the cytoplasm, while the other, cellular respiration, happens in the mitochondria. The role of the mitochondria makes sense when you compare cells, for those cells that have an extremely high energy need, such as muscle cells, have a larger quantity of mitochondria compared to other cells.

Prior to discussing glycolysis, I should say a little about adenosine triphosphate (ATP). ATP is the primary energy currency of the cell (see Figure 3.9). To understand the difference between that and glucose, think about going to get something to eat from a vending machine. Although you could use either a credit card or a check to pay in a local store, vending machines only take cash. In the same way, glucose is stored energy, which is converted to the cell's currency, or cash, ATP. Each glucose molecule broken down by the cell releases the equivalent of about 38 ATP's worth of energy, with the remaining energy (about 62 percent) released as heat.

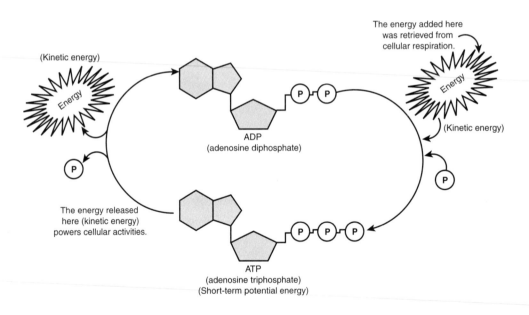

Figure 3.9

Diagram of ATP showing the release and gain of energy from the loss and gain of a phosphate group.

(©Michael J. Vieira Lazaroff)

What makes ATP useful as "cash" is related to its structure. As its name suggests, ATP is made of an adenosine molecule with three phosphates added. (See Chapter 2 for the structure of a nucleotide, for ATP is just a modified RNA nucleotide.) The ATP molecule releases energy when it loses a phosphate to become ADP, and it gains energy when it later gains a phosphate to become ATP again. This gaining of energy when a phosphate group is added is called *phosphorylation*. The beauty of such a setup is that both ATP and ADP and the phosphate group can be used over and over again. In this sense, you can think of ATP/ADP as a rechargeable battery.

ADP + Phosphate + Energy ↔ ATP

The first step in getting energy from glucose is *glycolysis* (*glyco* = sugar and *lysis* = to break), during which one glucose molecule in the cytoplasm, once again courtesy of the digestive and cardiovascular systems, is broken down into two molecules of pyruvic acid, which releases enough energy to make two molecules of ATP. If there is no oxygen (courtesy of the respiratory and cardiovascular systems), that is all of the energy that the cell can get from one molecule of glucose. Two ATP is not enough to power the cell, however, and I will explore that lack of oxygen in Chapter 9.

Following glycolysis is the Krebs cycle, which is a series of reactions in the mitochondria that provide energy, some of which is used in the next step, the electron transport chain. Instead of exploring the details of the Krebs cycle—which involves the juggling of various energy-carrying molecules, followed by the electron transport chain in and on the convoluted membranes of the mitochondria (whose shape maximizes the surface area to volume ratio)—I will talk about the importance of these reactions. The combination of the two reactions yields an additional 34 ATP, but those reactions require the presence of oxygen.

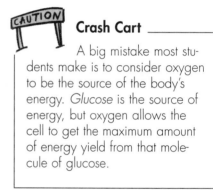

Crash Cart

A big mistake most students make is to consider oxygen to be the source of the body's energy. *Glucose* is the source of energy, but oxygen allows the cell to get the maximum amount of energy yield from that molecule of glucose.

What a difference breathing oxygen makes—2 ATP versus 38 ATP! If you think about it, this explains a lot. The brain dies when we stop breathing, not so much from the lack of oxygen, but from the lack of energy. When we exercise we need more energy; to get more energy means we need to get more oxygen—this is why we breathe more deeply when we exercise. To be out of breath? That means we can't get enough oxygen to meet our energy needs! This, as well as the extra energy we release in the form of heat when we burn sugar, will come up again when we talk about the muscular system, which is not only our major energy user but also our major source of body heat.

Disease Is Cellular in Origin

Given all the basic and specific functions of cells, it makes sense that diseases must arise out of a failure of cells to do their jobs properly. A problem with lysosomes is responsible for the effects of Tay-Sachs disease. A lack of specialized pancreatic beta cells is responsible for the effects of Type I diabetes. An extra chromosome 21 in every cell causes the effects of Down's syndrome. The list is seemingly endless.

With all that we have learned about the cell, and all that we are on the threshold of learning, medicine has made tremendous changes this past century, with many more changes on the horizon. Some of this has led to a field of study involving groups of similar cells (called tissues); this field of study, which is essential to understanding biopsies, is called *histology*.

Cell Junctions

Have you ever thought about how all of our cells stick together? I mean, without some sort of strong junctions, we would all fly apart every time we sneezed! To prevent this comically gruesome end, animal cells use three basic junctions: *anchoring*, *tight*, and *communicating junctions* (see Figure 3.10).

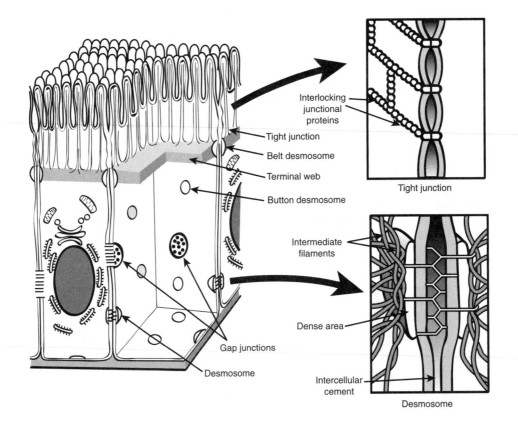

Figure 3.10

The types of cell junctions in a tissue are related to the function of that tissue.

Anchoring junctions simply hold cells together, but loosely. Strong protein fibers anchor the cells to one another, but just as an anchor connects a boat to the ocean floor via a chain, the cells are not in direct contact. Small spaces called *interstices* are filled with a fluid called *interstitial fluid*. This fluid is responsible for keeping the cells moist (necessary for gas exchange) and for the delivery and removal of materials to and from the cells.

This works fine for most organs, but it can be a mess if the cell's role is to hold fluids. Organs such as the urinary bladder and the gallbladder need a different type of junction, one that won't allow fluid to leak. These junctions, appropriately enough, are called *tight junctions*.

The last type of junction is called a *communicating junction*. This junction is so named because it is a tube-like opening between two cells that allows molecules through; these molecules carry instructions to the cells, thus communicating information to them. These junctions are very important in cardiac cells, allowing them to contract in unison.

Communicating junctions also tell cells that there are other cells nearby. This somewhat obvious piece of information is important in terms of mitosis, because a cell uses the junction to know that it must stop dividing. This is known as *contact inhibition*. Cancer cells don't have communicating junctions, so they keep dividing, creating pressure on neighboring tissues, causing pain and damage, and ultimately preventing the tissues and organs from doing their job. This effect of uncontrolled cell division, which is the true nature of cancer, is what makes it lethal.

> **The Big Picture**
>
> *Interstitial fluid* fills the spaces between our cells called *interstices*. This fluid, which leaves the capillaries (where it is called *plasma*) through diffusion and pinocytosis, circulates around the cells (delivering water, nutrients, and O_2, and removing wastes and CO_2) and is returned to the bloodstream via the lymphatic system (where it is called *lymph*). Interstitial fluid is the true contact between the blood and the cells.

The Least You Need to Know

- Cells are the basic unit of structure and function in our bodies, by virtue of their organelles.

- Cell membranes transport materials in and out of the cell, and also identify the cell based on the surface proteins.

- Molecules move in and out through either passive transport (diffusion and osmosis, which require no energy) or active transport (molecular pumps or endo- and exocytosis, which require energy).

- DNA is responsible for heredity because it can replicate and control protein synthesis.

- Cells get their energy by breaking down sugar (and other molecules) using oxygen; the energy yield is 19 times higher (38 ATP versus 2 ATP) when oxygen is available.

Chapter 4

Anyone Got a Tissue?

In This Chapter

- ◆ Tissues in medicine
- ◆ Epithelial tissues
- ◆ Connective tissues
- ◆ Muscular tissues
- ◆ Nervous tissues

As our knowledge of cells grew in the past century, so did our knowledge of tissues, which are simply groups of similar cells that work together performing the same task. This is the simplest form of teamwork in the body, and you will find that the sophistication of teamwork grows as you explore not only the connections between organs, but those between entire organ systems as well. At this larger level, the analogy of a team needs to be replaced with that of a symphony, with individual instruments playing in glorious counterpoint to produce a sound that exceeds the sum of the individual instruments.

In this chapter, I lay some of the groundwork beyond simple cells, to the configurations in which they are found. These configurations, or tissues, come in four distinct types and numerous subtypes, each of which is best suited to certain tasks in the body. In subsequent chapters you will see organs described in terms of their tissues, and it will become apparent that, with few exceptions, organs are composed of examples of all four basic tissue types: epithelial, connective, muscular, and nervous.

Biopsy Central

The field of *histology* is the study of tissues, and far from an obscure branch of medicine, it is a part of everyday diagnosis. Whenever a patient has a surgical biopsy, outpatient or inpatient, the tissue needs to be prepared for viewing under a microscope. The tissue is then examined and its structure is compared to known features of both normal and abnormal tissues in order to make a diagnosis. Diagnosis of cancer requires this use of histology, for tumors can be either benign (noncancerous, but not necessarily harmless, depending on its impact on neighboring tissues) or malignant (cancerous), and the difference can be seen under the microscope. A biopsy will not show only one tissue, for organs are made up of all four types of tissues (see Figure 4.1). One such example is skin. The part you see when you look in the mirror is epithelial tissue, but underneath there is a network of connective tissue and nerves. Muscles might surprise you in terms of their location: They attach to the hairs. The hairs? Of course! How else do you get goose bumps?

Figure 4.1

Organs, such as skin, contain all four of the tissue types: epithelial, connective, muscular, and nervous.

(LifeART©1989–2001, Lippincott Williams & Wilkins)

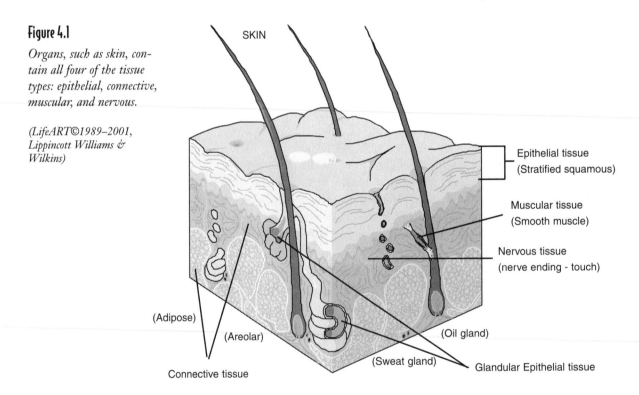

SKIN

Epithelial tissue
(Stratified squamous)

Muscular tissue
(Smooth muscle)

Nervous tissue
(nerve ending - touch)

(Adipose)

(Areolar)

(Oil gland)

(Sweat gland)

Glandular Epithelial tissue

Connective tissue

Microscopy was an enormous advance in medicine. The process of slide preparation takes approximately three days, which explains the delay for accurate biopsy results. What follows is a condensed version of the process. Whenever tissue is harvested, one of the first steps is to prevent decay. To that end, the tissues must be dehydrated in a series of either formaldehyde solutions (formalin) or alcohol solutions.

Next, the tissues are embedded in melted wax, which is then allowed to harden. Finally, the tissues are cut into *extremely* thin slices with a microtome. You can think of a microtome as the lab version of a deli slicer. Unlike a deli slicer, however, a microtome has a fixed blade, and it advances the tissue forward at a very precise rate to allow uniformly thin slices (as little as 3 microns, or micrometers).

Medical Records _____

This is all well and good, but three days is a long time to wait! If you are performing cancer surgery, it is imperative that you know you have removed all the cancer and that the borders of the tissue you have removed are "clean" (no cancer cells). If not, the cancer has spread past the tumor you have removed. If you don't find this out until after surgery, you have to schedule another surgery. A frozen section (using liquid nitrogen) can be done in under an hour, allowing the surgeons to wait for the results before continuing.

The tissues will be observed through a light microscope, and if the tissue sample is too thick the extra layers will block out the light that is needed to reach the eye. Individual cells are translucent (blocking out some of the light), but with too many layers they appear opaque. But why the wax? Have you ever tried to cut a tomato, only to have it crush because of the hollow area with the seeds? The wax permeates the microscopic crevices (interstices, or interstitial spaces) between the cells, supporting the tissue from the pressure of the microtome blade, thus preventing the tissue from deformation by the blade and allowing it to be seen as it appears in the body. Before the slide can be made, however, the thin tissue slice needs to be placed flat on the slide, which is not as easy as it sounds. By floating the tissue slice in warm water, you can scoop it up with a slide, and as long as there are no folds or wrinkles, which will prevent the cover slip from laying flat, you can continue. At this point the wax needs to be chemically dissolved, which will then allow the stain to permeate the tissue, highlighting the tissue for view under the microscope.

Paste a cover slip on, and you are almost done. The final step, once the slides are prepared, is perhaps the most difficult. At this point, the tissue must be examined for abnormalities, and this requires a trained eye and dedication, for abnormalities are often subtle.

Who's on Top?

Epithelial tissues, which make up 3 percent of your body weight, differ from the other three in some significant ways (see Figure 4.2). They don't move (unlike muscular tissue), they don't send messages (unlike nervous tissue), and their cells are all touching one another (unlike connective tissue). They are the most widely varied in terms of structure and function in the body.

Figure 4.2

Epithelial tissues have two surfaces, one free, and one attached to connective tissue through a basement membrane.

(©Michael J. Vieira Lazaroff)

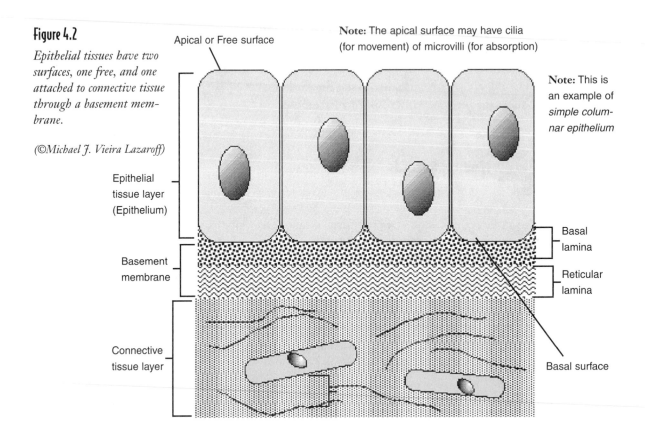

As to the basic layout of epithelial tissue, it can always be found on the inside of hollow organs, and the outside of *all* organs, above a connective tissue layer (*epi* = above), with a basement membrane in between.

The basement membrane consists of a basal lamina (with collagen and other substances secreted by the epithelial layer), and a reticular lamina (with reticular fibers, and other substances secreted by the connective tissue layer). Since epithelial tissue is avascular (meaning it contains no blood vessels), the connective tissue layer is essential so that the epithelial cells can get their nutrients from the blood vessels in the connective tissue beneath.

Simple, Stratified, and Pseudostratified

Connective tissue cells come in three basic shapes: *squamous, cuboidal,* and *columnar*. If you can't use the names here, you just aren't trying! With a name like *cuboidal*, you shouldn't be surprised that the cells look like *cubes! Columnar* should be just as easy, as the cells are longer, making them shaped like *columns!* The last shape is a bit harder. *Squamous* cells are like scales, or pancakes. If you can remember "being *squashed* like a *pancake*," then you'll remember the *pancake* shape of the *squamous* cells.

Shape, alone, is not enough, however, as cells of the same shape are organized in different ways. A single layer is called *simple*. Thus there are simple squamous, simple cuboidal, and simple columnar epithelial tissues (see Figure 4.3). Layers of cells are called *stratified* (see Figure 4.4). The skin, for instance, is stratified squamous, but there are also stratified cuboidal and stratified columnar epithelial tissues. Stratified columnar epithelial tissues are also divided into ciliated (with cilia) and unciliated varieties.

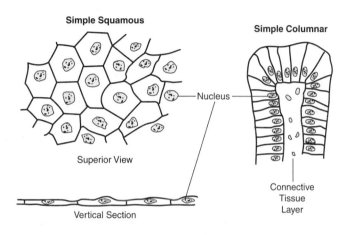

Figure 4.3

The three simple epithelial tissues.

(©2003 www.clipart.com)

Figure 4.4

Stratified epithelial tissues.

(©2003 www.clipart.com)

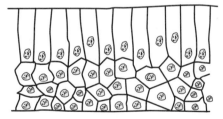

Sounds easy enough, so I might as well throw a wrench into the mix. There are two types of tissue that look like they can't make up their mind—one of them simple, one of them stratified. Transitional epithelium is a stratified tissue that looks like it can't make up its mind as to whether it is squamous or cuboidal. Stratified squamous epithelium starts off more rounded, and then starts to flatten toward the apical surface, but transitional epithelium fails to flatten, hence looking more cuboidal (see Figure 4.5).

Pseudostratified columnar epithelium looks like it has more than one layer because of the position of the nucleus. Simple columnar epithelial cells have their nuclei at close to the same level, but pseudostratified have them at widely differing levels, with the cell narrowing in the area without the nucleus. Under the microscope this gives the appearance of being stratified, hence pseudostratified (see Figure 4.5).

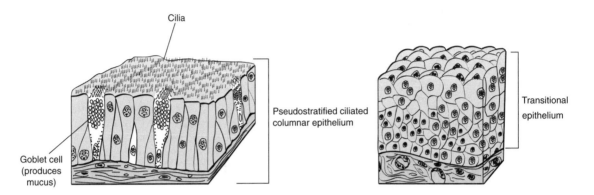

Figure 4.5

Pseudostratified and transitional epithelial tissues.

Membranes

Membranes are everywhere. The inside and outside linings of organs are membranes. These vary from the mucous membranes pumping out enzymes and mucus in the walls of the small intestine to the glistening serous membrane on the outside of the same organ, preventing its movement from creating so much friction that the organ would wear itself raw (SCRAPE!), not to mention the membrane you see when you look in the mirror—your skin! As the following table illustrates, these membrane types perform a wide variety of functions in the body.

Types of Epithelial Membranes

Type	Location	Function	Examples
Mucous	Lining of tubes	Moisten and protect from enzymes	Stomach, trachea, vagina
Serous	Outside organs	Lubrication	All thoracic, abdominal, and pelvic organs
Cutaneous	Body surface	Protection	Skin
Synovial	Synovial joints	Line and protect synovial cavity	Elbow, knee, and so on

Medical Records _____

There are three types of friction in this world: sliding, rolling, and fluid friction. Any type of friction (resistance against a moving object) involves the transfer of kinetic energy (motion) into heat. Too much heat can damage tissues. The greatest amount of friction is sliding friction. Dragging a box would ultimately produce enough friction to wear through the cardboard; an example would be the blister from a long hike in bad boots. Less heat is produced in rolling friction, but there is enough to increase the pressure in your tires after a long drive; the human body does not use rolling friction. The least amount is fluid friction, as in the oil that lubricates cars; this friction is used *all over the body!*

Glands

The body produces a large amount of fluid for various purposes: cooling, protection, lubrication, breaking down food, and feeding young. These fluids need to be released either into interstitial fluid (see Chapter 3) or directly into organs. Given that epithelial tissues line organs, they are the logical tissue for the makeup of glands. All this pumping and dripping and secreting happens in two basic types of glands: endocrine glands, which release fluids into the interstitial spaces, and exocrine glands, which release fluid, and cell parts (!), or both into ducts (see Figure 4.6). The types of glandular secretions can be found in the following table.

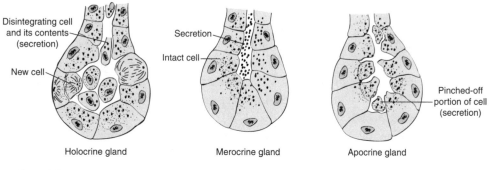

Figure 4.6

Glands differ from one another not only in terms of the contents of the secretions, but also in terms of the effect on the cell as it releases the secretions.

Types of Glandular Secretions

Type	Mechanism of Secretion	Examples
Endocrine	Into interstitial fluid	Hormones (for example, thyroid)
Exocrine	Into ducts	Merocrine, apocrine, holocrine
Merocrine	Exocytosis (from Golgi apparatus) cell intact	Mucus, sweat, tears
Apocrine	Apical part of the cell pinches off, regrows	Milk, axillary sweat
Holocrine	Entire cell ruptures	Sebaceous (oil) fluid

Muscle Beach

Muscle tissue, which makes up a full 50 percent of body weight, is a very special tissue in its ability to move. There are three types of muscle tissue: smooth, skeletal, and cardiac. The location of at least two of the three is obvious from the names—skeletal muscle attaches to bones to provide movement, and cardiac muscle is within the heart—but where is smooth muscle found? Have you ever gotten goose bumps? How did your hair stand on end? Have you ever leaned over a low drinking fountain to drink? How did the water make it uphill to your stomach? The muscles responsible for those, and hundreds of other small actions, are smooth muscles. The next question would be, how do you tell them apart, apart from the Realtor's credo of location, location, location?

Voluntary and Involuntary

The first division has to do with the conscious control of the muscles. On the one hand, if you put your hand on a hot burner, it is probably a good idea if you can respond before burning your skin! Muscles in charge of gross movement, also known as skeletal muscles, are under voluntary control. On the other hand, I don't want to have to think about when I should be releasing bile from my gallbladder. Those muscles, which are smooth, are involuntary. If you forgot about having your heart beat, you would be dead in minutes. Luckily, those cardiac tissues are also involuntary.

Striated and Smooth

The table and image illustrate how the three different types of muscle tissue are organized. Skeletal and cardiac tissues are easily seen as relatives because of their overall appearance; both types have evenly spaced dark stripes, or striations, that run perpendicular to the length of the muscle cell. These striations are actually the units of muscle contraction called *sarcomeres*, which I discuss in detail in Chapter 8. Cardiac muscle cells are short and branching, unlike the long cells in skeletal muscle.

Striated and smooth muscles both have the same contractile proteins, but the arrangement in striated muscles is so heavily organized that its contractions are much faster and stronger than that of smooth muscle. A quick look at the following table reveals that the strength and speed of the contraction go a long way toward explaining why certain muscle tissue types are found where they are. Figure 4.7 illustrates the three types of muscle tissue.

Characteristics of Muscle Tissues

Tissue	Nervous Control	Cell Appearance	Examples
Skeletal	Voluntary	*Striated*, long, multinucleated	Biceps brachii
Cardiac	Involuntary	*Striated*, short, branched	Heart
Smooth	Involuntary	*Smooth*, short, tapered ends	Gastrointestinal tract

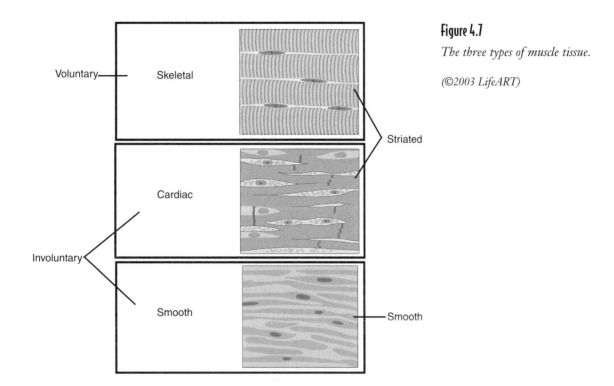

Figure 4.7

The three types of muscle tissue.

(©2003 LifeART)

Let's Get Connected

One of the most ironic things about connective tissue (which makes up a surprising 45 percent of body weight) is that the cells, with few exceptions, are not actually physically connected! The tissue gets its name because of its function: connective tissues in the body connect other tissues and body systems. This should make sense, because in the digestive tract (see Chapter 14) the connective tissue layer (submucosa) connects the epithelial mucosa to the muscularis (muscle layer) beneath. The cartilage in tendons connects muscle to bone (itself a connective tissue), and the cartilage in ligaments connects bone to bone. Blood vessels are imbedded in connective tissue throughout the body. Blood, which is itself a connective tissue, connects all of the body systems to one another (perhaps making it *the* connective tissue!).

The Matrix

With all these different types, it's probably a good idea to see what they have in common. Connective tissues are made of widely separated cells in a matrix. Although there are a number of different matrices, they can still be divided into three basic types—fluid, fibrous, and crystalline—and the type of matrix determines the function of the tissue.

Blood

A fluid matrix makes sense when you think of blood; the plasma in the blood (part of the cardiovascular system) links the blood cells (both red and white). This enables the tissue to be highly mobile and to squeeze through the blood vessels regardless of size. The fluid, which is mainly water (see Chapter 2), enables blood to carry an enormous variety of materials throughout the body (due to water's excellent ability as a solvent). The fluid matrix of blood makes circulation possible, making blood the epitome of all connective tissues (see Figure 4.8).

Figure 4.8

Blood is special in terms of its fluid matrix.

(©2003 LifeART)

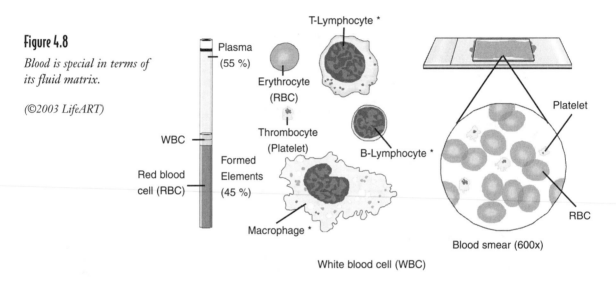

Bone

Bone is the opposite of blood in that its matrix is crystalline (see Figure 5.2). This hard, rigid matrix is crucial to the primary functions of bone: to provide support for the body, and attachments for muscles to pull, thus making movement possible. This rigidity is accomplished through a matrix of calcium crystals: tricalcium phosphate ($Ca_3(PO_4)^*$ $(OH)_2$)), and calcium carbonate ($CaCO_3$) crystals. This use of calcium and phosphorus has an added benefit, in that bone makes a perfect calcium reserve. This is important because, as you will learn in Chapter 8, calcium plays a crucial role in muscle contraction.

Don't Be Dense!

Cartilage, and other dense connective tissues, has a very different role from other connective tissues. The role of dense connective tissue is to be firm, but flexible. The body—with all of its bending, and flexing, and turning, and pulling—requires tissue with a certain amount of elasticity. Such tissues have a fibrous matrix, and the nature of the fibers determines how strong and how flexible such tissues are.

Hyaline cartilage, which you might know as gristle, is at the end of long bones, and provides flexibility at joints. It is known for its fine collagen fibers, which are not visible under a light microscope. Dense regular connective tissue (see Figure 4.9), which is in tendons, has strong, wavy, parallel fibers, which explain its strong elasticity, but in only one direction. Dense irregular connective tissue, such as the dermis of the skin, also has tough fibers, but the orientation varies. This is a great help in keeping skin extremely flexible, despite the multiple directions that it can be pulled! Elastic connective tissue, on the other hand, does not have as much organization in its fibers, making it less firm; this is very helpful in its role in the lungs, for if it were too firm, breathing would be too difficult.

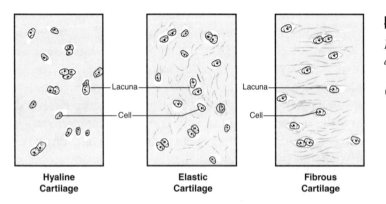

Figure 4.9

Forms of cartilage and other dense connective tissues.

(©2003 www.clipart.com)

Fat and Fascia

Adipose tissue (fat) is an interesting tissue because fat cells are basically all vacuole. The nucleus and cytoplasm are all around the edge of a large vacuole for the storage of fat molecules. You tend to deposit fat wherever you have fascia: under the skin, around the heart (the pericardium), along the intestinal mesentery, and so on. Some might call these fat cells evil because they don't disappear when you lose weight, but rather they simply release the contents of the vacuoles, content to wait until you lose your resolve, open that bag of potato chips, and rebuild the fat deposits!

Fascia, the friend of fat deposits (how alliterative!), appears in three different varieties, based on its location. Superficial fascia—also called *subcutaneous* (*sub* = under and *cutaneous* = skin) or *hypodermis* (*hypo* = under and *dermis* = skin)—is found under the skin, at the point where the skin attaches to the muscle. Using the directional terms discussed in Chapter 2, the deep fascia is called, uh, *deep fascia*! This fascia goes directly around the muscle and merges with the tendon, as you might have noticed when you eat a chicken drumstick! The last fascia is called subserous, which lies under the serous membrane, between it and the deep fascia.

Don't Be Nervous!

The body is so large that without the ability to communicate with itself, the right hand literally would not know what the left hand was doing! To accomplish this you have nervous, or neural tissue, which makes up only 2 percent of body weight, making it the least common tissue in the body. The chief ability of this tissue is the conduction of chemical impulses through polarizing and depolarizing membranes. (I will discuss this in more detail in Chapter 19.) But all the work of conduction means that these nerve cells also need help.

Neurons

Neurons are the glory of the nervous system, the cells you think about when you think about the brain. The basic shape consists of a roundish cell body with long, branching extensions on either end; these extensions are called axons and dendrites (see Figure 4.10). The three parts together can make the entire cell extremely long (up to 2 meters, or over 6 feet!). Although they look fairly similar, there is one significant difference: A dendrite carries nerve impulses *to* the cell body, while an axon carries the impulses *away from* the cell body. This difference means that each neuron can transmit in only one direction.

Figure 4.10

A neuron may do the "thinking," but it depends on the labor of the neuroglia to survive.

(©2003 LifeART)

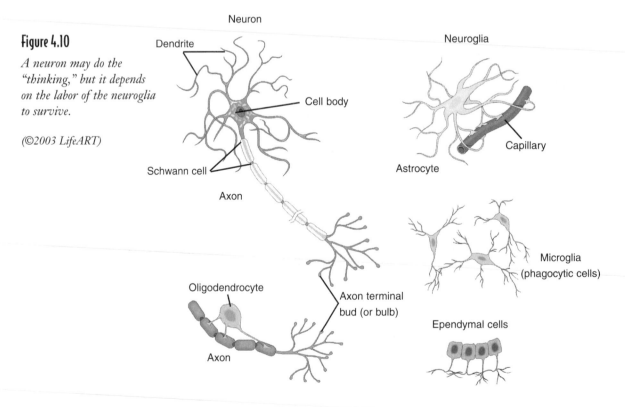

Support Staff

Have you ever heard the hackneyed expression, "you only use 10 percent of your brain"? Although that seems reasonable considering some people I've met, and certain political figures definitely come to mind, it isn't true. This is based in part on the fact that only a small percentage of neural tissue is actually neurons, which are doing the thinking, the control of the body, the sensing, and so on. The rest of the cells, a sort of neural proletariat, are support cells.

These support cells, called *neuroglia*, perform a variety of functions. They have a lot to do with maintaining the structure of neural tissues. In addition, in their supporting role, they provide nutrition to the neurons, and they clean up the tissues through phagocytosis. A special type of neuroglia called a *Schwann cell* is important in nerve cell transmission because it acts as insulation similar to the insulation on a piece of electrical wire.

The Least You Need to Know

- An understanding of normal and abnormal varieties of each tissue, the essence of histology, is important in the analysis of biopsied tissues.

- Epithelial tissues line organs, and are the tissues from which all glandular tissues are made.

- Epithelial tissues are classified according to their shape (squamous, cuboidal, columnar) and configuration (simple, stratified, pseudostratified).

- Connective tissues link other tissue types, and the function of each type is based on the nature of each one's matrix: fluid (blood), flexible (cartilage), and crystalline (bone).

- Muscular tissues are responsible for the body's movement, and they are classified by appearance (smooth or striated) and control (voluntary or involuntary).

- Nervous tissue consists of neurons, which transmit impulses, and glial cells, which support the neurons.

Part 2

If the Foundation Is Good

Many times people find that most of their organs seem to be working just fine. We are most likely to feel problems in our foundation: bones and muscles. If you have ever pulled a muscle or broken a bone, you know what I mean; if you haven't, you've got to get out more! As animals we have to move to get our food, unlike plants who sit there and make their own. Movement is based on a cool collection of 206 Halloween favorites (our skeleton), the connections between the bones (our joints), and the meaty means to move them (over 600 muscles).

So let's sharpen those knives, cut down to the bone, and get right to the meat of the problem!

Cut to the Bone

In This Chapter

- ◆ The functions of the skeleton
- ◆ The part of a long bone
- ◆ The types of bones
- ◆ Bone development, growth, remodeling, and repair
- ◆ The types of fractures

Skeleton is an interesting word; big fat weddings aside, the name comes from the Greek word *skeletos*, which means "dried up body." This might make sense when you see desiccated mummies with skeletal features due to the loss of moisture, but one look at the true nature of the skeleton and you will see that this is all wrong. Bone is actually very much alive, very moist (50 percent water), and ever-changing.

Some bones grow for your entire life, while others stop growing in your twenties. In order to do their job, bones come in an astonishing variety of shapes. The number of ways bones can break is just as varied. These incredible living organs have a remarkable capacity for fixing themselves, and the marks they leave bear silent witness to our fractured pasts.

Function Junction

In the Oscars there's always an interesting separation between Best Actor and Actress and Best *Supporting* Actor and Actress. Unlike glitzy awards shows, however, it should be known that bones do more than act in the shadows in a supporting role. The functions of the skeleton include …

♦ **Support.**   An organism our size would not be able to stand without the strong framework provided by bone.

♦ **Protection.**   As you will see in Chapter 6, there are numerous delicate structures—brain, spinal cord, and so on—that rely on the skeleton to protect them.

♦ **Motion.**   In providing a place for muscle attachment, for both the moving bone and for the stationary bone, the skeleton makes movement possible.

♦ **Mineral storage.**   As I discussed in Chapter 4, the crystal matrix (mostly calcium and phosphate crystals) of bone provides an ideal storage site, one which is drawn on and added to, as needed, to maintain homeostasis.

♦ **Energy storage.**   If you have ever seen someone crack open a chicken bone and slurp on the marrow, then you have seen, first hand, the fat-rich yellow marrow (rich in adipose tissue).

♦ **Hemopoiesis.**   Whaaaat? Wait a minute, *hemo*, as in hemoglobin? Yes. *Hemopoiesis* is the production of blood cells (both red and white). This occurs in the *red marrow*.

The Parts of a Bone

Most people envision bone as being uniformly solid, but nothing could be farther from the truth. For one thing, as you will see later in this chapter, bones come in many different shapes—long, short, flat, irregular, wormian, and sesamoid—which have much in common, despite their differences. A typical bone can be broken down into multiple parts, each with a particular function:

♦ **Epiphysis.**   This part is at the extreme ends of the bone (*epi* = above), where joints (articulations) form.

♦ **Articular cartilage.**   A layer of hyaline cartilage, called *articular cartilage*, exists to reduce friction and absorb shock at synovial joints (see Chapter 7).

♦ **Diaphysis.**   The shaft of a long bone, which is the direction at which the bone can withstand the most stress.

♦ **Metaphysis.**   The metaphysis is the place where the diaphysis meets the epiphysis. This is where major bone growth occurs, as well as where blood enters the bone.

♦ **Periosteum.**   A thin membrane that covers the outside of the bone, where tendons and ligaments attach to the bone. The outer fibrous layer is where blood vessels, nerves, and lymphatics connect to the bone, while the inner osteogenic layer has bone cells necessary for the growth and repair of bone.

♦ **Medullary (or marrow) cavity.**   This hollow cavity, in the diaphysis, is for the storage of yellow marrow.

♦ **Endosteum.**   This membrane lines the medullary cavity, and contains *osteoprogenitor* cells (unspecialized bone cells, as you will soon see).

Up, Down, and Middle

As you can see in Figure 5.1, the shaft of a long bone is called the diaphysis. The central, fat-storing marrow cavity is found inside the diaphysis. At each end of the bone, at the site of the synovial joint, is an area called epiphysis. At the juncture between the two is an area called the metaphysis.

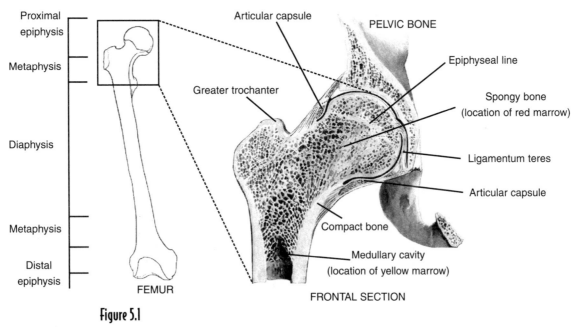

Figure 5.1

The many parts of a typical long bone. The example shown here is a femur.

(©2003 www.clipart.com)

Remember that organs, including bones, need three connections: blood vessels (both arteries and veins), lymphatics, and nerves. These structures enter the bone through little holes called *foramina*. A hole specifically for blood vessels is called a nutrient foramen (the singular form of *foramina*). Any student can tell if a skeleton is real by simply looking for foramina around the metaphysis. Another clue is the weight: Real bones are lighter than solid models, due to the openings for red and yellow marrow.

Beyond the entering and exiting nerves and vessels, the metaphysis is also the location of the *epiphyseal plates*, which are the primary growth centers of a long bone. There are four zones in the epiphyseal plate. The *zone of resting cartilage* is not involved in growth, but it does anchor the plate to the rest of the bone. The *zone of proliferating cartilage* and *zone of hypertrophic cartilage* are both involved in producing chondrocytes (cartilage cells, see Chapter 4), but the latter zone is where maturation of the cells occurs. The last zone, where the bone actually forms, is known as the *zone of calcified cartilage*.

As we age, the epiphyseal plates, which are less dense than bone and show up darker on an X-ray, will *ossify* (turn to bone), at which point they will appear as a light line (called the *epiphyseal line*). This marks the end of a bone's ability to grow longer; this ossification is usually complete by the early to mid twenties (although the sternum doesn't finish until after 30). The facial bones, and often the hands and feet, however, do not stop growing, which explains why a young Jimmy Stewart looked very different than he did as an old man.

The Big Picture

A certain pituitary disorder involves the overproduction of human growth hormone, or hGH. In a child, this results in gigantism, whereas too little hGH results in one form of dwarfism (other forms are caused by either extreme malnutrition or, in the case of *achondroplasia*, a dominant gene). As an adult, due to the formation of the epiphyseal line, the bones of the face, hands, and feet will enlarge dramatically. This condition, which is seen in certain movie villains, is called *acromegaly*.

The Harder They Come

Compact bone is notable for the wide spacing of the cells within a hard crystal matrix (see Figure 5.2). You may remember from Chapter 4 that both wide spacing and a matrix were characteristics of connective tissue. The main feature of compact bone is its strength. It provides protection for places outside a soft structure, such as in the flat bones of the skull. Compact bone also supports the stress placed on it. In a long bone, the stress is best absorbed along the longitudinal axis of the diaphysis. This arrangement is great for a bone like the femur, which absorbs stress in that direction, but the same cannot be said for the clavicle, which can be easily fractured if it receives a downward blow perpendicular to the diaphysis.

Microscopically, compact (or dense) bone is distinguished by its arrangement of osteocytes (bone cells) in concentric circles of matrix. Just as people settle around sources of water, these rings, or concentric lamellae, are arranged around a central haversian canal, which holds blood vessels. The combination of the concentric lamellae and the haversian canal is called an osteon, or haversian system. In addition to the haversian canal, there are perpendicular ones called perforating canals that connect haversian canals, and help to provide blood not only to the deeper haversian systems, but also to the marrow cavity.

The osteocytes look a little like ants because of the arrangement of little canals called canaliculi around each cell; these canaliculi, whose name always makes me think of an Italian dessert, are where the interstitial fluid is found (see Chapter 3). Canaliculi extend outward in every direction from the lacuna (see Chapter 4), which is the space where the osteocyte is found.

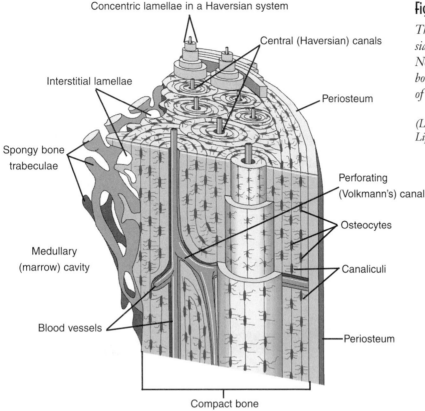

Concentric lamellae in a Haversian system

Central (Haversian) canals

Interstitial lamellae

Periosteum

Spongy bone trabeculae

Perforating (Volkmann's) canal

Osteocytes

Medullary (marrow) cavity

Canaliculi

Blood vessels

Periosteum

Compact bone

Figure 5.2

This is a diagram of haversian systems in compact bone. Note the organization of the bone is based on the location of blood vessels.

(LifeART©1989–2001, Lippincott Williams & Wilkins)

Not Just For Mopping Up Spills

Spongy or *cancellous bone* is very different in appearance. Rather than rigid concentric systems, spongy bone looks, well, spongy. The appearance is due to an irregular collection of overlapping and interconnected spokes called *trabeculae* (refer to Figure 5.2). To understand the function of spongy bone, note that it appears most commonly in the epiphysis, just under a protective compact layer. The compact layer provides firm attachment for that articular cartilage, both of which help to protect from the friction found in every synovial joint.

So why the spongy part? In terms of stress at the joint, imagine jumping in the air and landing hard on your feet while keeping your legs straight; a great deal of stress will be felt not only in your knees, but also where your femur articulates with your pelvis (see Chapter 6), not to mention in your back. You can easily reduce the stress by bending your knees and ankles; such bending absorbs the stress of the impact. Now do you know the reason for spongy bone? That's right, to absorb some of the shock of impact at synovial joints.

The screwy multidirectional trabeculae make it possible to absorb stress from multiple directions. In addition, the spaces between the trabeculae make spongy bone much lighter, thus making the skeleton as a whole much lighter. These spaces serve another purpose; they are filled with red bone marrow, the site of hemopoiesis.

That's the Long and Short of It

A lot of school kids learn that they have 206 bones, but most never learn that number of bones may vary from person to person! The troublemakers are two types of bones that grow in unusual places, as a result of our development. One type is called *sesamoid*, because they are flat, and usually small, bones shaped like sesame seeds; sesamoid bones develop in tendons around the hands, feet, and knees (in all, a total of 26 locations). The number of these varies, but we all have a pair in our knees: the kneecaps, or *patellae*.

The other type is called suttural bones (or wormian bones), because they form in the connective tissue when cranial bones form sutures. These bones form as little islands in the developing sutures. A quick look at a suture on a real skull (rather than a model) often reveals multiple small wormian bones. The rest of the more "garden variety" bones are listed in the following table (and in Figure 5.3).

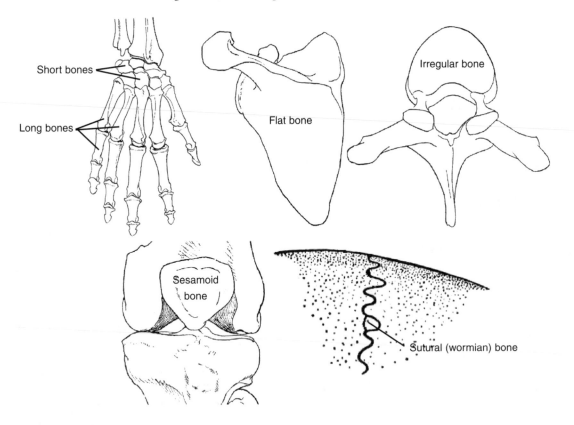

Figure 5.3

Examples of each of the bone types.

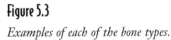

(©2003www.clipart.com)

The Four Major Bone Types

Bone Type	Description	Location
Long	Long, slender	Arms, legs, phalanges
Short	Short, box-like	Carpals, tarsals
Flat	Thin, parallel surfaces	Skull, ribs, scapulae
Irregular	Complex, with projections	Vertebrae and skull

The More Things Change ...

Bones can be very surprising. The simple fact that bones can *repair themselves* seems almost miraculous! Imagine, if you can, being able to pay a little more for a car that actually fixes itself! Simply put, bones are not the dried up sticks you thought they were, but rather complex and dynamic structures.

Bone Development

Bone development is a multistep process from hyaline cartilage to compact and spongy bone. The process is a bit like the opposite of certain hard candies, for bone starts out soft, with the hard part starting in the center:

1. Bone starts as hyaline cartilage. Chondrocytes in the diaphysis enlarge and then die, leaving cavities in their wake.

2. Perichondral cells (around the cartilage) turn into osteoblasts and begin to form a bony outer layer, and blood vessels start to grow around the edge of the developing bone.

3. Blood vessels start to grow in the central region of the bone, and fibroblasts become osteoblasts in the same region and start to build spongy bone in a region called the *primary ossification center*. This process spreads toward the epiphyseal regions.

4. The spongy bone in the center is broken down by osteoclasts, forming the marrow cavity. The bone grows longer and wider.

5. Osteoblasts and capillaries begin to infiltrate the epiphyses, producing *secondary ossification centers*.

6. The cartilage in the epiphyses is replaced by spongy bone, and the epiphyseal plate forms in the metaphysis. Eventually compact bone forms around the perimeter of the bone, leaving the epiphyseal plate as the area for bone growth.

Making and Breaking

The histology of bone is cool because people think of growth as a constructive process. Growth makes perfect sense because bones get longer, but what about the width? The width of a bone increases, but at the same time something has to happen to the medullary (marrow) cavity. If the medullary cavity walls all grew inward, the cavity would fill up during childhood. In order to have a medullary cavity the same relative size as the bone grows, the width of the cavity must increase. The only way to do that is to literally destroy the cells that line the walls of the cavity.

Medical Records _____

One of the coolest things about our fetal development is what happens to our tail! As you will see in Chapter 7, we do indeed have a tail! Our tail, known as the coccyx, is extremely short (between three to five fused vertebrae). When we are an embryo we grow a longer tail, only to destroy it during development! Why waste all that time and energy? The answer? Evolution! Our evolutionary ancestors had tails, and some of our development contains vestiges (vestigial structures) from our evolutionary past!

All of these changes require different cell types. Since bone is basically connective tissue, we should talk a little about the matrix that separates the cells. The matrix in bone is about $1/2$ mineral salts, $1/4$ protein fibers, and $1/4$ water. Water? Yes! Bone cells are alive, and to stay alive they need to have contact with the cardiovascular system. Each cell is bathed in interstitial fluid, which flows throughout the haversian system in all the canaliculi described previously.

There are four types of bone cells. First of all, connective tissues all derive from an un-differentiated form called *mesenchyme*. From this mesenchyme there is an undifferentiated bone cell called an *osteoprogenitor cell*. These cells are found in both the periosteum and the endosteum, as well as the various canals that carry blood throughout bone.

Osteoprogenitor cells divide (through mitosis, see Chapter 3) and can develop into bone building cells called *osteoblasts*. Although they cannot divide, they can secrete collagen necessary for bone building. As they build bone and become surrounded by the matrix, they eventually stop making collagen and become the main cells in bone: *osteocytes* (osteo = bone and *cyte* = cell).

The last type of bone cell is thought to come from a type of white blood cell called a *monocyte*. These cells migrate all over the body, eventually becoming macrophages (see Chapter 3). Some of these end up in bone, and become somewhat iconoclastic *bone-breaking* cells called *osteoclasts*. It is the osteoclasts in the endosteum that destroy the matrix of the bone lining the medullary cavity, thus making it wider as the bone grows larger.

Out with the Old ...

Osteoblasts and osteoclasts are busy little critters. Due to the need for a certain level of calcium in the body, especially in terms of muscle contraction (see Chapter 8), the matrix in bone needs to be continually broken down (to release calcium) and rebuilt (to deposit calcium). This continual activity is called *remodeling*.

The Big Picture

Two hormones are responsible for calcium homeostasis: parathyroid hormone (PTH) from the parathyroid glands increases blood calcium levels, and calcitonin from the thyroid decreases blood calcium levels. PTH *stimulates* osteoclasts to break down the matrix, thus releasing calcium. Calcitonin, on the other hand, *inhibits* osteoclasts. This leaves the osteoblasts free to deposit the calcium in bone. As such, which hormone would you expect to be prescribed to patients with osteoporosis? That's right, calcitonin!

Remodeling usually occurs at a balanced rate, which means for every osteon broken down, another is rebuilt. The rate of turnover varies according to the location of the bone, as well as the part of the bone, but about 20 percent of the bone tissue in a young adult is remodeled in a year. Areas of bone that receive a lot of stress, such as the spongy bone in the head of the femur can be replaced in anywhere from 4 to 6 months, but the compact bone in the diaphysis, which receives far less stress, is hardly changed at all.

Medical Records

Bones respond to stress by increasing osteoblast activity. A highly active person will have more pronounced bone landmarks, particularly where muscles attach, due to the increased stresses there. One unusual thing about the bone matrix crystals is that they produce a tiny electrical field when stressed, and osteoblasts are attracted to the field, and once there they start to produce bone. Doctors have used this to their advantage by stimulating bone growth electrically in particularly bad fractures.

Given the increase in bone density in response to stress, an active person can build stronger bone. Young women who are very active can actually reduce their risk of developing *osteoporosis* (the thinning of bone matrix, most commonly in older women) in their later years. One cause of osteoporosis is not having enough dietary calcium during pregnancy; the baby's need to build bone will cause the mother to draw calcium from her own bones, causing them to become more porous—osteoporosis means porous bone.

Another far less common way to reduce bone density is to become an astronaut! In order to mimic the effect of gravity on bone, an astronaut would have to exercise for about eight hours a day to prevent bone loss; as that is not possible, a typical astronaut is asked to exercise only two hours a day! For this very reason, physical therapy for women with osteoporosis involves a great deal of weight-bearing exercises; the greater the stress, the busier the osteoblasts!

Now with a Self-Repairing Option!

One of the things I have always found amusing about medicine is how far we have come in some areas, and how little things have changed in others. Although we are able to detect the extent of a brain tumor via a completely noninvasive MRI, when it comes to fixing broken bones, most of the time all we are doing is helping the body to do what it does naturally. We can find out the extent of the break with a simple X-ray, but at that point some old-fashioned horse sense takes over. The only thing left is to set the bone in the right position and immobilize it, often in the same type of plaster cast that my parents wore when they were little. The real wonder is what the body does next, completely without the doctor's help!

Them's the Breaks!

There are many different types of fractures, depending on the nature of the accident, as well as the age of the patient (see Figure 5.4)! If you ask around, you'll find that most people broke more bones in childhood than at any other time in their lives. I personally broke my left arm once, and my right arm twice, all before the age of 13, and I've broken nothing else in close to 30 years!

Figure 5.4

Some examples of typical fractures.

(LifeART©1989–2001, Lippincott Williams & Wilkins)

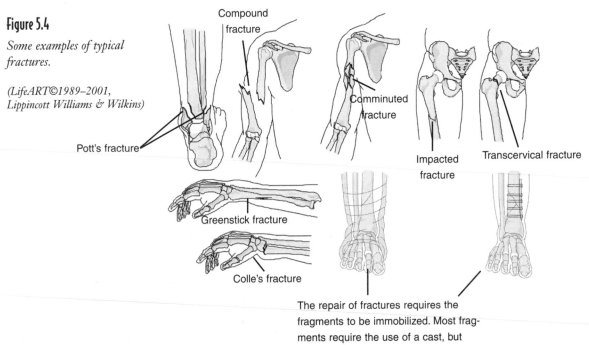

Compound fracture

Comminuted fracture

Impacted fracture

Transcervical fracture

Pott's fracture

Greenstick fracture

Colle's fracture

The repair of fractures requires the fragments to be immobilized. Most fragments require the use of a cast, but some fractures require the use of pins.

Part of that is simply due to the greater activity of the young, not to mention a little of the carelessness of youth! Luckily, the fractures of childhood are usually the least troublesome. These fractures are given the somewhat poetic name of *"greenstick"* fractures. This is best understood by a trip into the woods. If you take two sticks, a green one straight from a healthy tree, and a brown one from the ground, and then bend them to the point of breaking, you will see a startling difference.

The brown stick will snap in two, whereas the green one will bend far more, and when it breaks it will not break all the way through. The bones of the young are more pliable, and less likely to break all the way through, merely bending on the other side of the bone opposite the fracture. In terms of our evolutionary past, a young primate in the wild, learning the ropes, or the branches, as it were, did not have access to a veterinarian to set a fracture 30 million years ago. Greenstick fractures and self-healing bones are definitely a good evolutionary combination!

At the opposite end of the spectrum is the frailty of the bones in the elderly. Part of this, especially in women, is due to osteoporosis. When the bone is weakened it no longer can resist stresses that never used to be a problem. Fractures based on medical conditions, including, but not limited to osteoporosis, are called *pathologic* fractures.

Before looking at the following table of fracture types it is important to mention two other criteria that are used to describe a fracture. Fractures that only break part of the way across the bone are called *partial* fractures, whereas those that break the bone into two or more pieces are called *complete* fractures. When a complete fracture occurs, the anatomical alignment of the bone fragments is sometimes retained, and at other times the alignment is disturbed. The disturbed variety is called a *displaced* fracture, while the other variety is, for lack of a better term, called a *nondisplaced* fracture.

Lastly, fractures have a way of making sharp edges at the point of the break. If none of the pieces break through the skin, that fracture, regardless of any other complexity, is called a *simple* or *closed* fracture. On the more gruesome side, however, are those fractures where a sharp edge of the bone actually pierces the skin; these fractures are known as *compound* or *open* fractures.

Types of Fractures

Fracture	Description
Comminuted	Many small fragments exist at the break
Spiral	Bone is forcefully twisted apart
Stress	Due to repeated stress wearing down bone
Transverse	Break is perpendicular to longitudinal axis
Colle's	Break at distal end of the radius
Pott's	Break at distal articulation of the fibula

Fibula, Heal Thyself

Okay, so you've decided, gathering all the courage your four-year-old frame can muster, to ride your tricycle, with no brakes I might add, down your parent's steep driveway. You find yourself going too fast, the tricycle is getting out of control, and you fall over on your side, fracturing your right ulna. *Ouch!*

Now that the bone is broken, how does it *fix itself* (see Figure 5.5)? To start off, broken bones bleed! The clot that ultimately forms (after a few hours) and surrounds the bone fragments is called a *fracture hematoma*. An *internal callus* of cartilage later forms on the inside of the break, and an *external callus* of bone and cartilage stabilizes the outside of the bone. During this callus formation, spongy bone starts to form in the middle of the break.

Figure 5.5

The steps involved in the repair of fractures.

(LifeART©1989–2001, Lippincott Williams & Wilkins)

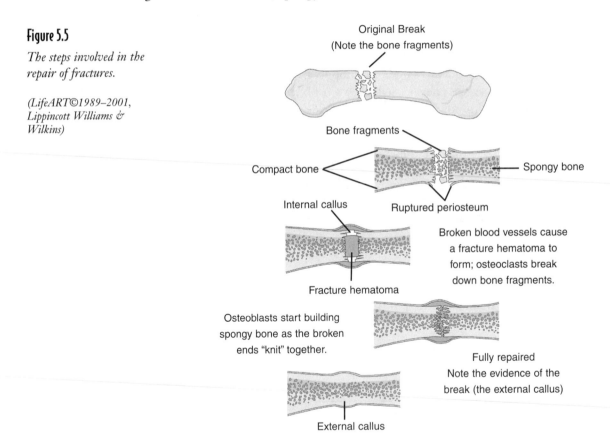

Original Break
(Note the bone fragments)

Bone fragments

Compact bone

Spongy bone

Ruptured periosteum

Internal callus

Broken blood vessels cause a fracture hematoma to form; osteoclasts break down bone fragments.

Fracture hematoma

Osteoblasts start building spongy bone as the broken ends "knit" together.

Fully repaired
Note the evidence of the break (the external callus)

External callus

As the trabeculae from each side of the break unite, or knit, and the external callus starts to ossify, the dead bone, the fragments, and the edges of the break are removed by osteoclasts and replaced by the activity of osteoblasts. After the fracture has healed, the bone is left with a widening at the point of the repair called an *external callus*; although remodeling may cause this to decrease in size over time, it is unlikely to disappear entirely.

Medical Records _____

One interesting thing about bones is their ability to record past injuries through the formation of an external callus. That knowledge is very useful in science and medicine. When wild primates are captured and x-rayed, they routinely are found to have broken bones; so much for the graceful swinging of monkeys! Next time you go to a natural history museum, see if you can find any broken bones in the skeletons (you may be surprised how often you can!). In medicine, a child with a broken arm may be x-rayed more thoroughly to see whether the "fall" was really just another in a long history of child abuse.

The Least You Need to Know

- The functions of bone go beyond support and muscle attachment, but also include storage of fat and minerals, as well as bone cell production.

- Bone develops from depositing calcium in cartilage.

- Bone has two main forms, spongy and compact, and two forms of marrow, yellow for fat storage, and red for bone cell production.

- Bones constantly break down and rebuild using their osteoclasts and osteoblasts, respectively, in a process known as remodeling.

- Bones rebuild themselves by forming a clot at the break, building first a calcium callus, then spongy bone, and then destroying the dead bone and ossifying the callus.

Chapter 6

Head, Shoulders, Knees, and Toes

In This Chapter

- The axial and appendicular skeletons
- The importance of bone landmarks
- Major structural support of the axial skeleton
- The range of movement of the appendicular skeleton

The 206 bones of the skeleton are marvels of structural design. Knowing all 206 is actually far easier than it seems at first! For one thing, having bilateral symmetry makes life a lot easier, because so many of the bones (except the ones exactly on the midline) have a mirror image on the other side! This knowledge is useful in reconstructing skeletons, for a trained eye can identify a bone as left or right in an instant.

In this chapter, I get to the bare bones, one body section at a time. It is not possible to cover all of the bone landmarks in a book of this size. If you want to pursue the matter in more detail, you can find such information in any college-level anatomy text.

Two Skeletons in One!

It might be weird to think of us having two skeletons, but, in a sense, we do! From evolutionary history, we know that the skull, and then the spine, evolved first. That portion of the skeleton is called the *axial* skeleton, which makes sense, given its central location as somewhat of a vertical axis. The axial skeleton is primarily the spine and its immediate offshoots: ribs, sternum, and, of course, the skull (see Figure 6.1).

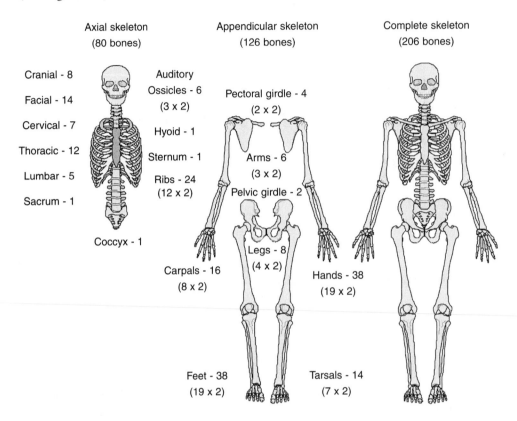

Axial skeleton
(80 bones)

Appendicular skeleton
(126 bones)

Complete skeleton
(206 bones)

Cranial - 8

Facial - 14

Cervical - 7

Thoracic - 12

Lumbar - 5

Sacrum - 1

Coccyx - 1

Auditory
Ossicles - 6
(3 x 2)

Hyoid - 1

Sternum - 1

Ribs - 24
(12 x 2)

Carpals - 16
(8 x 2)

Feet - 38
(19 x 2)

Pectoral girdle - 4
(2 x 2)

Arms - 6
(3 x 2)

Pelvic girdle - 2

Legs - 8
(4 x 2)

Hands - 38
(19 x 2)

Tarsals - 14
(7 x 2)

Figure 6.1

The division of the skeleton into axial and appendicular sections.

(LifeART©1989–2001, Lippincott Williams & Wilkins)

As our vertebrate ancestors continued to evolve, we had a greater and greater need to maneuver around our environment. By developing first fins, then legs, then arms, and then fingers, we have evolved a fantastically subtle arrangement of bones that form appendages, hence the name appendicular skeleton (refer to Figure 6.1). The appendicular skeleton is made of the pectoral girdle (scapula and clavicle), arms, pelvic girdle (pelvic bones), and the legs. A schematic view of the full skeleton can be seen in Figure 6.2.

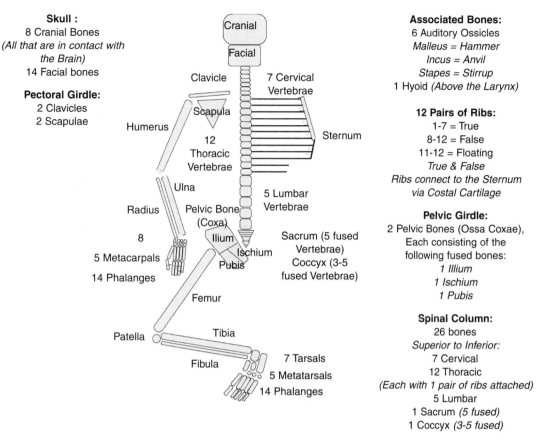

Skull :
8 Cranial Bones
(All that are in contact with the Brain)
14 Facial bones

Pectoral Girdle:
2 Clavicles
2 Scapulae

Cranial

Facial

Clavicle

Scapula

Humerus

12 Thoracic Vertebrae

Ulna

Radius

8

5 Metacarpals

14 Phalanges

Pelvic Bone (Coxa)

Ilium

Ischium

Pubis

7 Cervical Vertebrae

Sternum

5 Lumbar Vertebrae

Sacrum (5 fused Vertebrae)

Coccyx (3-5 fused Vertebrae)

Femur

Patella

Tibia

Fibula

7 Tarsals

5 Metatarsals

14 Phalanges

Associated Bones:
6 Auditory Ossicles
Malleus = Hammer
Incus = Anvil
Stapes = Stirrup
1 Hyoid *(Above the Larynx)*

12 Pairs of Ribs:
1-7 = True
8-12 = False
11-12 = Floating
True & False
Ribs connect to the Sternum via Costal Cartilage

Pelvic Girdle:
2 Pelvic Bones (Ossa Coxae),
Each consisting of the
following fused bones:
1 Illium
1 Ischium
1 Pubis

Spinal Column:
26 bones
Superior to Inferior:
7 Cervical
12 Thoracic
(Each with 1 pair of ribs attached)
5 Lumbar
1 Sacrum *(5 fused)*
1 Coccyx *(3-5 fused)*

Figure 6.2

Schematic diagram of the skeleton for easy reference.

(©Michael J. Vieira Lazaroff)

Landmarks and Markings

There is no bone in our skeleton that looks like the typical cartoon bone, with the butt-like shape at each end! The subtle and often bizarre shapes of true bones are directly related to their functions, and the need for muscle attachment. Muscles connect to bone via cartilage known as tendons, and bones connect to each other by cartilage known as ligaments. It can be hard to attach to a very smooth surface, so bones cooperate by making various crests and bumps to which the muscles attach. The larger the muscle, the bigger, and some might say uglier, the landmarks.

Remember which bones articulate with each bone you are studying, for the name of the articulating bone is often part of the landmark (for example, the radial notch is on the ulna, because the ulna articulates with the radius at the radial notch). You will also find directional terms invaluable here. Those bumps on your "ankle," which are actually on the leg directly above the ankle, have names based on their locations: the medial malleolus is on the medial side of the leg, and the one on the lateral side has to be the lateral malleolus). The following table provides an overview of the appearance and function of the various bone landmarks.

Bone Landmarks

Landmark Type	Appearance	Function
General Processes		
Process	Obvious projection	Muscle or joint
Condyle	Big, rounded projection	Articulation
Facet	Flat, smooth surface	Articulation
Head	Like a head, on a neck	Articulation
Special Processes for Connective Tissue Attachment		
Crest	Large ridge or border	Muscle attachment
Epicondyle	Bump "above a condyle"	Muscle or joint
Line (linea)	Small ridge or border	Muscle or joint
Spinous process	Thin, sharp process (spine)	Muscle or joint
Trochanter	Large projection (on femur)	Muscle attachment
Tubercle	Small, round process	Muscle attachment
Tuberosity	Large, rough process	Muscle attachment
Paranasal sinus	Air-filled cavity with canal	Humidity, sound
Openings and Depressions		
Canal	A narrow tube	Sound, blood
Fissure	A narrow cleft	Nerves, blood
Fontanel	Dense connective tissue between fetal cranial bones	To aid birth (flexibility)
Foramen	A hole, literally nutrient—for blood	Nerves, blood, and ligaments
Fossa	Shallow depression	Articulation
Groove or sulcus	A long, thin depression	Nerves, blood
Meatus	A larger canal	Sound

Oh, My Achin' Back!

The spine is far more than just the back. It is helpful here to think about the different sections of the spine. Each bone in the spine is called a *vertebra*: 7 cervical, 12 thoracic, 5 lumbar, 1 sacral, and 1 coccygeal, for a grand total of 26 vertebrae. The vertebrae in each section are numbered separately, starting from the cranial to the superior to inferior caudal (see Figure 6.3).

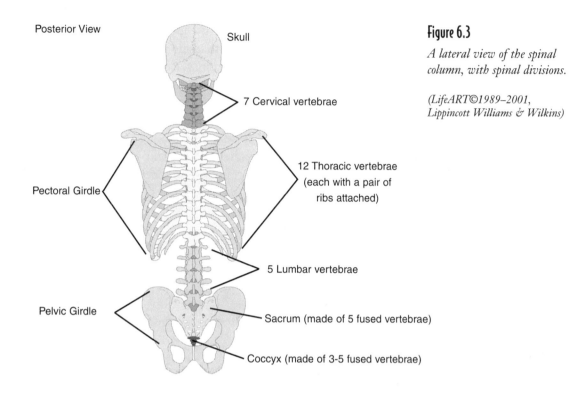

Posterior View

Skull

7 Cervical vertebrae

12 Thoracic vertebrae
(each with a pair of
ribs attached)

Pectoral Girdle

5 Lumbar vertebrae

Pelvic Girdle

Sacrum (made of 5 fused vertebrae)

Coccyx (made of 3-5 fused vertebrae)

Figure 6.3

*A lateral view of the spinal
column, with spinal divisions.*

*(LifeART©1989–2001,
Lippincott Williams & Wilkins)*

You have seven cervical vertebrae, which get progressively larger from one to seven. The top two vertebrae, because they are so distinct in shape and function, have special names. Those of you who liked mythology will get a kick out of the name of the first cervical: the atlas vertebra. Remember Atlas, carrying the world on his shoulders? Well, in this case the atlas vertebra carries a skull on its "shoulders."

The axis vertebra (the second cervical) is known for a vertical projection called the *dens*; the dens acts like an axis (hence the name of the vertebra) for the pivot joint that makes shaking your head possible. The shape of these two vertebrae is so unusual that they are truly unique. All other vertebrae share certain distinct features, which makes them all "variations on a theme."

The anterior part of every vertebra is called the *body* (except the atlas, which doesn't have one, and the axis, in which it is called the dens). The body bears the majority of the weight of the person. Immediately posterior to the body is a large opening called the *vertebral foramen*, for the spinal cord. If you connect all of the vertebral foramina together you form the vertebral cavity (see Chapter 1). Three processes extend out from the vertebral foramen. The posterior process is called the spinous process and can be felt along the midline of your back. The two lateral processes (think directional terms again) are called transverse processes, and they provide attachment not only for muscles, but in the case of the thoracic vertebrae, for the ribs.

Beneath the 7 cervical vertebrae (the same number as the giraffe!) are 12 thoracic vertebrae. The division is an easy one, for each of the 12 thoracic vertebrae has a pair of ribs (for a total of 24), one on each side, attached to the transverse process and the body, and no ribs are attached to either the 7 cervical vertebrae above or the 5 lumbar vertebrae below. Given the need for the spine to support your weight, being bipedal (walking on two legs), the vertebrae get progressively larger the farther down the spine. As such, the five lumbar vertebrae are the largest and the heaviest. The experienced eye can identify not only which group the vertebra is in, but which number in the group, but to those just learning, the first and second thoracic might be confused for a cervical, or the eleventh and twelfth thoracic might be confused for a lumbar (see Figure 6.4).

Figure 6.4

A comparison of cervical, thoracic, lumbar, sacral, and coccygeal vertebrae.

(LifeART©1989–2001, Lippincott Williams & Wilkins)

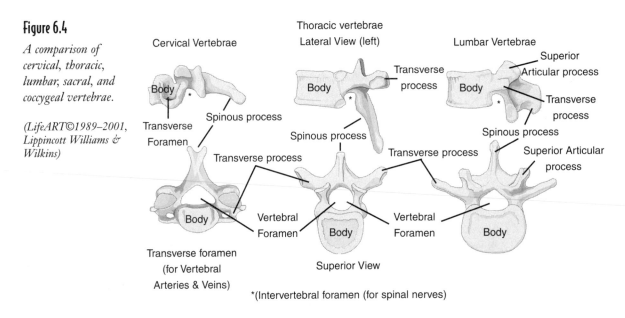

Flex Your Muscles

To make things easier, you can use a simple system of abbreviation when referring to vertebrae: C = cervical, T = thoracic, and L = Lumbar. Thus the third cervical vertebra would be C_3, and the last lumbar would be L_5. This system is also used for the spinal nerves, with a couple of modifications. For one thing there are eight cervical nerves (C_1–C_8), and the sacrum, being five fused vertebrae, has five spinal nerves called *sacral nerves* (S_1–S_5).

There are two unique vertebrae at the very end of the spine, beneath the lumbar: the sacrum and the coccyx. What makes them unique is that each of them is actually several vertebrae that are fused together. The sacrum, which articulates with a pelvic bone on each side to make the pelvis, is made of five fused vertebrae. The fusion of these vertebrae may mean a loss in flexibility, but it is more than made up for by the added stability. The last stop is the coccyx, which is made of anywhere from three to five fused vertebrae.

Medical Records _____

Is your tail longer than your neighbor's? You might not know this, but you do indeed have a tail. The tail of a cat or dog, for example, has many more vertebrae, which are called caudal vertebrae (caudal means tail). The large number allows for a great deal of flexibility in the tail. If you have ever seen a cat arch its back, then you have seen the great flexibility they have, which is due in part to the one extra thoracic and two extra lumbar vertebrae they have.

The Cage

When looking at a frog you might notice the sternum and clavicles providing protection for the heart and the surprisingly small lungs, but you will also notice the complete lack of ribs. Being cold-blooded (ectothermic), amphibians don't need to eat as much food as we do, but since we use the heat produced from burning sugar to warm our body (endothermic = warm blooded) we not only need to eat more, we need to *breathe* more. Because of that we have larger lungs, a greater need to protect them, and a more sophisticated way of bringing air in. All these differences are reflected in the size and shape of, not to mention the mere *existence* of, our ribcage.

Crash Cart _____

It's time to put a myth to rest. Every year in my anatomy class there are students who "learned" somewhere that men and women have a different number of ribs. Since the class has already talked about being bilateral critters, I always ask whether it makes sense to have one extra rib; students then decide that an extra pair makes more sense. At this point I have my students compare a male skeleton and a female skeleton. A simple count shows 24 ribs in each skeleton, organized as 12 pairs. Despite being childishly easy to disprove, this myth somehow refuses to die.

The thoracic cavity is defined by the ribcage. The overall shape is a bit like a flattened barrel that narrows at the top (see Figure 6.5). The narrowing is due to the small, flat, almost horizontal first pair of ribs. The parietal surface of the cavity (the inside wall) is the parietal pleura, which lubricates the lungs when you breathe (see Chapter 13). The bottom of the cavity is the diaphragm, which is crucial in your ability to breathe (see Chapter 13).

In addition to providing attachment for the diaphragm, the ribs also help breathing by providing attachment for the intercostal muscles. The external intercostals (superficial) are used when you inhale, and the internal intercostals (deep) are used when you forcibly exhale. When the external intercostals contract they pull the ribcage upward and outward;

the internal intercostals pull the ribcage downward and inward. The movement of each pair of ribs looks quite a bit like the handle of a bucket in the way that they move. The ribs can be divided into true and false ribs based on their attachment to the sternum. Ribs 1 through 7 (remember that we are talking about pairs 1 through 7) attach directly to the sternum via costal cartilage, and for that they are called true ribs. Ribs 8 through 12 are called false ribs, but they need to be further subdivided. Ribs 8 through 10 do attach to the sternum, but indirectly; these all have costal cartilage, but their cartilage attaches to the cartilage for rib 7, which makes the attachment indirect. Ribs 11 and 12 are called floating ribs, but despite their name, they do not float around because they do attach to T_{11} and T_{12}; the name "floating" is just because they do not attach to the sternum at all.

Figure 6.5

The ribcage, showing rib types and sternum.

(LifeART©1989–2001, Lippincott Williams & Wilkins)

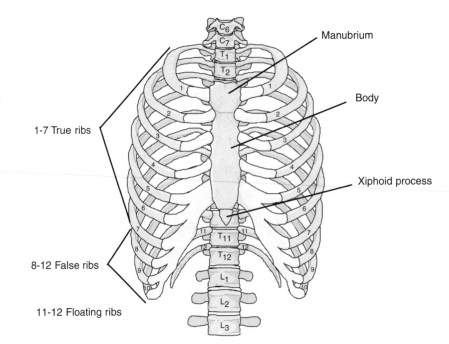

1-7 True ribs

8-12 False ribs

11-12 Floating ribs

Manubrium

Body

Xiphoid process

The *sternum*, often called the breastbone, is medial to the ribs, and provides attachment for ribs 1 through 10. It is considered one bone, but it has three distinct parts. The superior portion is called the *manubrium*, and it articulates with the clavicles and the first pair of ribs. The intermediate portion is called the *body* (a term often used for the largest portion of a complex structure, as in the body of a vertebra), and it articulates directly with ribs 1 through 7, and indirectly with ribs 8 through 10. The inferior portion is called the delicate *xiphoid process*.

The angle of the costal cartilage for the false ribs produces something of an arch, and the curve is continued in the dome-shaped diaphragm. For this reason, the ribcage protects more than just the lungs. The superior portion of the liver and the stomach are also partially protected. The heart is found directly above the diaphragm, directly behind the sternum.

Medical Records

The location of the heart behind the sternum is convenient for CPR (cardiopulmonary resuscitation). By pressing down on the body of the sternum (right above the xiphoid process) you can compress the heart, thus pumping blood from it. You must avoid the xiphoid process because it can easily break, but it is easy to find if you follow the curve of the costal cartilage medially to the top of the arch. The resilience of the ribcage is also very important, for after you press on the sternum (which compresses the ribs) the ribcage bounces back to its original shape. These repeated compressions keep the blood flowing.

Joggin' Your Noggin

Given how many times kids have seen diagrams that just say "skull," most people are surprised to learn that there are 22 bones in the skull, and a total of 29 bones in the head! Let's get the extra seven out of the way first. Although not specifically part of the skull, there is a bone at the top of the larynx (your voice box) called the *hyoid bone* (see Figure 6.6). If you feel the point where the bottom of your jaw meets your neck you may feel it. The other six are a bit hidden, and although they are deep within a bone, they are not part of the skull; they are the auditory ossicles, the three bones in each inner ear. I discuss these bones—the malleus (hammer), incus (anvil), and stapes (stirrup)—in Chapter 22.

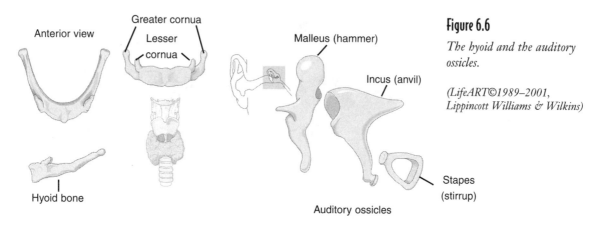

Figure 6.6

The hyoid and the auditory ossicles.

(LifeART©1989–2001, Lippincott Williams & Wilkins)

The other 22 bones are divided into two categories based on location: 8 cranial bones, which have direct contact with the brain, and 14 facial bones. Most of these bones come in pairs, except for two of the facial, and half of the cranial. Remember, the only excuse for not being in pairs is if the bone is found on the midline, and even then the bone is clearly bilaterally symmetrical.

Close Mental Contact

Cranial bones, which were the first to evolve, have as their primary function the protection of the brain. The development of these bones in the growing child is tied to the development of the brain. The growth is so tightly matched that the inside of the skull is a cast of the shape of the brain. If you examine the inside of the cranial bones you can also trace the flow of many of the major blood vessels. This information is useful, for it has allowed paleontologists to make casts of the brain case and trace the evolution of the brain—a nifty trick considering that soft tissue almost never fossilizes!

Most of the names you learn here will come in handy later, because the lobes of the brain share the same names as some of the cranial bones (see Figure 6.7). The hardest bone to remember by far is the *frontal bone*, because its name is no help in figuring out its location. Oh, wait, never mind! The frontal bone (which protects the two frontal lobes) is notable for its frontal sinus (behind your eyebrows). You'll learn more about the sinuses in the next section, and in Chapter 13.

In the very back is the occipital bone (protecting the two occipital lobes). Its most obvious feature is the *foramen magnum*, which literally means *big hole*! Why such a big hole? Why, for the spinal cord, of course! Another major feature is the pair of *occipital condyles* on either side of the foramen magnum; these condyles are the only place where the skull articulates with the spine (or with *any* part of the body).

Figure 6.7

Anterior and lateral view of the skull.

(©2003 www.clipart.com)

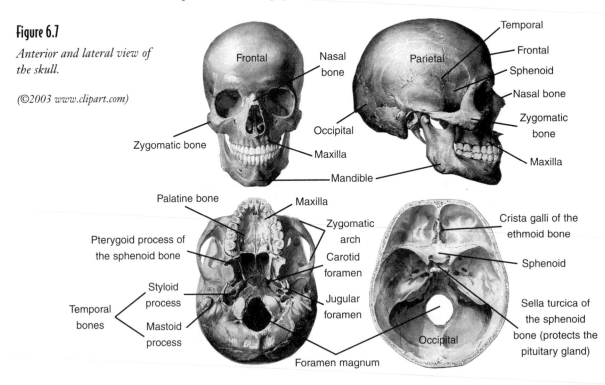

On both sides of the skull are the two parietal bones (which protect the two parietal lobes). These bones are the only pair of cranial bones that articulate with each other. Only a handful of bilateral bones in the body articulate with each other apart from the parietal bones: pelvic, palatine, maxillary, and nasal (the last three are all facial bones).

The temporal bones, by comparison, have a lot going on. For one thing, the external acoustic canal marks the entrance to the middle and inner ear, which reside within the temporal bone. In addition, both the carotid artery and the internal jugular vein enter the skull through the temporal bone.

The *ethmoid* and *sphenoid* bones, by contrast, are often overlooked simply because their location is a bit more hidden. The ethmoid forms the roof of the nasal cavity, and contains the olfactory nerve, which is responsible for smell. A small vertical projection called the *christa galli* has direct physical contact with the brain. The ethmoid is also known for its sinuses. Two pairs of projections on the bottom of the ethmoid called the superior and middle nasal conchae form four of the six turbinates in the nasal cavity (see Chapters 13 and 22).

Medical Records

One can often, but not always, determine the gender of a person by looking at the skull. The external occipital protuberance at the back of the occipital bone (can you feel yours?) is for muscle attachment, and as males are, on average, larger, and thus have larger muscles, this landmark is usually bigger in males.

The sphenoid, on the other hand, is more complex, as it extends from one temple to the other. This cool bone looks like a headless bird of prey! In addition to its sinuses, the sphenoid also contains the *sella turcica* (Turkish saddle), which protects the pituitary gland (see Chapter 18). Beneath the saddle the two optic nerves cross at the *optic chiasma*, after entering the sphenoid at the *optic foramina*.

The cranial bones were not always connected, but rather started life as separate bones. As the bones grow together they form the meandering junctions known as sutures. Before these sutures form, however, they are known as *fontanels*, often called *"soft spots"* on the baby's skull (see Figure 6.8). There are six in all, which may surprise some people. The fontanels greatly increase the malleability of the fetal skull, allowing the skull to fit through the cervix and the vaginal canal, which really are too narrow for it!

The largest fontanel is the best known, and may be the only palpable one; it is found at the juncture between the two halves of the frontal bone and the two parietal bones, which will eventually become the *coronal suture*. The anterior fontanel implies the existence of the *posterior* fontanel, which is where the two parietal bones join the occipital bone, the location of the future *lambdoid suture*. The other four fontanels appear in pairs on either side of the skull. They are the *anterolateral* or *sphenoid fontanel*, where the sphenoid, temporal, parietal, and frontal bones meet, and the *posterolateral* or *mastoid fontanel*, where the temporal and occipital bones meet. There are two other major sutures in the skull, but neither is associated with fontanels: the *sagittal suture*, between the two parietal bones, and the *squamous suture*, between the parietal bone and the temporal bone (which is the only major suture that appears as a pair).

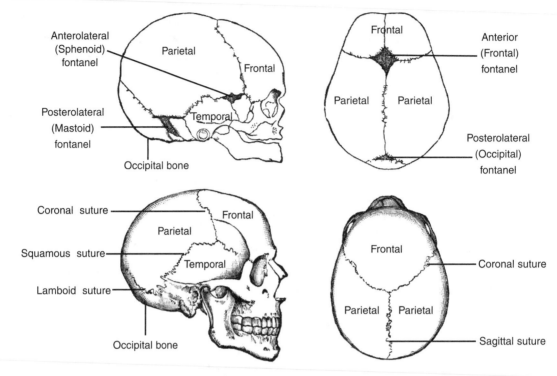

Note: Skulls are not to scale

Figure 6.8

The six fontanels are responsible for easing the passage of the baby's skull through the birth canal. These ultimately become the sutures in the skull.

(©2003 www.clipart.com)

Not Just a Pretty Face

The face, by contrast, is a composite of 14 facial bones. With the exception of the mandible, or lower jaw bone, and the *vomer* (Marge Simpson's favorite bone?), which forms the bulk of the nasal septum (the rest of the septum is from the *perpendicular plate* of the ethmoid), the facial bones appear in pairs. The largest pair are the maxillary bones, which make up the upper jaw. These also contain sinuses. The maxillary is the most complex facial bone, for it makes up the roof of the mouth, and the inferior and lateral borders of the nasal cavity.

The bridge of the nose is made by the tiny pair of nasal bones, with the bulk of what we call the nose being composed of cartilage. Inside the nose are two curled bones called the inferior nasal conchae. The very back of the nasal cavity, as well as the back of the hard palate, is from the pair of *palatine bones*.

The inferior border of the eye socket, or *orbit*, is part of the maxillary bone, and the superior border is part of the frontal bone, but the lateral border comes from another bone with a rather cool name, the *zygomatic* ("it dices, slices, chops …") bone. There is a posterior facing *temporal process*, which joins with the anterior facing *zygomatic process* of the temporal bone; together these form the *zygomatic arch*, also known as the *cheekbone*. The only pair of bones left are the pair of tiny *lacrimal bones*, which are within the orbit.

The Paranasal Sinuses

If you have ever had a sinus headache, you are intimately aware of the presence of the paranasal sinuses (see Figure 6.9). There are a pair of sinuses in the frontal, sphenoid, and maxillary bones, as well as a series of 3 to 18 *cells* in the ethmoid bone. The sinuses help to humidify the inhaled air by virtue of their mucous membranes. The sinuses also cause a unique resonance in your voice as the sound waves produced in the larynx bounce in and out.

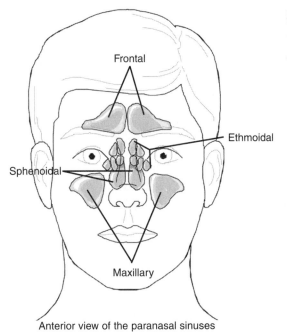

Figure 6.9

The paranasal sinuses, the curse of everyone with allergies.

(LifeART©1989–2001, Lippincott Williams & Wilkins)

Anterior view of the paranasal sinuses

Flex Your Muscles

In the presence of a sinus infection, the tiny drainage areas are plugged, and there is a painful buildup of pressure. Think about the last sinus headache you had. Did you feel pain above your eyes (frontal sinuses)? Did you feel pain in your cheeks (maxillary sinuses)? How about between your eyes (ethmoid sinuses)? Behind your eyes (sphenoid sinuses)?

The Other Girdle

When people hear the word *girdle* they usually think about a horrendous undergarment with about a hundred ties, seemingly designed to cut off all breathing. They don't usually think about the shoulders. Your shoulders, which are your introduction to the appendicular skeleton, are composed of only two pairs of bones, the two *scapulae* or *shoulder blades*, and the two *clavicles* or *collar bones* (see Figure 6.10).

Figure 6.10

The shoulders are made up of the scapulae and clavicles of the pectoral girdle.

(LifeART©1989–2001, Lippincott Williams & Wilkins)

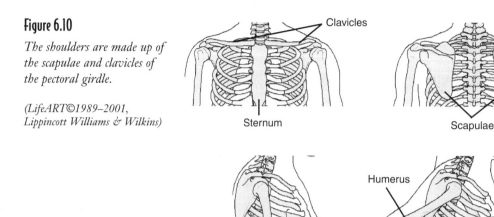

What is notable about these bones is their incredible range of motion. The clavicles are the least mobile, due to their attachment to the manubrium of the sternum and to the *acromion process* of the scapula, but each scapula is only connected to the appendicular skeleton, at the clavicle and at the highly mobile *humerus*, which articulates with the *glenoid cavity* of the scapula in a shallow ball-and-socket joint (see Chapter 7). The scapulae are held in place by deep muscles, which accounts in part for the wide range of motion, as well as the ease with which a shoulder can be dislocated (ouch!).

A Juggler in the Making

In buffalo chicken wings, the wing drumstick is the *humerus*, notable for the ball that attaches to the scapula, and the hinge joint at the distal end. When you bend your elbow you can feel a depression on the posterior face of the distal end, called the *olecranon fossa*, which articulates with the *olecranon process* on the proximal end of one of the bones of the lower area, the *ulna*.

The other part of that yummy chicken wing has a pair of bones, just like you: the *radius*, and the *ulna*. The aforementioned olecranon process is the tip of your elbow. The proximal end of the radius, the wheel-like *head*, pivots in the *radial notch* of the ulna, just as the *head* of the ulna pivots in the *ulnar notch* of the radius. These two articulations make the movement known as *pronation* possible (see Chapter 7).

At the distal end of the radius and ulna is a collection of eight bones called *carpals*, which make up the wrist, or *carpus* (see Figure 6.11). Each bone is capable of little movement against the other carpals, but the collection of all these gliding joints (see Chapter 7) makes for a relatively wide range of motion. These eight bones form two rows, proximal and distal. The proximal group, from medial to lateral are the *pisiform* (which you can feel in the corner of your hand, opposite the thumb), *triquetrum*, *lunate*, and the fracture-prone *scaphoid*. The distal group, which articulates with the five *metacarpals* that form the body of the hand, are, also from medial to lateral, the *hamate*, *capitate*, *trapezoid*, and *trapezium*.

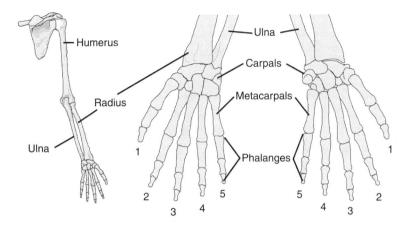

Figure 6.11

The bones of the arms and hands.

(LifeART©1989–2001, Lippincott Williams & Wilkins)

The shape of the metacarpals is similar to the bones of the fingers, or *phalanges*, the singular of which is *phalanx*. The similarity of shape makes the skeletal hand look as if it has incredibly long fingers. Each bone has a proximal *base*, an intermediate *shaft* (which is really just a short diaphysis; see Chapter 5), and a distal *head*. The head of the more proximal bone forms a hinge joint with the base of the next bone.

Flex Your Muscles

Each of the fingers, metacarpals, toes, and metatarsals has a number for quick reference, just like the vertebra. If you ever played the piano the numbers should sound familiar! The thumbs, as well as the big toes, are number one, the index finger is number two, up to the little finger, which is number five. Next time you see someone at a game holding up an index finger and shouting "we're number one," you should correct them and say they look like number two. Then again, you might get punched!

There are 14 phalanges on each hand, because the thumb, or *pollux*, has two, a proximal and distal phalanx, and each of the other fingers has three phalanges, proximal, middle, and distal. The arrangement of the bones in the toes, also called phalanges, is analogous to those in the hand, including the pair in the thumblike big toe, or *hallux*. Immediately proximal to the toes are the five *metatarsals*, which are analogous to the five metacarpals.

Hip Huggers

The pelvis, or pelvic girdle, is made of two bones, each called a *pelvic coxa*. Each is made of three bones that fuse during fetal development: the ilium, the ischium, and the pubis. The flared shape of the pelvis is very different from the more linear pelvis of our quadrupedal ancestors, but it makes perfect sense. The linear pelvis was longer, and helped to provide support for the abdominopelvic organs suspended beneath.

In your bipedal body, the abdominopelvic organs fall along your center of gravity, and the flared pelvis, especially the wide *iliac crest* along the outer rim, supports the weight of the organs from underneath. The placement of the acetabulum, the socket portion of the ball-and-socket joint with the head of the femur, is also placed differently in bipeds, to reflect the placement of the legs along the center of gravity.

The pelvis is the most recognizable difference between males and females (see Figure 6.12). The differences are many, and they are all related to the fact that the female pelvis must accommodate the birth of the baby. The upper rim, which forms part of the *pelvic inlet*, is more flared in the female, to support the added weight of the growing fetus and uterus. The opening in the bottom is called the *pelvic outlet*, and rightfully so, since the baby comes out through it! Other differences affect the outlet, making it larger, such as a wider angle beneath the pubic symphysis, and a straighter tilt to the sacrum, which articulates with the medial borders of the ilium, thus allowing more room for the baby.

Figure 6.12

The differences between the pelvis of the female and the male are related to a woman's ability to give birth.

(©2003 www.clipart.com)

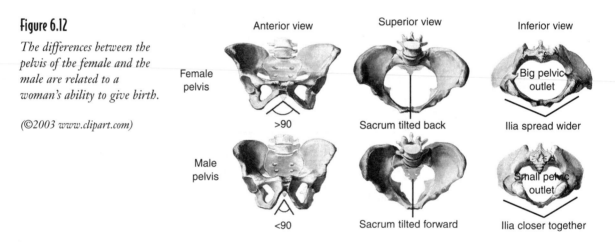

Anterior view

Superior view

Inferior view

Female pelvis

>90

Sacrum tilted back

Big pelvic outlet

Ilia spread wider

Male pelvis

<90

Sacrum tilted forward

Small pelvic outlet

Ilia closer together

These Legs Are Made for Walking

The form of the legs is very similar to that of the arms. There is a single femur with a ball-and-socket joint, just like the humerus, followed by two bones in the lower leg, tarsals (the ankle) instead of carpals, and feet that are very much like hands (see Figure 6.13). Nonetheless, the differences are revealing. The legs, unlike the supple arms, evolved for support. In that way there is a powerful *femur*, and an equally powerful *tibia*, that provide the bulk of the support. The much smaller fibula is more important to the ankle than to the knee.

Unlike the elbow, which has that nifty olecranon process and fossa combo to prevent hyper-extension (see Chapter 7), the knee only has a little, almost floating *patella*, or *kneecap*, to protect it. This explains why it is such a lousy joint, so prone to injury. The distal end of the lateral fibula is called the *lateral malleolus*, which you can feel as the bump on the outside of your leg, just above your ankle. The *medial malleolus* is the bump on the medial side of your leg, and it is on the distal end of the tibia.

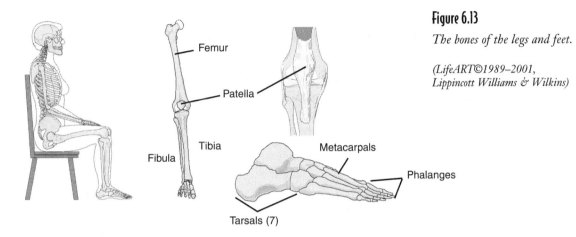

Figure 6.13

The bones of the legs and feet.

(LifeART©1989–2001, Lippincott Williams & Wilkins)

You looked at the metatarsals and phalanges when you looked at the hands. The *tarsals*, which make up the *ankle*, are interesting if you compare them to other animals, for we walk on ours, unlike pets, who walk on tiptoe! By walking on our ankles, the greater surface area of the foot made it easier to support the weight of the body. There are seven, rather than the eight in the wrist.

Flex Your Muscles

Have you ever noticed how horses, dogs, and cats have backward knees? Well, guess what? They *don't!* What you think is their knee is actually their ankle! They are walking on tiptoe! If you look higher up the leg you will see a very normal looking knee. Why walk on tiptoe? Well, if you have ever sprinted, then you know why. For speed! That's why you can never win in a race with your dog!

The weight of the leg, and the rest of the body for that matter, rests upon the *talus*, which is immediately superior to the *calcaneus*, better known as the heel of your foot, even though it is really your ankle! Farther forward are the *cuboid*, which articulates with the fourth and fifth metatarsals, and the *navicular*, which is immediately posterior to the three *cuneiforms*, each of which in turn articulates with a separate metatarsal. The three cuneiforms, called first, second, and third, are also termed by location as medial, intermediate, and lateral.

The Least You Need to Know

- The skeleton is divided into two parts, axial (skull, ribcage, and spine) and appendicular (shoulder, arms, hips, and legs).

- The vertebrae are divided into 7 cervical (neck), 12 thoracic (chest, each attached to a pair of ribs), lumbar (lower back), sacral (part of the pelvis), and coccyx (your vestigial tail).

- There are seven pairs of true ribs, connected directly to the sternum, and five pairs of false ribs; the last two pairs do not connect with the sternum at all, and are called floating ribs.

- The skull consists of 8 cranial bones, which protect the brain, and 14 facial bones.

- The pelvis of the female is wider, especially in the pelvic outlet, to allow for the birth of the baby.

- The form of the arms and legs, hands and feet, are very similar, but the structure of the bones differs based on their role for either manipulation or support.

Chapter 7

Oil Can! Oil Can!

In This Chapter

◆ The different types of joints

◆ The structure of a synovial joint

◆ The six types of synovial joints

◆ The types of movements at synovial joints

Now that you've learned all 206 bones, it's time to get a handle on how they are physically put together. A joint is, quite simply, the junction between two bones. Some people automatically think about movement here, but if you think about furniture, the place where two fixed pieces of wood join is also called a joint (remember, the word *join* is the root of the word *joint*). Joints are classified based on the range of motion possible, from a fixed joint to the most highly flexible joint known as a ball-and-socket joint.

There are also names for the incredibly wide range of motions that are possible in the body. You will find that these movements, like the directional terms, are also found in pairs (as are the muscles, as you will see in Chapter 10). Also like the directional terms, you will find these movement terms important in the naming of muscles.

Types of Joints

Any place where two bones meet is called an articulation (or joint). It might help you to remember that a person is articulate when she/he can *string words together* (making joints!) well. Don't forget that bones are used for more than movement, so the range of motion is going to vary a great deal, with the least amount of motion in those bones that are mainly for protection. There are two basic ways to classify joints (as seen in the following table): by virtue of structure and by virtue of function. (There's that structure and function duality again!)

Classification of Joints

Joint Type	Characteristics	Location
Structural Classification		
Fibrous	No joint cavity, held by fibrous connective tissue	Skull sutures, tooth and mandible
Cartilaginous	No joint cavity, held by cartilage	Pubic symphysis, epiphyseal plate
Synovial	Joint cavity, bones are surrounded by articular capsule, and often accessory ligaments	Knee, elbow, phalanges, and so on
Functional Classification		
Synarthrosis	Immovable joint	Skull sutures, tooth and mandible, epiphyseal plate
Amphiarthrosis	Slightly movable joint	Pubic symphysis, tibia-fibula (distal end)
Diarthrosis	Freely movable* joint	Knee, elbow, phalanges, and so on

Diarthrosis joints are "freely movable," because they can move freely, but only within a limited range of motion based on the type of diarthrosis.

I Can't Move!

We have all learned so well that bones help us to move (which explains why some texts refer to a musculoskeletal system) that we forget how often bones are used for simple structural support and protection. One very useful way to accomplish these functions is to be immobile. There are three types of immobile joints in the body. *Sutures* in the skull are one example; their principal function is to protect the brain. These fibrous joints, which form as the skull develops, resemble the meandering pattern of a river, and their pattern is unique in every skull. Remember the fontanels from the last chapter? Fontanels, which are basically preossified sutures, enable the skull to be flexible during delivery.

A *gomphosis* is a peg-and-socket joint. This type of joint makes perfect sense when you consider your teeth. The root of each tooth is the peg shape, and in both the inferior surface of the maxilla and the superior surface of the mandible, there is a line of sockets for each of the teeth. At the tip of the root there is a periodontal ligament that connects the root to the jaw, which is what makes pulling teeth so difficult.

Lastly, a *synchondrosis* is an example of a joint where you don't think a joint would be, simply because it is *within* the bone. As I mentioned in Chapter 5, at the epiphysis of every long bone is an epiphyseal plate of hyaline cartilage where the bone growth occurs, at least until adulthood when it ossifies (turns to bone), at which point it is called a *synostosis*.

A Little Wiggle Room

There are two types of slightly movable joints (*amphiarthrosis*): *syndesmosis* and *symphysis*. A syndesmosis is similar to a suture, complete with the fibrous connective tissue, but it is more flexible. Such a joint is useful if the body needs to link two bones, but allow a little flexibility. A perfect example can be found between the tibia and the fibula. The proximal joint involves only those two bones, but the distal end of each bone also articulates with the talus, which is part of the ankle. The joint between the tibia and fibula thus needs to be both rigid and flexible.

The other amphiarthrosis is called the symphysis, and it is characterized by a broad, flat piece of fibrocartilage, which both cushions the joint and allows some movement. There are two examples of this in the body: the intervertebral disks and the pubic symphysis. The intervertebral disks wear down and flatten over time, which is one of the reasons why people get shorter when they age! The pubic symphysis is the only place where the two pelvic bones articulate (the other articulations are with the femur and the sacrum). This joint is loosened by the hormone relaxin during the ninth month of pregnancy, which eases delivery.

Make Room!

A diarthrosis is also called a synovial joint, named after the structures that make them so freely movable. The defining characteristic of a synovial joint is what is called the synovial joint cavity, filled with synovial fluid, which separates the bones in the joint. This cavity, and the structures in and around it, are more complex than other joints, but they are necessary in terms of providing so much motion.

At the end of each articulating bone is a cap of hyaline cartilage called *articular cartilage*, which serves to protect the bones, but does not connect the two bones. You might remember having seen it the last time you ate a chicken leg. Around the joint as a whole is an articular capsule, which encases the synovial joint cavity. The articular capsule has two layers: a tough outer layer called the fibrous capsule and an inner synovial membrane.

Medical Records

With all these terms ending in -*arthrosis*, is it any wonder that an inflammation of a joint is called arthritis? There are many forms of arthritis, with a wide variety of causes. Perhaps the most debilitating is osteoarthritis, in which the cartilaginous parts of the joint actually ossify (turn into bone), thus permanently eliminating movement in that joint.

The fibrous capsule, which is made of dense, irregular connective tissue, has two distinct functions. One is to provide enough flexibility to allow movement, and the other is to provide enough strength to prevent dislocating the joint. The greater the flexibility in a joint, the easier it is to dislocate the joint. The ball-and-socket joint allows the greatest range of motion, but the shoulder has far more flexibility than the hip. This is presumably related to our evolutionary past, in terms of needing the pelvis to provide adequate support for our bipedal nature, and a holdout of our ancestors' heavy use of brachiation (arm over arm swinging—think monkey bars). As such, it is understandable that the shoulders get dislocated far more often than the hips.

The inner synovial membrane is a bit of a mish-mash, with areolar connective tissue, adipose tissue, and some epithelial tissue with the cells uncharacteristically widely spaced. This membrane secretes a viscous *synovial fluid* the consistency of raw egg white that lubricates the joint. Wear and tear on the joint is also cleaned up via macrophages in the fluid. Lastly, since cartilage is avascular, the fluid aids in the distribution of nutrients to the cartilage around the joint.

Medical Records

Have you ever heard people crack their knuckles and wondered what that gross sound was? By pulling on a knuckle you increase the volume of the cavity, thus decreasing the pressure (see Boyle's law in Chapter 14); this can cause some of the synovial fluid to evaporate, and thus make a bubble. When the pressure in the capsule exceeds that in the bubble, the bubble implodes—thus the pop! Any small bubbles left need to fully dissolve before the joint can be cracked again (23 to 30 minutes).

Most synovial joints also have accessory ligaments. Ligaments outside the capsule are called *extracapsular ligaments*, with those inside the capsule called *intracapsular ligaments*. A joint with a lot of use will also have a pad of fibrocartilage called an *articular disk*, which is also called a meniscus. All of this motion might also cause a lot of friction against the muscle and skin. An extra measure of protection are the bursa, which are similar to articular capsules, complete with fluid, but without the bones. When these capsules get inflamed, that condition is known as bursitis.

Hinges, Pivots, and Saddles ... Oh My!

Despite all their similarities, there are six basic types of synovial joints (see Figure 7.1). What makes these types different is the type of motion that is possible at each. What follows is a brief description of each type, with examples, of course! Keep these joints in mind, as they will help you to learn the movements later.

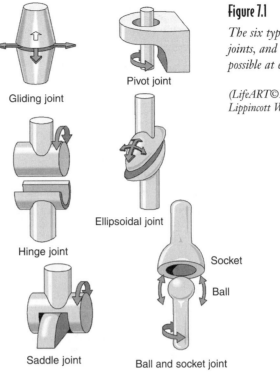

Figure 7.1

The six types of synovial joints, and the type of motion possible at each.

(LifeART©1989–2001, Lippincott Williams & Wilkins)

Slipping and Sliding

A gliding (or arthrodial) joint, which is the simplest type of diarthrosis possible, only allows side-to-side motion. This is obviously very limited, but it definitely has its uses. These joints are found between the carpals and between the tarsals. Individually they may not provide much movement, but together they provide great flexibility. Gliding joints can also be found at both ends of the clavicles (scapula and sternum), and at both ends of the ribs (sternum and vertbra).

Like Opening a Door

A far greater amount of movement is possible in a hinge (or ginglymus) joint, but it is a limited movement nonetheless. Hinge joints act like, well, hinges! The motion is called angular, but the limitation is that hinge joints allow motion along only one plane. Hinge joints are found all over the body. Most people think of the elbow and the knee, but hinge joints are also found between the phalanges of the fingers and toes, in the ankle, and between the occipital bone and the atlas vertebra (a.k.a. C_1).

Side to Side

A pivot (or trochoid) joint also allows a lot of movement, this time called rotational movement; but once again it is limited to a single plane. One characteristic of the bones at such joints is that they have a rounded surface. One example is the proximal end of the radius (which is shaped like a wheel), and of the ulna (shaped like a wheel well), allowing you to flip your hand over. Another example is between the atlas vertebra and the dens (shaped like a round peg) of the axis vertebra (see Chapter 6); without this you wouldn't be able to shake your head and say "No!"

Ellipsoidal

An ellipsoidal joint has an advantage over hinge joints in that they can move on two planes (side to side, and back and forth), rather than just one. These joints are also called condyloid because they are a combination of an elliptical cavity and a rounded condyle. One example of this is the articulation between the cavity at the distal end of the radius, and two of the carpals (scaphoid and lunate). Ellipsoidal joints are also found where the proximal phalanges meet the metacarpals and metatarsals.

Back in the Saddle Again

A saddle joint is a modified ellipsoidal joint, but with a pair of bones in a rider and saddle shape. The first metacarpal (of the thumb) and the trapezium (on of the carpals) form just such a joint. A saddle joint has a much wider range of flexibility, including movement on two axes (biaxial). One form of movement, called circumduction, should be well known to you if you've ever twiddled your thumbs.

Sock It to Me, Lucille!

Of all the joints in the body, this is, by far, the most flexible. If you have ever had a G.I. Joe doll, or a Barbie action figure, and you have ever changed the clothes of these exceedingly anatomically *incorrect* toys—no one is really shaped that way, especially in terms of *proportion*—you have seen this type of joint firsthand. In both the toy hip and toy shoulder, as in humans, there is a ball-like shape at the proximal end of the arm and leg that fits into the plastic shoulder and hip.

In humans, the ball is the head of the humerus and the head of the femur; the head of the humerus fits nicely into the rounded glenoid cavity of the scapula, whereas the head of the femur fits into the even more rounded acetabulum of the pelvic coxa, or pelvis bone. Imagine swinging your arm to throw the third strike in the bottom of the ninth, clinching your victory in game seven of the World Series, and you will have used a movement that is a wonder of the ball-and-socket joint: circumduction. A quick look at the intense number of tendons and ligaments at each of these joints, but especially in the hip, will explain why people are far more likely to dislocate their shoulders.

The Dancer's Alphabet

This title is an interesting analogy. No matter how glorious a dance is, it can be broken down into simpler parts. Many of a dancer's movements are combinations of many of the simple movements to follow. Given that all of the glory of literature in English, from Shakespeare, to Poe, to Swift, to Austen, to Dickens, to Dickinson, to Hiaasen, to, well, you name it, is all based on a brilliant juggling of the same 26 letters, I think it is safe to say that the analogy holds.

Flex Your Muscles

If you are trying to learn motions by memorizing a book, you are missing the point! People are visually, auditory, or kinesthetic learners. The last one, kinesthetic learners, learn by *doing*. I know, however, that every learner can learn this way. Take your book and stand up in front of a mirror. When you read about one of the motions in the book, *do it! In front of the mirror!* What's more, *say the name of the movement while you do it!* In essence, the best thing to do—I am not kidding!—is to choreograph a simple dance! This way you will see the movement (visual), hear the name (auditory), and feel the movement (kinesthetic). It works for my students every year. Trust me, you can't lose!

As you learn these movements, keep thinking in terms of opposites. The muscles that control these movements are found in what are called antagonistic pairs, and so it is only natural that the movements fall into pairs as well. Given all of the directional terms you learned in Chapter 1, this should be second nature to you. Don't plan on memorizing and forgetting these movements, by the way, because you will be using them, along with the directional terms and the names of the bones, whenever you learn muscle names. So? Let's get movin'! The following table provides an overview of the movements at synovial joints.

The Movements at Synovial Joints

Movement	Description	Examples
Angular	Hinge, ellipsoidal, saddle, ball-and-socket joints	See the following 5 movements
Flexion	Decreases the angle of a joint, moving away from anatomical position	Elbow, knee, phalanges
Extension	Increases the angle of a joint, returning to anatomical position	Elbow, knee, phalanges
Hyperextension	Continues extension, but moves beyond anatomical position (more than 180 degrees)	Neck, leg, wrist

continues

The Movements at Synovial Joints   (continued)

Movement	Description	Examples
Abduction	Decreasing angle laterally, moving away from anatomical position	Arm, leg, phalanges
Adduction	Increasing angle laterally, returning to anatomical position	Arm, leg, phalanges
Rotational	Ellipsoidal, saddle, pivot, ball-and-socket joints	See the 6 following movements
Pronation	Turns palm prone (dorsally)	Wrist
Supination	Turns palm supine (ventrally)	Wrist
Internal rot. (or medial)	Turns arm/leg toward trunk (or medially)	Arm, leg
External rot. (or lateral)	Turns arm/leg a way from trunk (or laterally)	Arm, leg
Left rotation	Turns head to the left	Head and neck
Right rotation	Turns head to the right	Head and neck
Special	Gliding, ellipsoidal, saddle, ball-and-socket joints	See the following 10 movements
Inversion	Faces sole of foot medially	Foot, ankle
Eversion	Faces sole of foot laterally	Foot, ankle
Dorsiflexion	Faces sole of foot medially	Foot, ankle
Plantar flex.	Faces sole of foot medially	Foot, ankle
Protraction	Moves jaw forward (out)	Mandible
Retraction	Pulls jaw backward (in)	Mandible
Depression	Pulls jaw open (down)	Mandible
Elevation	Pulls jaw closed (up)	Mandible
Lateral flex.	Angles head toward the shoulder	Head and neck
Opposition	"Opposable thumb," touches thumb to the tips of the other fingers	Thumb

A New Angle

One of the things I have to *unteach* every year is the idea that flexing and contracting is the same thing! Flexing is only one in a number of movements described as angular. Angular motion is, quite simply, motion in which one end of the bone is stationary, but the diaphysis, or shaft, of the bone changes its angle. First of all, we need to refer to the body in anatomical position (see Chapter 1), with the palms facing forward, and the arms, legs, back, and neck straight. Why? Because angular motion, including flexion, is motion relative to anatomical position.

When you stand in anatomical position, the bending of any joint that decreases the angle of that joint is known as flexion (see Figure 7.2). For example, in anatomical position, your arm, as a straight line, is a 180-degree angle; if you bend your elbow, to make your forearm parallel to the ground, you have a 90-degree angle. That is an example of flexion, as is the bending of the knee in the Hollywood foot-poppin' kiss. Don't forget fingers and toes!

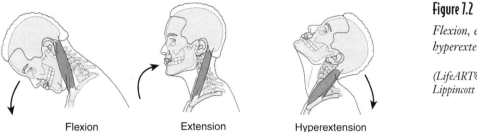

Flexion   Extension   Hyperextension

Figure 7.2

Flexion, extension, and hyperextension.

(LifeART©1989–2001, Lippincott Williams & Wilkins)

Extension, on the other hand, is simply the return of the joint to anatomical position. An extreme version of this is called hyperextension, when movement in the direction of the extension is continued, but beyond anatomical position. In some joints, such as the elbow and knee, such movement involves injury; just thinking about landing badly on the knee and having it bend forward instead of back is a creepy thought! Other joints, however, do this freely, as you do with your head when you look up at the sky. You hyperextend your leg at the hip every time you walk; you do the same thing with your arm if you swing your arms while you walk!

Abduction and adduction are similar, except the movement is along an entirely different plane (see Figure 7.3). Flexion and extension act on a sagittal plane, with the movement always in the anterior/posterior direction. Abduction and adduction, on the other hand, act along a frontal plane, with the movement in a lateral/medial direction. Abduction and adduction are the domains of ball-and-socket joints at the pelvis and shoulder, and the saddle joints at the base of each finger.

Crash Cart

Don't confuse flexion and extension of the elbow and knee with flexion and extension of the entire arm or leg (in which the elbow or knee may remain straight)! Flexion and extension are the only movements possible at a hinge joint, but they are also two of the many possible movements at ball-and-socket joints, for example.

It is easy to remember the difference if you look at the root of the words. *Abduction* comes from the root *abduct*, which means to "take away." The movement of your arms up and your legs wide at the start of a jumping jack is a classic example of abduction. *Adduction*, on the other hand is the opposite motion, to *add* the arms and legs to the body, as when a, soldier comes to attention. *Circumduction*, as its name implies, involves a circular motion, as you would when you swing your arms, such as in the windup to throw a ball (see Figure 7.3). This movement, in a limited fashion, happens at ellipsoidal and saddle joints, but it is best done at ball-and-socket joints.

Figure 7.3

*Abduction, adduction, and
circumduction.*

*(LifeART©1989–2001,
Lippincott Williams &
Wilkins)*

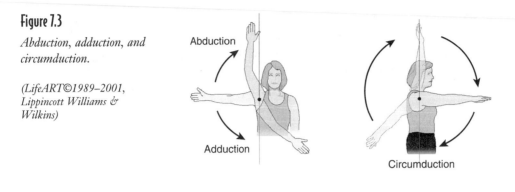

Don't Wait. Rotate!

Rotational motion is motion where the tip of the shaft of the bone stays at the same angle, but it rotates along its longitudinal (or long) axis, like a pencil rolling between your fingers.

Remember the directional terms *prone* (face down) and *supine* (face up) from Chapter 1? If you think about anatomical position, the hands are always facing forward (toward the ventral surface); the "back of your hand" is, like the rest of your body, dorsal. If you keep those in mind, *pronation* and *supination* shouldn't be a problem. Try this. Dangle your arm down to one side and try turning your palm so it faces backward, but *don't* move your shoulder! That movement is called *pronation*.

The opposite motion, from the palm facing backward to the palm facing forward is *supination*. It is important to note that at no point is there any movement at the shoulder. All of the movement is in the lower arm. In pronation the head of the radius pivots in the radial notch of the ulna, and the distal end of the radius ends up crossed over the ulna. This is the reason why anatomical position has the palms facing forward, for in this position the radius and ulna appear parallel. Figure 7.4 compares these different types of motions.

 Crash Cart _____

Don't confuse pronation and supination with medial and lateral rotation. Both pairs are rotational movements, but pronation and supination are specialized. In internal (or medial) rotation and external (or lateral) rotation of the arm, the radius and ulna stay parallel, and all the movement is at the shoulder (or at the hip in the case of rotation of the leg). Pronation and supination (of the arm only) are entirely due to the radius and ulna, with no movement at the shoulder.

Internal rotation (sometimes called medial rotation) and external rotation (sometimes called lateral rotation) differ from pronation and supination by the location of the rotation; unlike pronation and supination, there is no movement in the lower arm. In medial rotation all the movement is in the shoulder, and the arm rotates so that, once again, the palm faced backward, but this time the radius and ulna don't cross. Think of that movement as rotating the arm toward the midline of the body, hence the name medial rotation. Reverse that motion, moving the shoulder to return to anatomical position, and you have just done lateral rotation. Once again, think of rotating the arm toward the side, putting the lateral in lateral rotation. Also note that medial rotation and lateral rotation are also possible in the legs.

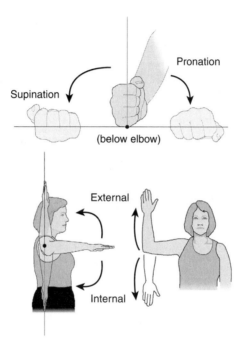

Figure 7.4

Pronation, supination, medial (internal) rotation, and lateral (external) rotation.

(LifeART©1989–2001, Lippincott Williams & Wilkins)

Right rotation and left rotation of the head are fairly straightforward, as it were (see Figure 7.5). Why do I specify right and left? I have to! With your head being right on the midline, I can't use the words *medial* and *lateral*, now can I?

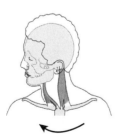

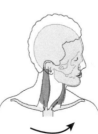

Right Rotation   Anatomical Position   Left Rotation

Figure 7.5

Left rotation and right rotation.

(LifeART©1989–2001, Lippincott Williams & Wilkins)

Special Movements

Special movements are called that simply because they don't easily fit in the other categories. You can, however, divide them according to the part of the body involved. For those of you foot fetishists, let's start from the ground up, as it were. If you stand up with both feet flat, and then swivel your foot so the soles are facing each other, inward, in other words, that is called *inversion*. If you move your feet so that the soles are facing outward—in an imitation of the Jerry Lewis walk?—you are doing *eversion*.

Now imagine you have a cat who spends half the night hacking up fur balls, and after you wake up to your alarm which was set, as usual, freakishly early, you stumble right into a puddle of vomit and hair (Eeeeewwww). Right after you put your bare foot into the gunk, you stop cold and lift your toes off the ground, with your heel still on the floor. That disgusting scenario is a perfect example of *dorsiflexion*, and it is a scenario that my daughter's favorite cat, Jimmy, has left for me far too many a morning!

At least *plantar flexion* is a little less gross. Stand up, and then stand on tiptoe. That's it, plantar flexion. Presumably they are both called flexion because anatomical position starts out with the foot at a 90-degree angle. If you stand up on tiptoe you can also feel very clearly which muscle is being used: the gastrocnemius. Figure 7.6 compares these different movements.

Figure 7.6

Inversion, eversion, dorsi-flexion, plantar flexion.

(©*Michael J. Vieira Lazaroff*)

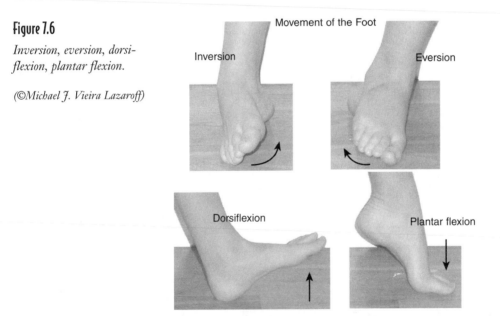

A little higher up and you can focus on some movements that involve the jaw. Have you ever been so surprised that your jaw drops? That movement is called depression, because to "depress" is to move down. If that's hard to swallow, the opposite makes perfect sense: To raise the jaw is called elevation (see me in Figure 7.7).

Movement of the Jaw

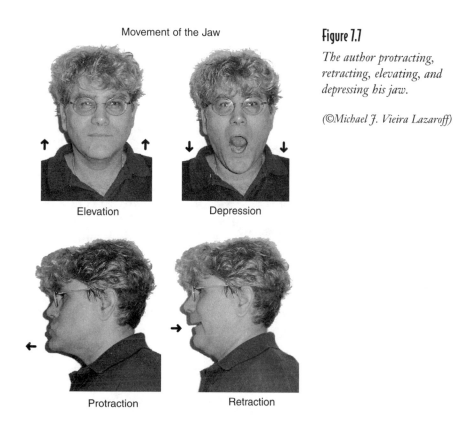

Elevation   Depression

Protraction   Retraction

Figure 7.7

The author protracting, retracting, elevating, and depressing his jaw.

(©Michael J. Vieira Lazaroff)

The other two movements are kind of funny looking, and Jim Carey has made a good career out of using these two. Protraction is when you stick your jaw forward, as if you are trying to impersonate a cave man! Now imagine you are just offered a strawberry and used kitty litter pie … I know, that's gross, but that's the point! Don't you pull your jaw back in disgust? At least I hope you do! That movement is called *retraction* (see me in Figure 7.7).

There are just two movements left in the dancer's vocabulary. One involves the head. If you tilt your head from side to side, that is another form of flexion, and since it's to the side, it makes sense that it be called *lateral flexion* (see Figure 7.8). Just as you had to specify with rotation of the head, you need to specify right and left.

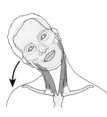  

Right Lateral Flexion   Anatomical Position   Left Lateral Flexion

Figure 7.8

Lateral flexion.

(LifeART©1989–2001, Lippincott Williams & Wilkins)

Saving the best for last, I have just one more movement. This movement is only possible with the special gift called an opposable thumb, which is why it's called *opposition*. Look at your hands, and, with one hand, touch the tip of your thumb, in quick succession, with the tip of each of your other fingers. That little movement makes something called the precision grip possible.

When my daughter was very young (I'm talking a few months old) she used to pick up Cheerios off her high chair by the handful. When her hand got sticky enough she would just put her wet hand on the pile of Cheerios and eat them off her hand. As she grew older, and her brain trained her thumb and forefinger to pick up Cheerios one by one, she would do it that way. It is a far less efficient way to eat cereal, but that is a crucial step in survival, that precision grip, which led humans to excel in terms of using tools. Even in the details of eating Cheerios, there is evidence of evolution!

The Least You Need to Know

♦ There are many different types of articulations (or joints) that are classified according to both structure and function.

♦ The three functional types of joints are the immobile, slightly movable, and freely movable.

♦ Freely movable synovial joints come in six different types based on the movements possible at each joint.

♦ The various movements of which the body is capable are divided according to angular, rotational, and special movements.

♦ Movements, just like directional terms and muscles, are found in pairs, with each motion in a pair producing the opposite effect of the other.

Chapter 8

A Meaty Idea

In This Chapter

- The structure of a muscles and muscle cells
- The neuromuscular junction's control of muscle cells
- How to interpret a myogram
- How a sarcomere works to cause a muscle to contract
- The development and breakdown of rigor mortis

You might have noticed a running trend in this book so far: The body is more complex than it seems. It is oh so easy to assume that problems within the body have a single cause, that structures and systems have a single function. With a little time, however, it becomes apparent that nothing about the body is that simple. Even a seemingly simple system such as the muscular system has multiple functions.

Amid this complexity there is, nonetheless, common ground. The same processes of cell transport used elsewhere, particularly in the nervous system, are used here as well. The muscular system is one of my favorites because a number of small details (integrating the concepts of active transport, facilitated diffusion, exocytosis, oxygen and energy usage, and so on, into a coherent whole) suddenly make sense when you apply them to the big picture, which is when and how muscles contract.

Function Junction

True to form, muscles weren't content doing one thing. Muscle tissue actually has four functions in the body:

- **Motion.**   This refers not only to getting from place to place, but also to movements within the body, such as peristalsis (see Chapter 14), or even developing "goose bumps"!

- **Stability and posture.**   If you could cause all of your muscles to relax you would collapse in a heap! It is surprising how many muscles relate to posture.

- **Controlling organ volume.**   Hollow organs, such as the gallbladder, need to release their contents from time to time, and they need muscles to do it.

- **Generating body heat, or thermogenesis.**   (Break it down: *thermo* = heat and *genesis* = creation, so thermogenesis = creation of heat!) The breakdown of glucose produces 36 ATP (with O_2 present), and releases 62 percent of the energy as heat! Most body heat is produced through the contraction of muscles.

All these glorious functions are related to four rather cool characteristics of muscle tissue: excitability, contractility, extensibility, and elasticity. Excitability, or (as I prefer to call it) irritability, refers to the ability of muscles to be stimulated by nerves. Contractility makes sense, because, after all, muscles contract. Extensibility is a little more difficult, because muscles don't lengthen on their own, but the fact that they can do so without causing *damage* is crucial. Last is the idea of elasticity, which means that, after any shortening or lengthening, the muscle will return to its original shape.

All Parts, Big and Small

To understand how a muscle contracts, it is helpful to think of those old Roman ships with the galley of slaves rowing the oars in unison. Each slave had only so much strength, but all their strength combined, not to mention the ability to row together in unison, gave the ship's oars incredible power.

A muscle not only has multiple units, but also needs that connect it with the rest of the body. Being irritable, muscles need to have a connection to motor neurons. All the need for ATP means the need for O_2 and CO_2 exchange, not to mention H_2O, waste, and glucose transport. These needs all require the presence of blood, lymph, and the vessels to carry them. Nerves, blood vessels, and lymphatics are held in connective tissue around the muscles, tissues known as fascia.

Fascia Galore!

Fascia refers to connective tissue that surrounds muscle, which provides both protection and flexibility. It is helpful if you have ever cooked chicken and had a chance to handle both raw and cooked meat, then you'll know fascia. Meat is, after all, muscle! The act of cooking causes the fascia to break down, which makes the comparison of raw and cooked meat very useful. You might never be able to look at meat the same way again, but you are, after all, what you eat!

Superficial fascia is so called because it is the closest to the skin, being immediately deep to the skin (which is why it is also called *subcutaneous fascia*). If you have ever eaten cooked chicken you probably noticed that the skin comes off easily. On the other hand, removing the skin from raw chicken is not so easy! To do so you need to pull up on the skin and gently cut the *superficial fascia* that holds the skin to the muscle. This fascia breaks down during cooking, which is why it's so easy to take the skin off after cooking.

This layer of fascia also has adipose tissue (fat), and it is where most excess fat is stored when you exceed your ideal body weight. In addition to storing fat and water, this layer provides insulation from heat loss, a cushion-like protection from trauma, and a place for vessels and nerves in and out of muscles. Figure 8.1 shows a view of the fascia.

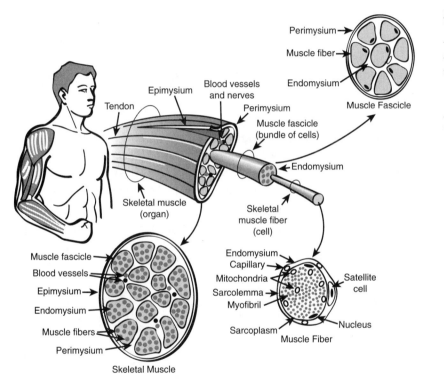

Figure 8.1

Fascia appears at several levels in and around every muscle.

(©2004 Pearson Education, Inc., publishing as Benjamin Cummings)

Since muscles come in groups, these groups are held together, with division to separate them by a deeper layer of fascia called, what else, *deep fascia*. This layer, by providing a division between muscles, helps the muscles to contract without interference. Each muscle in the group has its own layer of outer fascia called *epimysium*. Thus each muscle in the group has its outer *epimysium*, all of which is surrounded by *deep fascia*.

Fascicles

Have you ever noticed that meat that is overcooked tends to be stringy? That illustrates an interesting point—that muscles are made of many parallel fibers that contract in the same direction. Individual muscle cells are called myofibers, and these myofibers are arranged in groups of around 10 to 100, and these groups are called *fascicles*. Between these myofibers of every fascicle is a connective tissue called *endomysium*. Finally, around every fascicle is another layer of fascia called *perimysium* (*peri* meaning around, in this case around the fascicle). When you overcook meat the perimysium breaks down, thus highlighting the individual fascicles, making the meat look stringy!

Myofibers

This layer upon layer effect continues in each muscle cell or myofiber. In case you haven't noticed, *myo* means muscle; so when you read *myofiber* think muscle fiber. These cells have a few terms of their own. For instance, using another prefix, *sarco-*, which also means muscle, the cytoplasm becomes the sarcoplasm, and the E.R. (*endo*plasmic reticulum) becomes the S.R. (*sarco*plasmic reticulum).

Given the importance of the membrane in receiving the message to contract, the myofiber membrane becomes the *sarcolemma*. Not to be outdone, the neuron's cell membrane is called the *neurilemma*. The arrangement of proteins in a muscle cell is its crowning glory, and the thread-like proteins in the sarcoplasm are called *myofibrils*. These myofibrils form striations (see Figure 8.2) in skeletal and cardiac muscle of alternating myofilaments (thin and thick filaments), which I will explore when I discuss the unit of muscle contraction, the sarcomere.

Figure 8.2

Two views of a myofiber.

(©2003 www.clipart.com and ©Michael J. Vieira Lazaroff)

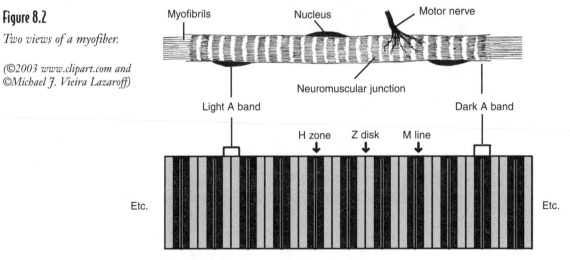

Note: The M line is in the middle of the H zone.

The Neuromuscular Junction

A myofiber is an extraordinarily sophisticated piece of cellular machinery, but, in the end, it still only does what it is told! Each muscle cell that contracts is connected to a motor neuron. Muscles, when they contract, go whole hog, which means each cell is in an all-or-nothing mode. In order for a large muscle to have a strong contraction means that every cell in that muscle must be told to contract; weaker contractions mean fewer cells contract. So how does a nerve cell tell a muscle cell to contract?

First, the nerve and muscle cells must make contact, yet the two cells don't actually touch. The junction between a neuron and a muscle fiber is called the *neuromuscular junction* (NMJ) (see Figure 8.3). The junction, just as in the junction between neurons, is called a chemical synapse, and there is always a space between the cells called a synaptic cleft.

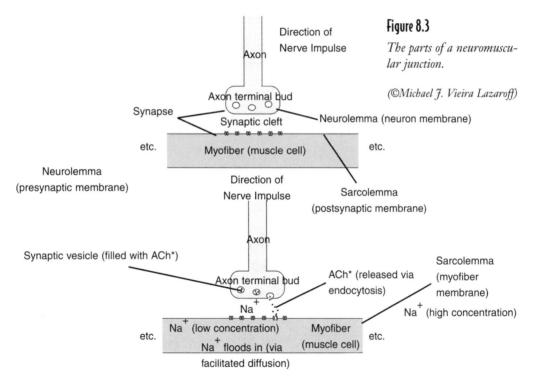

Figure 8.3

The parts of a neuromuscular junction.

(©*Michael J. Vieira Lazaroff*)

*ACh is short for the neurotransmitter acetylcholine.

The membranes of the two cells in a synapse are named because of the direction of the nerve impulse: the presynaptic membrane (the neurolemma) and the postsynaptic membrane (the sarcolemma). The message transmitted from the neuron to the myofiber is a chemical one called a *neurotransmitter*, and they work by changing the permeability of the postsynaptic membrane. This type of synapse, with a synaptic cleft, is a chemical synapse; heart cells, and many nerve cells, have electrical synapses, in which the cells actually touch and travel through communicating junctions called *gap junctions*.

Neurotransmitters and Exocytosis

Neurotransmitters are chemical messengers that travel across the synaptic cleft between neurons and neurons, or neurons and myofibers. A neurotransmitter called *acetylcholine* (ACh) is used at the neuromuscular junction. Acetylcholine is initially produced in the Golgi bodies (in the cell body), and then travels down the axon to the axon terminal bud in synaptic vesicles; since there is no E.R., and no Golgi bodies, in the terminal bud, the synaptic vesicles are recycled using both endocytosis, exocytosis, and mitochondria to recycle the ACh.

So what does it mean to change "the permeability of the postsynaptic membrane?" To start off, have another look again at active transport. You might remember that active transport is the movement of molecules against the concentration gradient, from a low concentration to a high concentration. In this case, using a nifty little engine called the Na^+/K^+ pump, sodium ions are kept at a high concentration in the synaptic cleft. The sarcolemma, when in this state, is considered polarized. Now remember, keeping up this membrane polarized takes energy, in the form of ATP. Ironically, this energy is used to keep the muscle relaxed.

At this point it is a good idea to look at what the word *relaxed* actually means. Most people envision relaxed as being somewhat akin to a teenager crashed on a couch, but that is nothing like relaxed muscle. Survival depends on the ability to react quickly in emergency situations, and a quick mobilization of muscles is crucial to that ability. In that sense, a relaxed muscle is a lot like a bow and arrow, with the arrow pulled back ready to be released. The bow and arrow, even though they are not moving, are primed to move quickly; just like a muscle, this state also requires energy.

The big advantage to active transport, in this situation, is that it creates a situation in which the sarcolemma can become depolarized all the more quickly. Diffusion alone is not the fastest way to move the sodium back to a state of equilibrium. The fastest way, without a doubt, is through facilitated diffusion. Facilitated diffusion requires a channel for the sodium to pass through, but if the channel were always open it would be awfully hard to maintain active transport. There has to be a way to open and close the channel when needed. When the muscle is relaxed, the channel is closed; it is only opened when the muscle cell needs to contract. It makes sense that such a channel should be kept under lock and key. The neurotransmitter is the chemical message that tells the cell to contract; the neurotransmitter, a cool molecule called acetylcholine (ACh), is the key. The ACh is kept in synaptic vesicles in the axon terminal bud. When the axon receives the message, via a similar change in the polarization of the neurilemma, the synaptic vesicle fuses with the neurilemma and releases its contents (the ACh) into the synaptic cleft through exocytosis. The key has been released; all the key needs now is a lock.

ACh, Receptors, and Facilitated Diffusion

All along the postsynaptic membrane, the sarcolemma, are receptors for the acetylcholine (ACh). The ACh binds with the ACh receptors, which open the channel, allowing the sodium ions to flood through the postsynaptic membrane via facilitated diffusion. Once again, the combination of active transport and facilitated diffusion makes this and depolarization incredibly fast (only two milliseconds!).

The only problem with this scenario is how the muscle gets relaxed again. This is easier to understand when you think about the enzyme acetylcholinesterase (ACh-esterase). When the ACh is released into the synaptic cleft, it is broken down very quickly by the acetylcholinesterase. In order to maintain a sustained contraction, the motor neuron needs to release an almost continuous supply of ACh—luckily there's a constant recycling of the neurotransmitter, or it would not be possible to contract a muscle for more than a few milliseconds!

My, Oh, Myogram!

Have you ever heard of a muscle twitch? A myogram is a graphic representation of the speed and strength of a muscle contraction (see Figure 8.4). To understand a myogram, let's start by looking at a muscle twitch. In this graph, the X-axis indicates time, and the Y-axis indicates the strength of the muscle contraction. If you look at the following figure, you will notice that a flat line indicates the relaxed muscle, and that the quick contraction and relaxation are indicated by the bell curve.

What you may not notice right away is the location of the stimulus by the motor neuron. If you look carefully you'll notice that the contraction does not start right after the stimulus, but that there is a delay called the latent period, before the muscle contraction actually starts. What is happening here, during the latent period? Think about it, for there are several steps, in order: exocytosis of the ACh, ACh binds to the ACh receptors, the channels open, and Na^+ ions flood through the newly opened channels (facilitated diffusion).

The rise of the curve is called the *contraction period*, and the decline is called the *relaxation period*. The relaxation period, however, is a reversal of the latent period: The ACh is broken down by the acetylcholinesterase, the ACh receptor channels close, and active transport returns the Na^+ ions to the synaptic cleft. So why is this period so much longer than the latent period? Remember, since active transport has to move the ions *upstream*, it does take longer, but also remember that facilitated diffusion is so much faster when it is preceded by active transport!

You've Fallen into My Treppe!

Each muscle twitch is a brief event. Because of the rapid breakdown of ACh, the myofiber starts to relax almost as soon as it starts to contract! Such a contraction cannot be full strength. If a second stimulus happens in the middle of the relaxation period, the strength of the resulting contraction will be a combination of the strength of the contraction at the point of the second stimulus, as well as the new contraction. This summative effect is called the staircase effect, or *treppe*, because of the staircase-like appearance of the myogram (*treppe* means staircase in German).

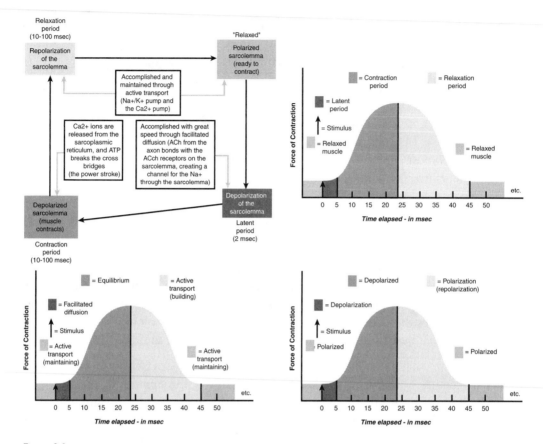

Figure 8.4

If you compare the flow chart of membrane transport with the various labeled myograms, you will see that they are representative of different versions of the same event.

(©Michael J. Vieira Lazaroff)

If you have ever heard of, or hopefully *had*, a tetanus shot, you probably think of tetanus as a bacterium. The bacterium, however, got its name—*Clostridium tetani*—from the sustained muscle contraction known as tetanus. In incomplete tetanus the muscle reaches peak strength, but it cannot maintain it (the relaxation period is allowed to start) due to the frequency of the stimuli from the motor neuron. When there are more frequent stimuli the relaxation period is not allowed to start, which results in a sustained contraction known as tetanus. (An infection with the tetanus bacterium affects muscle cells, causing sustained contractions where they are not wanted. For this reason tetanus, which is sometimes called *lockjaw*, can be fatal if untreated.)

Ca²⁺, Not Just for Healthy Bones!

Following all of this influx of Na^+ ions, a whole series of events occurs. The endoplasmic reticulum in muscle cells not only has a special name (the sarcoplasmic reticulum or S.R.), but it also has a special function. The S.R. acts as a calcium storage device (see Figure 8.5). Similar to the sarcolemma, the S.R. maintains a high concentration through active transport. One difference in the S.R. is the presence of a protein, with a really cool name, by the way, that helps to store the calcium: calsequestrin. Like a sequestered jury, the calcium, which is in its ionic form, Ca^{2+}, the Ca^{2+} is held in place by that protein.

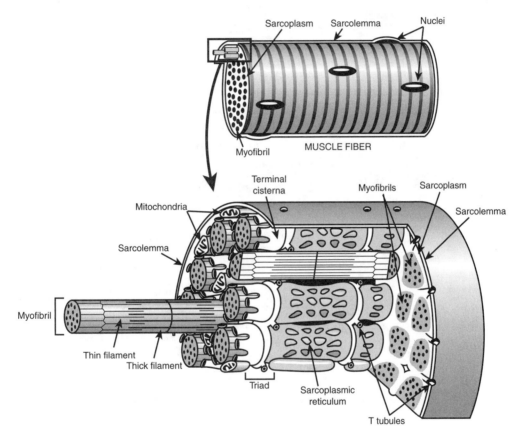

Figure 8.5

The conjunction of transverse tubules, and the layout of the sarcoplasmic reticulum, speed up the release of Ca^{2+} ions into the sarcomere.

(©2004 Pearson Education, Inc., publishing as Benjamin Cummings)

It's a good thing, too, because that Ca^{2+} can't be trusted. Its release, also through facilitated diffusion, triggers the actual contraction phase of the muscle. It turns out that the exocytosis of acetylcholine through the presynaptic membrane starts a chain reaction—first the influx of Na^+ through the postsynaptic membrane, then the exodus of Ca^{2+} ions out of the S.R., and then, well, you'll find out in the moving proteins section!

Sliding Home with Sarcomeres

At this point you need to take a good look at the place where the actual contraction happens. A number of fascinating proteins work together to make contraction possible. I will focus on striated muscles (skeletal and cardiac) here, because the proteins are highly organized into parallel stripes (striations) that literally get closer together when the muscle contracts. On a smaller scale, these striations are formed by sarcomeres, which are the unit of muscle contraction.

It is helpful here to take a look at the appearance of a striated muscle cell, for the parallel stripes fall into a clear pattern. Each part of the stripes has a specific name that is identical to the corresponding proteins in the sarcomere. Remember that each striation represents hundred of parallel sarcomeres. In the myofiber as a whole you first see the division into dark A bands, and light I bands, but closer examination reveals a series of other details such as the Z disk, H zone, and M line. Take a moment and try to find these same structures in the diagram of the sarcomere (see Figure 8.6).

Figure 8.6

The sarcomere is the unit of muscle contraction. Compare the parts of the sarcomere to the parts of the previous myofiber diagram.

(©Michael J. Vieira Lazaroff)

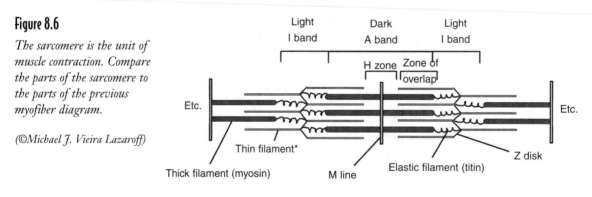

*The thin filament is made of three different proteins: actin, troponin, and tropomyosin. The troponin and tropomyosin form a helix, the T-T complex, around the actin.

The biggest details in the sarcomere are probably the thick and thin filaments. The thick filament, as you can imagine, corresponds to the dark A band, and the thin filament corresponds to the light I band. Much of the rest of the sarcomere can also be seen on the myofiber: Z disk, H zone, M line. The only structure that is not visible on the myofiber is the elastic filament (made of a protein called titin), which connects the thick filament to the Z disk.

Shrinking Stripes

When you take a much closer look, you see that there are more details to both the thin and thick filaments, and understanding those details is vital to understanding how a sarcomere works. The thick filament is made of a protein called *myosin*. Myosin is arranged as a large (thick) filament with numerous tails coming off in all directions, and at the end of each *myosin tail* is the *myosin head*. Also on the myosin head is an enzyme called *ATPase*, which is responsible for converting ATP into ADP, a free phosphate, and energy (see Chapter 3). The thin filament, however, is far more complex. The principal thin filament protein involved in contraction is called *actin*, and a helical chain of actin molecules forms the core of the thick filament. On the actin molecule is a *myosin binding site*, where the myosin head attaches; the attachment of myosin head and actin molecule is called a *cross bridge*. When left alone, actin and myosin bind together and stay there. A muscle in such a situation would be rigid, which is hardly the state of a relaxed muscle. It is therefore important that the cell have a way to lock and unlock the actin.

Surrounding the actin, locking it as a cable lock locks up a bicycle, is a pair of protein molecules called *troponin*, and *tropomyosin*. Because the molecules go together they are referred to as the *troponin-tropomyosin complex* (or *T-T complex* for short). The helical T-T complex wraps around the helical actin molecules, with the tropomyosin acting as the cable, and the troponin acting as the lock. When the T-T complex is locked, the tropomyosin covers up the myosin binding sites on the actin, thus making the muscle relaxed, and allowing it to be stretched by other muscles. (I will look at antagonistic pairs in more detail in Chapter 9.)

When unlocked, the tropomyosin moves aside, thus exposing the myosin binding sites and allowing the actin-myosin cross bridges to form. The filaments then contract, which means that the thin filaments slide over the thick filaments (hence the *sliding filament* name) in each sarcomere, thus making the muscle contract (see Figure 8.7). All of this raises two questions: How does the T-T complex get unlocked, and after forming the first cross bridge, how does the muscle avoid staying rigid and actually move? To answer those questions you need to look at two substances: Ca^{2+} and ATP.

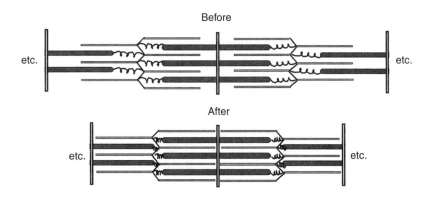

Figure 8.7

The shrinking of the space between the Z-discs, due to the sliding of thin filaments over thick filaments, makes muscle contraction possible.

(©Michael J. Vieira Lazaroff)

Moving Proteins

The calcium ions act as a key for the troponin, with the keyhole being a place ion the troponin molecule called, conveniently enough, the *calcium-binding site.* As you should remember, the calcium is released from the S.R. only upon the influx of sodium ions, which was in turn triggered by the ACh stimulus from the motor neuron. At this point, the muscle is free to contract. There is just one problem: the rigid actin-myosin cross-bridge.

To break each cross bridge requires the energy from one ATP molecule. Luckily the ATP-ase can be found at the end of each myosin head. It is helpful here to understand that forming the cross bridge the myosin head will swivel toward the M line (which acts to anchor the thick filament). By using ATP to break the bond, the swivel action pulls the thin filament (inertia also plays a role here, as an object in motion will stay in motion until acted upon by a force) toward the M line in an action called the *power stroke.*

By repeatedly forming and breaking the cross bridges, and thus doing repeated power strokes, the thin filaments as a whole will slide closer to the M line, thus making the whole sarcomere shorter. Even though ATP is needed for the active transport that keeps the muscle relaxed, far more ATP is needed to power the repeated power strokes that make contraction possible. All this is summarized in Figure 8.8.

Out of Breath?

This ATP use requires the presence of numerous mitochondria. In Chapter 3, I mentioned that each glucose molecule yielded enough energy to convert 38 ADP to 38 ATP, and I also explained that an additional 62 percent of the energy in the glucose was released as heat. This should explain why you sweat so much when you exercise, what with all that excess heat!

In addition to the glucose being delivered by the blood, muscle cells also store glucose in the form of the polymer glycogen (see Chapter 3). Glucose utilization also requires oxygen. In addition to the oxygen delivered by the blood, muscle cells can also hoard oxygen due to a hemoglobin-like molecule called myoglobin that can be found in the sarcoplasm, and which gives raw meat its color.

Despite all these precautions, sometimes the muscle cells demand more energy than the body can provide. Having enough glucose usually isn't the problem, but to get the maximum amount of ATP your cells also need oxygen. As you exercise, and your oxygen needs increase, you start to breathe more quickly, and more deeply, in order to get the oxygen, and thus the ATP, to all those sarcomeres. If you can't get enough oxygen, you get the sensation of being "out of breath," which is the body's way of saying *slow down!*

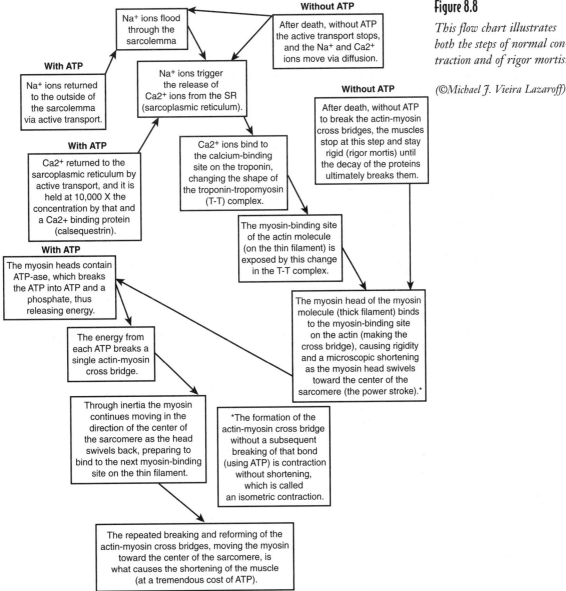

With ATP

Na+ ions returned to the outside of the sarcolemma via active transport.

Na+ ions flood through the sarcolemma

Without ATP

After death, without ATP the active transport stops, and the Na+ and Ca2+ ions move via diffusion.

Figure 8.8

This flow chart illustrates both the steps of normal contraction and of rigor mortis.

(©Michael J. Vieira Lazaroff)

Na+ ions trigger the release of Ca2+ ions from the SR (sarcoplasmic reticulum).

With ATP

Ca2+ returned to the sarcoplasmic reticulum by active transport, and it is held at 10,000 X the concentration by that and a Ca2+ binding protein (calsequestrin).

Ca2+ ions bind to the calcium-binding site on the troponin, changing the shape of the troponin-tropomyosin (T-T) complex.

Without ATP

After death, without ATP to break the actin-myosin cross bridges, the muscles stop at this step and stay rigid (rigor mortis) until the decay of the proteins ultimately breaks them.

The myosin-binding site of the actin molecule (on the thin filament) is exposed by this change in the T-T complex.

With ATP

The myosin heads contain ATP-ase, which breaks the ATP into ATP and a phosphate, thus releasing energy.

The energy from each ATP breaks a single actin-myosin cross bridge.

The myosin head of the myosin molecule (thick filament) binds to the myosin-binding site on the actin (making the cross bridge), causing rigidity and a microscopic shortening as the myosin head swivels toward the center of the sarcomere (the power stroke).*

Through inertia the myosin continues moving in the direction of the center of the sarcomere as the head swivels back, preparing to bind to the next myosin-binding site on the thin filament.

*The formation of the actin-myosin cross bridge without a subsequent breaking of that bond (using ATP) is contraction without shortening, which is called an isometric contraction.

The repeated breaking and reforming of the actin-myosin cross bridges, moving the myosin toward the center of the sarcomere, is what causes the shortening of the muscle (at a tremendous cost of ATP).

If you continue to exercise, the muscle cells will start to process glucose in the absence of oxygen. The first step, glycolysis, is the same, producing only 2 ATP. The pyruvic acid (see Chapter 3) is then converted, because of the lack of oxygen, into lactic acid (in a process known as lactic acid fermentation). The lactic acid buildup, which means a drop in the pH of the sarcoplasm, is painful. Pain is, of course, the body's way of telling you to stop what you are doing. When your muscles are feeling this pain, they are working too hard in the absence of oxygen. No oxygen means you are only getting $1/19$ the amount of energy, and you are in danger of *running out of energy!* That seems like a good point to stop!

Rigor Mortis

What will happen when you die? Rather than getting into a lengthy discussion of the soul, let's concentrate on what we know happens to the body. Life is spent in a relentless search for food (energy); this search doesn't end when you are unconscious or in a coma, for you will start to break down your own tissues, just to get your precious requirements of ATP. When you die, some of this breakdown will continue automatically, for longer in the big muscles than in the small ones, but what happens when you ultimately run out of ATP?

Medical Records

Have you ever wondered why dissection specimens are so stiff? If you think about it, if rigor mortis follows the loss of ATP, and the only way to reverse it (or rather continue forward to a loosening of the cross-bridges, as it is impossible to go back) is to have the muscles decay, and the specimen has been preserved to prevent decay, you have effectively trapped the specimen in a state of permanent rigor mortis!

Let's break it down, as it were: (1) the active transport stops as soon as the ATP runs out, (2) the Na^+ and Ca^{2+} flood into the sarcoplasm, (3) the Ca^{2+} unlocks the T-T complex, (4) the myosin-binding site on the actin forms a cross-bridge with the myosin heads on the thick filament, but (5) without any more ATP those cross-bridges cannot be broken. The muscle gets stiff, but it doesn't contract. This is the essence of rigor mortis. The only thing that ultimately breaks those cross-bridges in the absence of ATP is the breakdown of the proteins when they decay.

The Least You Need to Know

- Striated muscle cells have clearly defined patterns of proteins that make up sarcomeres, the unit of muscle contraction.

- Motor neurons stimulate muscle cells to contract by causing Na+ ions outside the polarized membrane to flood through, thus making the membrane depolarized.

- The repeated formation and breaking of actin-myosin cross-bridges produces multiple power strokes, thus making the thin filaments slide over the thick filaments.

- Muscle cells have a way of both locking the sarcomere (keeping the muscle relaxed), and of providing enough energy (ATP) to power the contractions.

- Rigor mortis is best understood as a continuation of all the other processes of stimulation and contraction, but without ATP.

Chapter 9

Muscle Bound

In This Chapter

- ◆ Parts of a skeletal muscle
- ◆ Types of skeletal muscle
- ◆ Types of levers
- ◆ How muscles are named
- ◆ Specific muscles

Now that you've learned the hard part, how the muscle cells actually work, it's time to learn the muscle names. There are about 700 muscles, including both superficial and deep, too many to cover in this book. Any good college-level text will provide numerous tables that provide specifics about origin, insertion, action, and motor nerves, should you need more specific information, but they can be awfully hard to interpret.

My job here is to make it all easier. All the muscle names can become a poly-syllabic sea that can leave the student at sea. The names, however, are actually very simple, once you learn the principles of muscle naming, because the names often provide incredible clues as to their location and action. This chapter, like Chapter 6 was for the skeleton, is a roadmap to help you to explore the muscular system.

Characteristics of Muscles

There are a few characteristics of muscles that I have yet to cover. Chapter 8 dealt mainly with the microscopic; this chapter deals with the larger characteristics, the macroscopic. These details are important in understanding the larger function of skeletal muscles, whereas the last chapter dealt, in general, with characteristics of all types of muscles.

Origin and Insertion

Skeletal muscle, true to its name, attaches to the bones of the skeleton. Such attachment is accomplished via tendons to at least two bones. But is the attachment the same for each bone? All the physical characteristics may be the same, but the function of each attachment is very different. For one thing, in order to provide controlled movement, it is important not only that the muscle *moves* a bone, but that the muscle is *anchored* while it moves the bone. Given that, there are three parts to a typical skeletal muscle: *origin, belly,* and *insertion* (see Figure 9.1).

Figure 9.1

Every skeletal muscle consists of tendons on either end, with an origin connected to one end, the insertion at the other, and the belly in the middle.

(LifeART©1989–2001, Lippincott Williams & Wilkins)

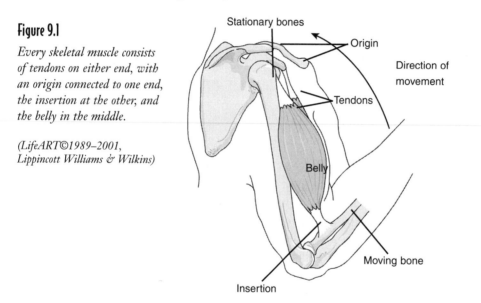

Stationary bones
Origin
Direction of movement
Tendons
Belly
Moving bone
Insertion

The widest part of a muscle is called the belly. The origin and insertion deal with the tendon attachments. First of all, the bone needs to be anchored; the attachment to the unmoving bone is called the origin. Next, the contraction of the belly pulls on the tendons, and the other bone moves; the attachment to the *moving* bone is called the insertion.

The Three Classes of Levers

The origin and the insertion bring up the concept of leverage. Synovial joints not only allow for movement, but they can act as part of a lever. There are three basic parts to any lever: fulcrum (F), effort (E), and resistance (R). The fulcrum (F) is the fixed point around which the lever moves. The effort (E) is a force that causes the lever to move. Last, the resistance is the weight that is acting against the movement; this resistance includes the weight of the bone being moved, plus any object that is being carried or moved.

Levers are divided into three classes, based on the location of the three parts in relation to each other, as you can see in Figure 9.2. Their relationship to each other says quite a bit about the lever. Levers can be used to increase or decrease the amount of work done for a particular effort. The definition of work is force multiplied by distance (W = FD).

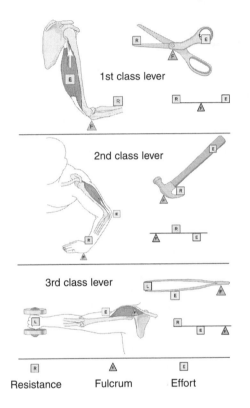

Figure 9.2

The three types of levers, both in the body and in common objects.

(LifeART©1989–2001, Lippincott Williams & Wilkins)

If you have ever been on a seesaw, you have been on a first-class lever (with the fulcrum in the middle). In the body, a first-class lever is used to lift your head. The effort is provided by the trapezius muscle pulling on the occipital bone, the weight of the face acts as the resistance, and the fulcrum is the joint between the occipital bone and the atlas vertebra.

Second- and third-class levers both have the fulcrum at one end; in both of these the direction of the effort is the same as the movement of the resistance. A second-class lever places the resistance in the middle; a typical example of this is a wheelbarrow. Second-class levers are the rarest in the body; one example is when you stand on tiptoe (using your gastrocnemius muscle to do plantar flexion), your toes and the ball of your foot make up the fulcrum, the effort is from the gastrocnemius, and the resistance (the weight of your body) is in the middle.

A third-class lever places the effort in the middle of the fulcrum and the resistance. These are the most common levers in the body, simply because the insertion of the muscle is proximal to the joint; were the insertion distal, there would need to be a lot of extra tissue needed to cover the muscle. (Imagine a straight muscle extending from the scapula to the distal part of the radius, and the inside of your elbow would disappear!) The disadvantage (mechanical disadvantage) of third-class levers is that quite a bit more strength is needed (effort) to pull on the bone (resistance).

Flex Your Muscles

To compare the amount of effort needed in a second- and a third-class lever, place your elbow on a table and relax your forearm completely. Now pinch your skin on the back of your hand and lift your forearm. By placing the effort far from the fulcrum (your elbow), you have made a second-class lever. Now pinch your skin close to the inside of your elbow and lift your forearm ... not so easy, is it? You have just switched the location of the effort and the resistance, making a third-class lever.

Arrangement of Fascicles

Fasciculi, or fascicles (see Chapter 8), can be arranged in very different ways. In parallel muscles the fascicles are, well, parallel, and the belly is the same width as the tendons. Most people think of muscles as being *fusiform*, which means that the fascicles are almost parallel, but that the belly is wider, and the muscle tapes toward both the origin and the insertion. Both these types have short tendons on each end.

Pennate muscles have shorter fascicles and much longer tendons, sometimes almost as long as the whole muscle. The location of the fascicles, and the number of tendons, determines which *type* of pinnate muscle it is. If the fascicles are on one side, it's called *unipennate*, and it's *bipennate* if the fascicles are on both sides. *Multipennate* muscles have multiple tendons, with the fascicles on both sides of the tendons. Figure 9.3 shows the various possible arrangements of fascicles in muscles.

The only muscle arrangement left is circular. These muscles have the fascicles in roughly concentric circles around an opening. The orbicularis oris, around the mouth (pucker up!), and the orbicularis oculi, around the eyes (Wink! Wink!), are examples of this type.

Acting in Concert

With almost 700 muscles, both superficial and deep, it makes sense that some of them might work together. If you remember the structure and function of the sarcomere (see Chapter 8), you may remember that muscles only work in one direction: contraction. After a contraction the muscle is at the mercy of other muscles, since it cannot elongate on its own. When the muscle is relaxed, however, the sarcomeres, and thus the muscle as a whole, can be stretched.

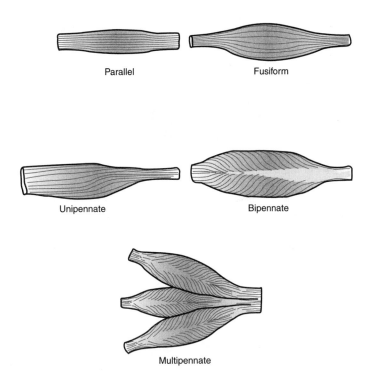

Figure 9.3

Fascicles come in many different arrangements in muscles.

That stretching, however, requires the contraction of other muscles. For that reason, we evolved muscles on opposite sides of the bones. Remember flexion and extension, or abduction and adduction, from Chapter 7? The pair of muscles that make these movements are called *antagonistic pairs*. The muscle doing the initial movement (of the pair of movements, flexion, for example) is called the *agonist*, or *prime mover*. The muscle doing the opposite movement, in this case extension, is called the *antagonist*.

 Crash Cart _____

An easy mistake to make is to assume that one muscle is always the agonist, and the other is always the antagonist. In truth, the agonist is always the muscle that is contracting at that moment (let's say the *biceps brachii*), and the antagonist is the muscle that is being stretched (let's say the *triceps brachii*). As soon as the triceps start to contract, it becomes the agonist, and the biceps become the antagonist.

With about 700 muscles, it makes sense that some muscles may not work alone, especially with both superficial and deep muscles. These helper muscles, called *synergists*, work to stabilize a movement, and also to increase the efficiency of the prime mover. Another type of muscle is called a *fixator*, and its job is to stabilize the origin of the prime mover. This is especially important in terms of the scapula.

When you compare the articulations of the scapula to those of the clavicles or the pelvic bones, you will notice something interesting. The clavicle articulates with the axial skeleton at the sternum, and the pelvic bones articulate at the sacrum, but the scapula doesn't articulate with the axial skeleton at all. The only thing holding the scapula in place is the action of the fixators; without them, the scapula would be pulled away from the body whenever we did something as mundane as flexing our forearm.

A Muscle by Any Other Name ...

With so many muscles, anatomists decided to name muscles using a few simple principles. Understanding these principles makes muscle names a snap. We have covered most of the essential principals already when we learned directional terms (Chapter 1), bone names (Chapter 6), and specific muscle movements (Chapter 7).

To give you an example of how to use the names, consider the tibialis anterior. From the name you know two things: It's not only near the *tibia* (tibialis), but it's *in front* of it (anterior). Another example is the *flexor carpi ulnaris*: It connects the *carpals* (carpi) and the *ulna* (ulnaris), and the movement it does is *flexion* (flexor). Not all the muscles are so easy, but you'll be surprised at how many muscles follow these simple naming rules. There are only a few loose ends, such as muscle shape and the direction of the muscle fiber, that we will need to get out of the way before you become an expert at muscle names. The seven characteristics used to name muscles can be seen in the following table.

Naming Muscles

Characteristic	Description	Example
Action	Muscle movement (Ch. 7)	*Supinator*
Fiber direction	Angle from midline	*Rectus abdominus*
Location	Bone, directional terms	*Tibialis anterior*
Number of origins	If more than one	*Biceps femoris*
Origin and insertion	Location of tendons	*Sternocleido-mastoid*
Shape	General muscle shape	*Deltoid*
Size	Big, small, long, short	*Adductor brevis*

All Shapes and Sizes

Muscle shape is often a factor in muscle names. Many of these are based on geometric shapes: The *deltoid* is shaped like a triangle, the rhomboideus major is rhomboid (or diamond) shaped, and the trapezius is trapezoid shaped. A nongeometric shape name is the *serratus anterior*, which is saw-toothed in shaped (think serrated knife).

The number of origins is also used. Most muscles have only one origin, so the exceptions usually have that fact as part of their name. *Biceps*, means two origins, *triceps* means three, and *quadriceps* means four origins.

In terms of muscle size, the words *maximus* (big), *medius* (medium), and *minimus* (small) are often used. Think of extending your leg (you can feel this when you climb stairs), and you will be thinking of the large *gluteus maximus*, the mid-size *gluteus medius*, and the smaller *gluteus minimus*. These muscles, as you will often find, are organized from superficial (maximus) to deep (minimus). The words *major* and *minor* are also used to mean big and small, as in the *zygomaticus major* and the *zygomaticus minor*. Remember, think in terms of groups when you hear certain muscle names: If there's a major, look for a minor; if there's a maximus, look for a minimus!

The length of a muscle is sometimes part of the muscle name. The word *longus*, as you would imagine, means long, as in the *adductor longus*. Short muscles use the word *brevis*, meaning short (as in brevity), as in the *peroneus brevis*. Well, that's the long and short of it!

Flex Your Muscles

The name of the muscle includes more than just biceps. Were you aware that you have two biceps? In your arm you have the *biceps brachii* (brachii means arm), and in your leg, close to your femur, you have the *biceps femoris*.

Medical Records

The gluteal muscles are used in extending your leg, which is very useful in walking. Other animals have such muscles, but none as big as yours! We are bipeds, which means the muscles of the legs are used extensively as we walk. What is the implication of this fact? Humans have big butts!

Fibers Every Which Way

Muscles are sometimes named by virtue of the direction of the muscle fibers. There are only three angles that the muscle fibers can go when compared to the midline: parallel, perpendicular, and oblique. Abdominal muscles easily illustrate all three names. The term *rectus* means that the fibers are parallel; the *rectus abdominus* is, guess where, on the abdomen, and its fibers are parallel to the *linea alba* (which is on the midline). Just as in the section name, perpendicular fibers are called *transverse*; the *transversus abdominus* is a deep muscle with perpendicular fibers. Lastly, the *external and internal obliques* are both superficial to the transversus abdominus, but their fibers run at oblique angles (see Figure 9.4).

Figure 9.4

The direction of the muscle fibers, which indicates the direction of the contraction, is referred to by the angle from the midline.

(LifeART©1989–2001, Lippincott Williams & Wilkins)

Fiber direction in relation to the midline

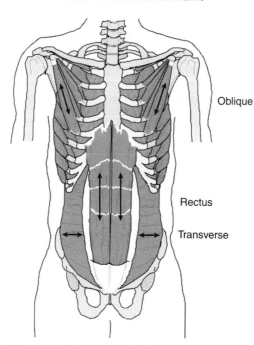

Oblique

Rectus

Transverse

Location, Location, Location

You will find that learning the bone names will be invaluable here. Since muscle attaches to bone, it makes a lot of sense to put the name of the bone it attaches to into the name of the muscle. There are too many examples to name them all here, and we already mentioned a couple before (*tibialis anterior* and *flexor carpi ulnaris*), so I'll just give you one more: The *frontalis* is attached to the frontal bone (you use it to raise your eyebrows).

Sometimes muscle names will use structures other than bones, such as the *buccinator* (buccal means mouth) attaching to the mouth. Another example is the *levator labii superioris*, which attaches to the lips (labii refers to lips). The last example also has within it another clue as to its location. If you were planning on looking *under* the mouth for the muscle, then you just aren't paying attention to the directional term *superior!*

You will often find a directional term, or a muscle movement, added to the bone name. Use this as a clue in remembering the muscle's location. The "dorsi" in *latissimus dorsi* tells you that the muscle is dorsal. You wouldn't look for the *vastus lateralis* on the medial side of the leg, but you would find it on the lateral side.

Remember the origin and the insertion? They can also find their way into muscle names. One example is the oddly named *sternocleidomastoid*. There are two origins on this muscle—the sternum (*sterno*) and the clavicle (*cleido-*)—and one insertion—the mastoid process of the temporal bone (*mastoid*).

Lights, Camera, Action!

Muscles are, after all, all about movements. As such, there are two things you should do when you are learning the muscles. The first thing is to look for any movement in the name, such as elevation in the *levator scapulae.* You should also know where to look for it, not only near the scapula, but *superior* to the scapula (in order to pull it up, after all). You can find plenty of examples, such as the *flexor digitorum brevis* in the foot, or the *extensor carpi ulnaris.*

The other thing to do is to pay attention to the muscle's location in order to figure out what movement the muscle produces. You will often be asked to know the movements of the muscles. If the movement is *not* given in the name, you have to problem solve, to use some logic, in order to figure it out. Take the gastrocnemius, which has its origin on the distal, posterior end of the femur, and its insertion on the posterior surface of the *calcaneus* bone by the *calcaneal* or *Achilles tendon.* What does the muscle do? If you said plantar flexion … good for you!

Flex Your Muscles

Which is more important in terms of predicting a muscle's movement, its origin, or its insertion? Well, since the origin anchors the muscle to the unmoving bone, and the insertion attaches to the *moving* bone, I'd put my money on the insertion! Since muscle movements come in pairs, and antagonistic pairs perform them, be sure to pay close attention to which side of the bone the insertion is on, because the agonist and antagonists insertions will be on opposite sides.

A Quick Look

The muscles in a college-level text are usually put into tables with five columns. The columns will include, in addition to the muscle name, the origin, the insertion, the muscle action (or movement), and the motor nerve that provides the stimulus for the muscle to contract (see Chapter 8). As I said earlier, it is not possible to include all that information in a book of this size, but hopefully what we have covered in this chapter will help you to crack those tables.

As you look at the diagrams of the muscles, see what you can interpret about the movements of the muscles from the name, and from the location. You should also see how many clues you can use to help you to remember the names of the muscles. As you are doing all this, think about how much you have already learned can help you to learn the muscles. Figure 9.5 shows the major superficial muscles.

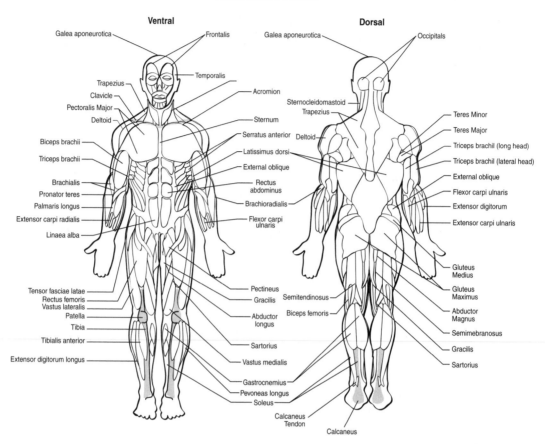

Figure 9.5

The superficial muscles of the body, as seen from both anterior and posterior views.

(LifeART©1989–2001, Lippincott Williams & Wilkins)

The Least You Need to Know

◆ The three part of skeletal muscle are the origin, belly, and insertion.

◆ Muscles act together, either as synergistic muscles, or as antagonistic pairs.

◆ Muscles and bones form each of the three classes of levers in the body.

◆ Muscle names are based on several factors: action, direction of fascicles, location, the number of origins, names of the origin and insertion, shape, and size.

◆ With most muscles names it is possible to learn not only the action, but the location of the muscle.

Part 3

Let's Make a Connection

I know what you're thinking. "Enough a'ready with the connections!" But connecting ideas and concepts is the key to understanding the physical connections between body systems! Think about your house or apartment. Do you have electricity? Unless you have solar power, or a gas generator, you are connected to a town/city electric company. Unless you have a well, you get town/city water coming in, and waste either connected to a sewer, or a personal septic system.

In the macroworld we are all connected, and the same goes for the microworld of our tissues! The means of most of those connections is the circulation of the cardiovascular and lymphatic systems. Food and water (from the digestive system) are delivered by the blood, but so are the wastes (from the tissues to the excretory system). Gases hitch a ride to and from the lungs (respiratory system) via the blood, too. In addition to carrying all that to every tissue in the body, hormones are carried from our endocrine glands to their target organs. Beyond that, blood carries nutrients either directly to the fetus, or indirectly to the baby via the mammary gland (reproductive system).

Chapter 10

The Red Sea

In This Chapter

- ◆ Functions of blood
- ◆ Chemical and cellular components of blood
- ◆ Connection between plasma, interstitial fluid, and lymph
- ◆ How clots form

You might remember from Chapter 4 that blood is a form of connective tissue (widely spaced cells in a matrix, in this case a fluid matrix). In this chapter, you will start to understand how blood might better be called *the* connective tissue. Most people grow up thinking of blood as part of the "circulatory" system, but as you shall see, there are in fact two systems involved in circulation: the cardiovascular system and the lymphatic system. In terms of transport, the cardiovascular system takes the top spot, but in terms of defending against bacteria and viruses, the lymphatic system gets top billing. In this chapter, I explore the connection between these two parallel systems.

Blood is the giver of life, the provider of food, water, and air, the waste remover (with the help of the kidneys; see Chapter 15), but it can also be the harbinger of death, if an infection makes its way into the blood and we become septic. Adult females have an average of four to five liters of blood, and adult males average about five to six liters. How the blood manages to get from place to place is the subject of Chapters 11 and 12. Here I discuss what blood actually *does*.

Function Junction

With apologies to Tina Turner, what's blood got to do with it? Blood does many wondrous things:

- Blood (plus vessels and the heart) is the primary transportation system for materials (the lymphatic system, as you will see, is the secondary transportation system): nutrients, water, wastes, O_2, and CO_2.

- Blood, which is slightly alkaline (7.35 to 7.45), regulates pH levels by absorbing acids from the interstitial fluid (see Chapter 4) and neutralizing them.

- Blood regulates body temperature by transferring (via the plasma) heat generated by muscles (hemopoiesis; see Chapter 9) to tissues throughout the body; heat can also be retained or lost through the constriction and dilation of vessels in the skin. As such, the average temperature of blood is higher than body temperature (38°C, or 100.4°F).

- Blood protects from fluid loss by clotting at the site of injuries.

- Blood protects from toxins and pathogens, through the action of white blood cells and antibodies.

Fluids Galore!

Blood makes up about 8 percent of the body's weight. Over half of its weight (about 55 percent) is the liquid matrix known as plasma. With exception of the oxygen, and a little of the carbon dioxide, all other materials in need of transport make their way via the plasma. The plasma, more so than the cells, links the two circulatory systems—cardiovascular and lymphatic—together in a continuum: plasma to interstitial fluid to lymph and back to plasma.

Plasma Before Delivery

Blood plasma is almost 92 percent water. The fact that water is such an excellent solvent (see Chapter 2) makes it possible for blood to carry so many dissolved substances (solutes). The majority of these substances are plasma proteins. The smallest of these proteins (which nonetheless make up about 60 percent of all blood proteins) are albumins, which are important in terms of osmosis, helping to move more water out of the capillaries. These albumins also help to transport steroid hormones.

About 35 percent of the proteins are called globulins. These include antibodies, which play an essential role in fighting infection. Two other globulins, alpha and beta, transport fat, fat-soluble vitamins, and iron. About 7 percent of plasma protein is fibrinogen, which is made in the liver, and is an essential part of the clotting process. The last 1 percent consists of regulatory proteins, such as proenzymes, enzymes, and hormones. Given that the endocrine system is powerless without the bloodstream, it's a bit sad that so *little* of the plasma is made up of hormones.

The remaining 1.5 percent of the plasma is made of other solutes: electrolytes, gases, nutrients, regulatory substances, vitamins, and waste products. Once again, these are awfully small amounts, but they are, nonetheless, very important! The electrolytes alone are quite varied: Na^+, K^+, Ca^{2+}, Mg^{2+}, Cl^-, HCO_3^-, HPO_4^{2-}, and HSO_4^{2-}. Sodium and calcium ions (Na^+ and Ca^{2+}), for example, are essential for muscle contraction (see Chapter 8), and bicarbonate ions (HCO_3^-) are essential in the transport of CO_2 to the lungs (see Chapter 13).

> ### The Big Picture
>
> The gases, nutrients, wastes, and regulatory substances all have their own body systems associated with them: O_2 is picked up and CO_2 is dropped off at the lungs (respiratory system), food and water are picked up at the small intestine (digestive system), wastes are filtered out at the kidneys (excretory system), and hormones are picked up and delivered (endocrine system). The blood needs to pick up and deliver materials to *every* system of the body; don't forget that every living cell needs food, water, etc. to be delivered and waste and so on to be removed by the blood! Now can you see why blood is a *connective* tissue?

When clotting occurs, fibrinogen molecules will interact to form fibrin molecules (which are fiber-like), which will also trap the blood cells. The fluid that is left is called *serum*, and it will have other differences (aside from no fibrinogen, in solute concentration compared to plasma), such as a lower level of calcium ions.

Fluid Between the Cracks

Why am I talking about interstitial fluid if this chapter is on blood? Well, blood is mainly about transporting materials to and from the body tissues; the means of actually getting the materials into the cells outside the blood vessels is by first transferring it to the fluid that surround those cells. That fluid is called interstitial fluid (see Chapter 3).

Solutes in that fluid, as well as some of the water itself, enter the cells, and solutes and water enter the interstitial fluid from the cells. As such, the concentration of the various substances in the interstitial fluid will change (for example, O_2 enters the cells, and CO_2 exits). In the same way, the concentration of the solutes in the blood are different in the blood vessels before the capillaries, than in the blood vessels after the capillaries. I discuss how the materials leave the capillaries in Chapter 12.

Draining the Cracks

Ultimately, this interstitial fluid is drained into the lymphatic capillaries (or lymphatics), where it is called *lymph*. It is important to get the concept that these fluids—blood, interstitial fluid, and lymph—are essentially the same, despite any differences in concentration; the main difference between the three is *location*. (In the same way, the only difference between magma and lava is location: magma is below the ground, and lava is above.) Lymphatic capillaries are open-ended vessels with periodic gaps in their walls, which are the entry point for the drainage of the interstitial fluid. The lymph fluid is ultimately emptied into to the plasma via the thoracic duct to the subclavian vein (see Chapter 12).

Cells, Bells!

My first introduction to white blood cells was that old sci-fi classic, *Fantastic Voyage*. (I recommend it, but to have the most fun, you should keep an eye out for the *mistakes!*) One of the final scenes involves the bad guy being eaten by a white blood cell! You may not have learned, however, that while there is only one type of red blood cell (RBC), that there are *many* white blood cells (WBCs), not to mention platelets! All of these cells are called *formed elements*, and they make up about 45 percent of blood volume.

In Chapter 5, I mentioned that the red bone marrow was the site of blood cell production in a process called *hemopoiesis*, which is summarized in Figure 10.1. In truth, not all blood cells are made in the red bone marrow, because while the B-lymphocytes are made in the bone (B for bone), the T-lymphocytes are made in the thymus (T for thymus). All of the blood cells come from the same stem cell— the *pluripotent hematopoietic stem cell*, its friends call it a *hemocytoblast*—in a model of differentiation (see Chapter 3).

Medical Records

Note that any percentages given for substances in the blood, either within the plasma, or as part of the formed elements, are averages. Your health may affect the amount of each substance. A drop in the number of erythrocytes (RBCs), for example, is one form of anemia. For this reason, blood tests can be very revealing. As a diabetic, I test my blood glucose level with a handheld meter up to eight times a day.

This figure may be a bit frightening, but try to look for the commonalities. From the one stem cell (the hemocytoblast) there are five cells into which it can develop. Note that the first cells all have the suffix "-blast," and that the three of the five cells have, in their name, a clue as to the final cell (for example, a lymphoblast will ultimately become a lymphocyte). The second step, with the exception of the early erythrocyte, all have similar names (for example, promonocytes ultimately become monocytes). Look for such strands as you follow the diagram, and you will see that it is not so frightening after all!

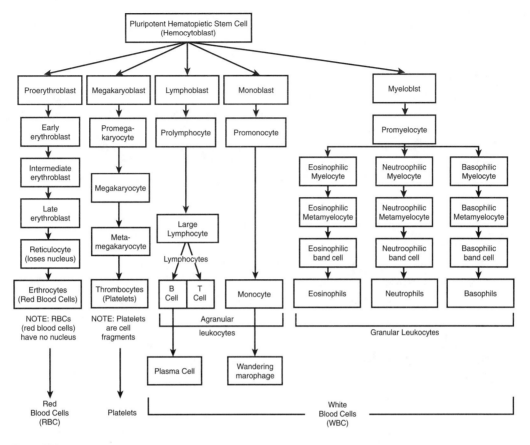

Figure 10.1

This diagram shows the development of all types of blood cells from the original hemocytoblast.

(©Michael J. Vieira Lazaroff)

Now I Know My RBCs

RBCs make up 99.9 percent of all formed elements, that's a nearly a 1000 to 1 ratio! RBCs give blood its red color. I always tell my students that *erythrocytes*, or *red blood cells* (*RBCs*), are on a suicide mission. From the earlier diagram you may remember that the last step of RBC formation is the loss of the nucleus (as well as ribosomes, mitochondria, and most of the other cell organelles). That it is a very important event, for it says a tremendous amount about that cell.

Without a nucleus (and its DNA cargo) or ribosomes, an RBC cannot make protein, and so it is trapped with its current complement of proteins. That means it can do very little to react to its environment, and it is completely incapable of repairing itself. The loss of mitochondria means it can process glucose only without oxygen (an irony, considering how much oxygen is in the RBC): Producing only 2 ATP versus 36 (see Chapter 3), the RBC must have *very low* energy needs.

Crash Cart

Guess what makes deoxygenated blood blue? Give up? *Nothing!* Once and for all, *blood is never blue!* Not even among aristocrats! Deoxygenated blood is *dark red*, and oxygenated blood is *bright red*, but it's always *red!* In terms of bleeding, blood cannot immediately turn red when exposed to the air, simply because of the low SA/V (see Chapter 3). Arteries are too deep to see, but veins look blue simply because the red is refracted by the tissue layers the light must pass through before reaching your eyes. As my wife says, "Blood never gets the blues!"

An RBC's role is basically as a floating sack of *hemoglobin*, which is the protein that carries oxygen. The iron in the hemoglobin causes the cells to turn more red in the presence of oxygen, just the way rusting iron turns red! Without a nucleus, RBCs are incapable of dividing, so they are on a suicide mission, working until they are destroyed by phagocytic cells (see Chapter 3), or in the spleen.

RBCs are a beautiful example of differentiation. Here we have a cell that does only one thing, lives for 120 days before it is broken down, and must constantly be replaced. The single-minded purpose of the cell is linked to the *hemoglobin*, or *Hb*, which makes up 95 percent of the proteins in the cell. The lack of the nucleus allows room for more hemoglobin molecules, which means that the cell can carry more oxygen.

So what's the big deal? After all, why can't we just carry the O_2 in the plasma, as we do the other molecules? The answer is due in part to the fact that O_2 is very poorly soluble in water (1.5 percent), and in part to the structure of the hemoglobin molecule itself. In Chapter 2, I discussed the quaternary structure of proteins. A hemoglobin molecule is roughly x-shaped, with two alpha chains (each at a diagonal), and two beta chains (each also at a diagonal); each of these four chains is a tertiary structure (see Figure 10.2).

In the midst of each chain is a heme group, which contains an iron ion (Fe^{2+}). Each heme group can hold one O_2 molecule, so each Hb molecule therefore carries four oxygen molecules; all this adds up to a tremendous amount considering we have about 280 million Hb molecules in every RBC! This arrangement allows the hemoglobin in the RBC to be far more efficient at carrying oxygen than plasma. In addition to carrying about 98.5 percent of the O_2, hemoglobin carries about 23 percent of the CO_2; the combination of CO_2 and Hb is called carbaminohemoglobin. The remaining CO_2 is carried as ions (see Chapter 14) in the plasma, or the cytoplasm of the RBC.

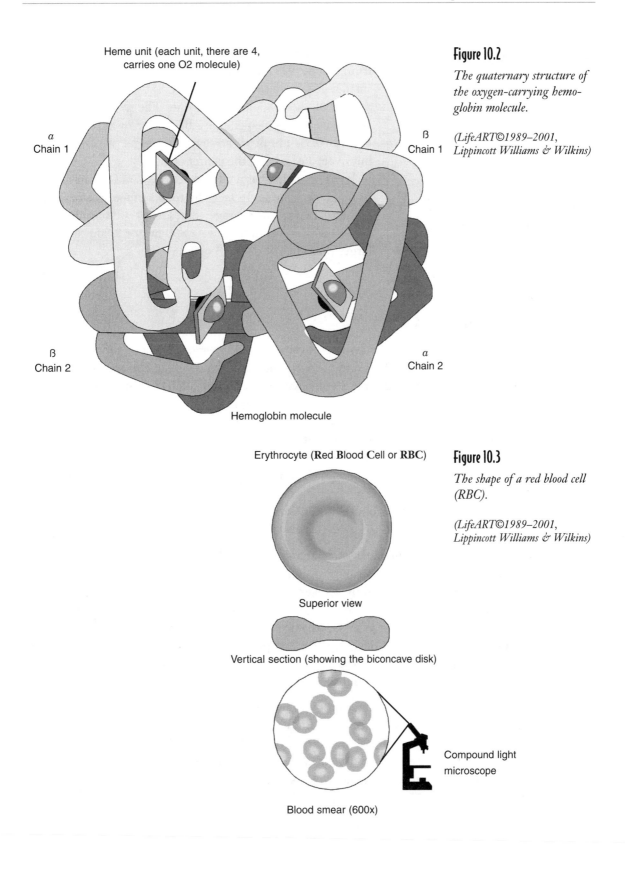

Heme unit (each unit, there are 4, carries one O2 molecule)

α Chain 1

β Chain 1

β Chain 2

α Chain 2

Hemoglobin molecule

Figure 10.2

The quaternary structure of the oxygen-carrying hemoglobin molecule.

(LifeART©1989–2001, Lippincott Williams & Wilkins)

Erythrocyte (**R**ed **B**lood **C**ell or **RBC**)

Superior view

Vertical section (showing the biconcave disk)

Compound light microscope

Blood smear (600x)

Figure 10.3

The shape of a red blood cell (RBC).

(LifeART©1989–2001, Lippincott Williams & Wilkins)

The Big Picture

RBCs are mostly concerned with carrying oxygen. The need for O_2 is connected to energy capture from glucose (see Chapter 3). With O_2 the energy available is 19 times higher (38 ATP instead of 2). It is clear that O_2 is essential for our survival, which explains the special connection between the cardiovascular and respiratory systems, as you shall see in Chapter 11.

Do you remember surface area to volume ratio (SA/V) from Chapter 3? The important idea to remember is that a smaller cell has a larger SA/V. RBC is the star of SA/V, with a SA/V of about 1.5 million to one! The shape of an RBC is a bit weird; it's shaped like a biconcave disk (refer to Figure 10.3). Think of the shape of a bagel or doughnut, but instead of a hole, the center is merely pinched in (like a bialy). This shape has a far greater SA/V ratio than a sphere of the same volume. In fact, the total surface area of all 25 trillion RBCs in the average adult's body—260 million in a single drop!—is equivalent to 3,800 m^2, or 2,000 times larger than that of our skin (about 1.9 m^2)!

Blood Types and Genetics

Before I talk about blood type, I should probably back up and talk about genetics. In Chapter 3, I talked about DNA and protein synthesis, as well as the chromosomes involved in mitosis. Don't forget that, unless you have a chromosomal defect such as Down's syndrome (Trisomy 21), each of your chromosomes comes as a homologous pair with one chromosome from each parent. These homologous chromosomes have the same type of gene on them, but the alleles are different. A gene is section of DNA that codes for a particular protein, but an *allele* is a variation of a gene. Since genes are on chromosomes, and chromosomes come in pairs, our genes usually come in pairs of alleles (those on the sex chromosomes—the twenty-third pair, that determine gender—might not, because females are XX and males are XY, or one X chromosome and one Y chromosome).

Alleles can be either *dominant* or *recessive*, so the combination of alleles in the pair determine the trait. A dominant allele, indicated by a capital letter, is one that is expressed as a trait (such as a widow's peak hairline) with either one or two copies of the allele: WW or Ww. A recessive trait, indicated by a lowercase letter, is one that is *only* expressed (such as a straight hairline) with two copies of the allele: ww. The combination of alleles—WW, Ww, and ww—is called a *genotype*: WW is *homozygous dominant*, Ww is *heterozygous*, and ww is *homozygous recessive*. The physical trait that is expressed is the *phenotype*, such as the type of hairline, and it is determined by the genotype: WW and Ww are both dominant (widow's peak), while ww is recessive (straight hairline).

That's a lot to get out of the way, but it will be useful when I talk about reproduction in Chapters 23. Now, on to blood type. Blood type refers to the antigens (in this case glycoproteins; see Chapter 3) on the membrane of the RBC, and it is determined genetically by two genes: the ABO gene and the Rh gene. Let's start off with the Rh gene, which gets its weird name from its initial discovery in the blood of the Rhesus monkey.

Crash Cart

Not all of genetics is so simple. Although some traits (can you roll your tongue?), including some disorders (sickle cell anemia, cystic fibrosis), are *monogenic*, or controlled by one gene, others are *polygenic*, or controlled by multiple genes. Polygenic traits include such features as eye color, hair color, skin color, and blood type. Some cancers, or rather the tendency to get some cancers (the impact of the environment on the genes is important in terms of actually getting cancer), are polygenic.

If the allele for the glycoprotein (indicated as Rh +) is inherited, the glycoprotein will appear on the cell membrane; there is another allele, which means that no antigen is displayed (indicated as Rh -). The only way to have a cell with no Rh antigen is if both chromosomes in the homologous pair are Rh negative (--); Rh negative is therefore recessive. You will be Rh positive if you inherit at least one Rh + allele (that is, ++ or +-).

The ABO gene, which gets its odd name from the three alleles, determines the other blood type: A, B, and O, of course. This gene illustrates another interesting thing that sometimes happens in inheritance: some genes are *codominant*. Codominant genes are ones in which two alleles, such as A (I^A) and B (I^B), are both dominant, and a heterozygous genotype (AB) produces a phenotype where both traits are expressed (the A and B glycoproteins).

Crash Cart

Don't confuse codominance with *incomplete dominance*, in which a heterozygous pair of alleles is a blending of two traits. Wavy hair is a heterozygous cross between a homozygous straight hair and homozygous curly hair; medium-set eyes are a heterozygous cross between homozygous close-set eyes and homozygous eyes set far apart. Another confusion involves polygenic traits and *pleiotropy*: polygenic traits involve *many* genes determining *one* trait, and pleiotropy involves *one* gene controlling many traits. Sickle-cell anemia is a homozygous recessive condition in which the recessive form of the hemoglobin molecule causes a number of effects in the body, which makes it a form of pleiotropy.

Similar to the Rh--allele, the O (I^O) allele can *only* produce the O phenotype if *both* inherited alleles are O (OO), which makes the O allele recessive. All of this is important because emergency procedures often involve loss of blood, and the only way to keep a patient alive is to replace some of the volume lost; without enough blood, tissues will not be able to receive the materials, especially O_2, needed to survive (the brain, in particular, dies very quickly in the absence of O_2). A simple idea is to take a little of the blood from one person and put it into another, a process called *transfusion*.

Compatibility of blood types is based on the fact that incompatible blood sample will *agglutinate*, or clump, when mixed. This is due to the presence of antibodies that attack foreign surface antigens (A, B or Rh +, but not O or Rh -). For this reason, these surface antigens are sometimes called *agglutinogens*, and the antibodies are called agglutinins. The three antibodies are called, and rightfully so, anti-A, anti-B, and anti-Rh. Anti-A and anti-B are always found in the blood of specific blood types (see the following table of blood types), even without prior exposure to the foreign antigen, but anti-Rh is only produced after exposure.

A quick look at the table of blood types illustrates some interesting points. Any incompatible blood type must have an antigen that is foreign to the recipient. AB+ blood contains *all three* antigens (A, B, and Rh +), and *none* of the antibodies; as such, there is no blood type that is incompatible to AB+, the *universal recipient*.

Blood Types

Type	Genotype(s)	Antibodies	Incompatible
Rh Blood Types			
Rh +	++ or +-	none	none
Rh -	--	anti-Rh	Rh +
ABO Blood Types			
A	AA or AO	anti-B	B, AB
B	BB or BO	anti-A	A, AB
AB	AB	none	none
O	OO	anti-A, anti-B	A, B, AB
ABO and Rh Combined			
A+	AA++, AA+- AO++, AO+-	anti-B	B+, B-, AB+, AB-
A-	AA--, AO--	anti-B, anti-Rh	B+, B-, AB+, AB-, O+
B+	BB++, BB+- BO++, BO+-	anti-A	A+, A-, AB+, AB-
B-	BB--, BO--	anti-A, anti-Rh	A+, A-, AB+, AB-, O+
AB+	AB++, AB+-	none	none
AB-	AB--	anti-Rh	A+, B+, AB+
O+	OO++, OO+-	anti-A, anti-B	A+, A-, B+, B- AB+, AB-
O-	OO--	anti-A, anti-B anti-Rh	A+, A-, B+, B- AB+, AB-, O+

Most people learned somewhere that type O is the universal donor, meaning anyone is able to receive it. That is only partly true. Do they mean O+ or O-? In order to be compatible to everyone, there can be *no foreign antigens*. Look at the table again. O+ has an Rh + antigen that is foreign to anyone with an Rh--(A-, B-, AB-, O-). O-, however, has *no foreign antigens*, making it the *universal donor.*

Now I Know My WBCs

Although there are many types of white blood cells (see Figure 10.4), they all share a similar function, that of protection from foreign substances, pathogens, and dead cells (they need to be cleared away, after all). Despite their similarities in function, there are differences; let's take a quick look at what they do.

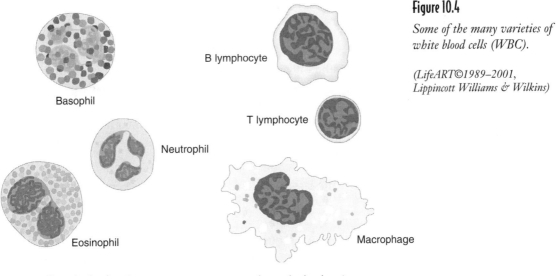

Figure 10.4

Some of the many varieties of white blood cells (WBC).

(LifeART©1989–2001, Lippincott Williams & Wilkins)

Basophil

B lymphocyte

T lymphocyte

Neutrophil

Eosinophil

Macrophage

Granular Leukocytes

Agranular Leukocytes

Monocytes do their work outside the blood vessels, although they can move in and out. They survive for a few months, wandering around (fittingly enough) as wandering macrophages, engulfing foreign bodies through phagocytosis (see Chapter 3). An elevation in the number of these in the blood is a sign of mononucleosis. One cool thing about these guys is what they do when the object they need to engulf is too big for only one to do the job: they call for backup. A bunch of them join up to make a *phagocytic giant cell*, which then does the job!

Lymphocytes have the longest lifespan of any white cell, living from months to decades! The

Flex Your Muscles

A *sign* is an objectively measurable indication of an illness, such as fever, swelling, and so on. Don't confuse this with a *symptom*, which is subjective, such as pain. As of this writing there is no objective measurement for pain.

long lifespan is useful in terms of the B cells, which make antibodies, and can store a memory of the antigen for decades, a fact that makes vaccines possible. I will discuss all of this in more detail when I explain immunity in Chapter 17.

If you have ever had any swelling, which acts to cushion and protect damaged tissue, basophils were responsible. Basophils respond by leaving the blood vessels nearest the injury and releasing chemicals, including histamines, which induce swelling. Some medicines used to reduce swelling are therefore antihistamines. Neutrophils and eosinophils are similar in that they both act through phagocytosis, and they both release cytotoxic (cell-killing) enzymes, but eosinophils in particular respond by attacking cells and pathogens that have antibodies attached. Although some antibodies can actually destroy a pathogen, most simply label the cells as food for the hungry eosinophils ("Come and get 'em!"). These same eosinophils are also important in getting rid of parasites, and are markers for allergies.

Clots in Store

The last formed elements are the only ones that are not truly cells. *Thrombocytes*, or *platelets*, are cell fragments that are essential to the clotting process, as discussed in the next section. The term *platelets* is more commonly used when referring to human blood, because they are only fragments; vertebrates (craniates?) other then mammals have truly nucleated cells performing this function, and thus making them more deserving of the name thrombocytes.

During blood cell formation, the largest cell is the metamegakaryocyte (sometimes called a megakaryocyte) in the red bone marrow. These cells continuously release cytoplasmic fragments filled with substances essential to blood clotting. The production of so many small platelets in mammals (over the larger thrombocytes of our nonmammalian ancestors) was an increase in clotting efficiency that went hand in hand with the higher metabolism of being mammals. Evolution strikes again!

The contents of platelets are held within a membrane, because if they were released they could trigger part of the clotting process. How the platelets' life-preserving contents—make no mistake here, because without their contents we would quickly lose our life-giving blood with the smallest injury—are actually released, as well as the rest of the clotting process, the sum of which is called *hemostasis* (prevention of blood loss), is the subject of our next section.

Clots Stem the Tide

Exactly how do clots form, and how are they helpful in the healing and repair process? I mentioned the role of the fracture hematoma in the healing of bone in Chapter 5 but clots also form scabs, which keep the injured tissues underneath protected while new cells are made.

First, many chemicals are involved in clotting. Twelve of them are called clotting factors; they are numbered I through XIII because factor VI was found to be the same as factor V. This list of 12 factors is nonetheless incomplete, for other substances, such as ADP and serotonin, are also involved. Rather than tracking down *all* of these and explaining their function, I will explore the highlights of the three phases of blood clotting: vascular phase, platelet phase, coagulation phase, and clot retraction.

Hemostasis begins with the vascular phase, a 30-minute period that starts when a blood vessel wall gets damaged. A vascular spasm constricts the vessel walls (*vasoconstriction*), which will reduce blood loss in larger vessels, but can completely stop capillary blood loss. The initial damage to the walls, coupled with the constriction, causes the exposure of the basement membrane (see Chapter 19), to which platelets attach; the walls also become, for lack of a better word, sticky, which not only helps platelet attachment, but also helps seal off smaller vessels. It is the release of chemicals (including local hormones) by the vessel walls, however, that initiates the second phase, the platelet phase.

Medical Records

Hemophilia is rare genetic disorder in which the cells cannot produce factor VIII. Although only one small part of the clotting process, the absence of that one factor makes the formation of the clot impossible. A chain, after all, is only as strong as its weakest link. Genetically engineered factor VIII has made it possible for hemophiliacs to live fairly normal lives by injecting factor VIII—previously isolated from whole blood—without the risk of blood-borne pathogens.

The platelet phase starts the attachment of the platelets to the exposed collagen fibers in the damaged vessel walls. Chemicals in the plasma and those released when the platelets themselves break open, stimulate further vasoconstriction, platelet aggregation, as well as the mitosis (see Chapter 3) necessary to, well, repair the walls! Don't forget that all repairs involve mitosis.

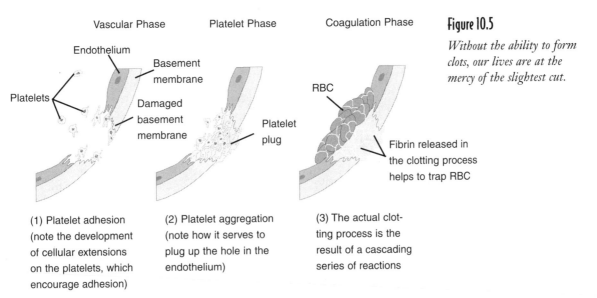

Figure 10.5

Without the ability to form clots, our lives are at the mercy of the slightest cut.

Vascular Phase — Endothelium, Basement membrane, Platelets, Damaged basement membrane

Platelet Phase — Platelet plug

Coagulation Phase — RBC, Fibrin released in the clotting process helps to trap RBC

(1) Platelet adhesion (note the development of cellular extensions on the platelets, which encourage adhesion)

(2) Platelet aggregation (note how it serves to plug up the hole in the endothelium)

(3) The actual clotting process is the result of a cascading series of reactions

The third phase is called the coagulation phase. It is called the third phase because it lasts longer than the platelet phase, but in reality they both start within 30 seconds of the damage to the vessel walls. This phase involves a cascade reaction involving multiple clotting factors, rather like a set of dominoes falling. The most important thing about this step is the chemical conversion (thanks to those clotting factors) of the dissolved fibrinogen in the plasma to a network of fibrin fibers. These fibers trap the RBCs and WBCs and thus stop the blood flow. Refer to Figure 10.5 for an overview of these processes.

Medical Records

Not all clots function normally. Sometimes a clot continues to grow, extending into the vessel, and ultimately blocking all blood flow through the vessel, instead of just stopping blood flow out. This type of clot is called a thrombus. A clots which breaks loose and gets lodged in a narrower vessel downstream is called an *embolus*. In both cases, any blood flow beyond either of these abnormal clots is stopped, and the subsequent tissues can die. The severity depends on the area affected.

The last phase, or *clot retraction*, pulls the broken edges of the vessel together, which prevents any further bleeding. This retraction makes the damaged area smaller, which makes the mitosis and final repair easier. The only thing left is *fibrinolysis*, which is the dissolving and breakdown of the clot after the repair is completed.

The Least You Need to Know

- Blood is our medium for transporting materials from one specialized part of the body (lung, small intestine, and so on) to another.

- Blood plasma, interstitial fluid, and lymph are extensions of each other, and are the fluids that carry materials throughout the body.

- Red blood cells are ideal for the transfer of gases, particularly oxygen, and are necessary due to the low solubility of oxygen in plasma.

- Blood types refer to the antigens on the surface of the RBCs, and they are used to safely transfuse blood from one person to another.

- The many types of white blood cells are essential to the body's defense against pathogens.

- A series of physical and chemical reactions involving platelets are responsible for blood vessels' ability to clot and stop blood loss.

Chapter 11

And the Beat Flows On

In This Chapter

♦ The anatomy of the four-chambered heart

♦ The pulmonary, systemic, and cardiac circuits

♦ Regulation of the heartbeat

♦ Electrocardiogram and the cardiac cycle

Valentine's Day, at least the greeting card variety, has forever done the anatomist a disservice. We have all grown up seeing the heart as an upside-down butt shape with a point at the opposite end—and don't even get me started on Cupid, with his wings coming out of his back (like an insect), rather than as arms (like a bird or bat)—but the heart looks nothing like that! Luckily, for people with a bit of a sick sense of humor like me, it is now possible to buy life-size, anatomically correct, milk chocolate hearts!

The heart is really a pretty cool organ, with an elegant shape that mirrors its function. Not only amazingly strong, but incredibly sturdy, the heart beats about 100,000 times a day, 36,500,000 times a year (not to mention an extra 25,000 times each leap year, and a total of about 2,739,375,000 times in an average lifetime (approximately 75 years, give or take). Resting for 24 hours will save your heart 20,000 beats, which is why your momma told you to stay in bed and rest when you were sick; just think of all the energy you save! Every day the heart pumps about 8,000 liters of blood, or enough, over the same average lifetime, to fill about $58\frac{1}{2}$ Olympic-size swimming pools (a total of 219,150,000 liters in a lifetime).

What I love about the heart, despite it not being the seat of the emotions (for that you need to look at the limbic system in the brain; see Chapter 20), is its subtlety. How the heart coordinates the efficient pumping of four chambers to each of the three circuits simultaneously is the epitome of evolutionary elegance.

Anatomy of the Heart

The four-chambered heart is the result of millions of years of evolution. The evolution of a system to circulate fluid was first, but the heart itself took much longer to reach full flower in vertebrates. Fish have a two-chambered heart: one *atrium* and one *ventricle*. An atrium (plural, atria) receives blood from the body and pumps it, using the muscular walls of the atrium, to the ventricle. The ventricle receives the blood from the atrium and pumps the blood out to the body.

Given that the ventricle must pump a greater distance than the atrium, it makes sense when you see a heart, that the walls of the ventricles are much thicker than those of the atria. The only other issue is how to guarantee that the pumping of the heart always goes in the same direction, because backflow would mean that oxygen and food would not be able to reach their target. The solution to that problem is simplicity itself: valves!

The shape of a valve allows it to open when the flow occurs in one direction, and close when the flow goes the opposite direction. In animals the valve shape is based on flaps (see Figure 11.1), and if you think about it, the free ends of the flaps point, just like an arrow, in the direction of the blood flow. As you shall see, the placement of those valves is crucial to the operation of the four-chambered heart.

Figure 11.1

As you can see, the shape of a valve makes the flow of fluids possible in only one direction.

(LifeART©1989–2001, Lippincott Williams & Wilkins)

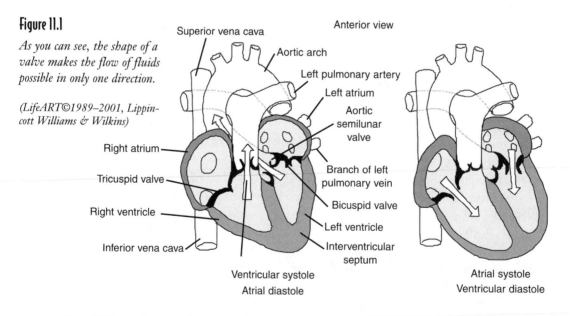

In addition, the coordination of the heartbeat makes the valves function more easily. The heart contracts from top to bottom, the two upper chambers, or atria, contract (systole) first, while the lower chambers, or ventricles, relax (diastole), and vice versa.

This alternation of systole and diastole allows, for example, the relaxed ventricles to be filled by the contracting atria.

But the heart is just the start. In Chapter 10, you learned what the blood was for, but I never really explained to you where the blood was going. All this pumping of the heart must be sending the blood somewhere, here and there ... but where?

To and Fro

In order to understand the cardiovascular system you need to grasp the vessel component of a closed system (see Chapter 10); there is, after all, a *there* there. As a closed system, despite some of the plasma leaving the vessels (see Chapters 10 and 12) the blood never leaves the vessels unless they are damaged and you start to hemorrhage (bleed).

The pathway of the blood is called a *circuit*, for the blood that leaves the heart it returned to it. There are six parts to every blood circuit, in order: the pump-like heart arteries carrying blood pumped out of the heart; smaller arterioles; capillaries, where all exchanges happen; venules, which start the return to the heart; and lastly the veins, which ultimately return connect to the heart.

The blood starts when it is pumped from the heart—don't forget here that the blood flow within the vessels travels in one direction! The second part is called an *artery*, which is easy to remember because the blood flows **a**way from the heart. In labeling the vessels, it is important to understand that many smaller arteries branch off from one major artery (think of a tree trunk dividing into smaller and smaller branches); each smaller artery has a specific name.

These artery names and locations are the same for everyone. On a smaller scale, the arteries continue to branch, but the branching becomes very small and starts to differ from person to person. These smaller branches, the third part of the circuit, are called arterioles. The smallest, but perhaps the most important part of every circuit, is the capillary, the fourth part of the circuit.

Capillaries are important because they are exceedingly small, with the lumen, or hollow center, of the vessel only wide enough to allow RBCs through one at a time. This small size is important, for the fluids that nourish the peripheral tissues, the wastes that are excreted, and the oxygen, water, and food that are absorbed all enter and leave the vessels here (to learn how, see Chapter 12). The fact that capillaries are so small, and that there are so many of them, means that there is an enormous surface area to volume ratio. This is an enormously important part of the circuit, for all the other portions are merely there to get the blood to and from the capillaries. Even the heart takes second stage, because, physiologically, it makes the most sense to say that all circuits start with the capillaries!

The fifth part of the circuit is analogous to the third part. Just as the arterioles are smaller versions of arteries, the venules are really just smaller versions of veins. The venules, which are once again as unique to the individual as the arterioles, drain the capillaries, and are ultimately drained into the veins (like smaller roads merging into larger highways). The sixth and last part of the circuit is analogous to the second part, for the veins are the largest vessels that take the blood to the heart. Similar to the arteries, there are many small veins that drain into the largest veins.

In the fish example, with the two-chambered heart, there is really only one circuit, called the systemic, that travels to and from the other systems of the body. Amphibians and reptiles developed a three-chambered heart to match the addition of two other circuits. Most students only know one of these circuits, the pulmonary, which is responsible for gas exchange. The third circuit, which is really a small branch that comes immediately off the systemic, is called the coronary, because its job is to nourish the heart; when a part of this circuit fails, and the blood flow stops, part of the heart might die, which is called a heart attack.

The Four Chambers

So why do you need four chambers if three worked just fine for frogs and lizards? Humans, and indeed all mammals (not to mention birds!), are *endothermic* (warm blooded). Warm bloodedness requires a great deal of oxygen, for the oxygen is used to generate both ATP and heat. A four-chambered heart is an enormous evolutionary advantage over a three-chambered heart. To understand this, you need to look at the chambers and the circuits together.

Remember the fish, with an atrium to receive blood from the body, and a ventricle to pump it out again? Well, with a three-chambered heart there are two ventricles and one atrium. The two atria emphasize a higher degree of separation between two of the circuits: the pulmonary circuit and the systemic circuit. At this point you need to start thinking of the heart in terms of left and right. The *right atrium* receives *deoxygenated* blood (low in O_2, and high in CO_2) from the systemic circuit, and the *left atrium* receives *oxygenated* blood (high in O_2, and low in CO_2) from the pulmonary circuit.

Medical Records

Some babies are born with a *ventricular septal defect,* which means an opening between the left and right ventricles, which means that their hearts are acting like three-chambered hearts. Surgery to correct the defect is necessary in order for the child to live a normal life.

This advance was only so good, however, because both atria pump the blood to the single ventricle. In a three-chambered heart the blood pumped out of the ventricle is a mixture of both oxygenated and deoxygenated blood. This blood is pumped out to both the pulmonary and the systemic circuit (in truth, because it is pumped right back to the tissues of the heart, it really goes to all three circuits). For ectothermic (cold blooded) animals that is plenty of oxygen, but it's just not enough for *you.*

Birds and mammals evolved a ventricular septum, turning one ventricle into two. The result is the evolution of entirely separate pulmonary and systemic circuits (see Figure 11.2). The blood sent to the lungs is completely deoxygenated, and the blood pumped out to the rest of the body is fully oxygenated. The evolution of two ventricles, making a four-chambered heart, doubled the amount of O_2 being sent to the tissues. The amount of food and waste in the blood going to the systemic circuit is not so cut and dried (see Chapter 12).

In the human heart the right atrium sends deoxygenated blood from the body to the right ventricle, which then pumps it to the lungs (pulmonary circuit). The left atrium sends oxygenated blood from the lungs to the left ventricle, which then pumps it to the body (systemic circuit).

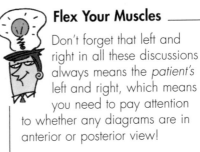

Flex Your Muscles

Don't forget that left and right in all these discussions always means the *patient's* left and right, which means you need to pay attention to whether any diagrams are in anterior or posterior view!

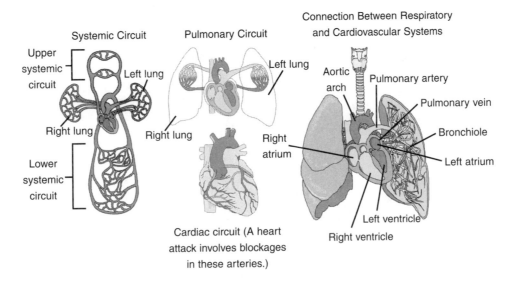

Figure 11.2

The human heart has four chambers, which equally separate the right and left sides of the heart, maximizing the oxygen content of the blood being sent to the systemic circuit.

(LifeART©1989–2001, Lippincott Williams & Wilkins)

Blood Vessels and Chambers

When you look at the orientation of the heart at the bottom of the thoracic cavity (see Chapter 13 to learn about the pericardium) you will see that, rather than being straight up and down, the heart is at an angle, and a bit twisted (kind of like me!). This is due in part to making room for the liver, and in part to the location of the many blood vessels that attach to the heart.

Figure 11.2 shows the blood vessels connected to the heart, but you may find the flow-chart in Figure 11.3 a bit easier to understand. Don't forget that the blood flow in the pulmonary and systemic circuits is continuous, meaning that blood from one circuit moves on immediately to the other circuit. Next, the central location of the heart means that blood going to the lungs needs to be pumped both left and right, and blood going to the body needs to be pumped both up and down. Thinking in terms of opposites will help you to remember the vessels.

Figure 11.3

This flowchart illustrates the flow of blood, in terms of opposite directions, to and from both the systemic and pulmonary circuit.

(©Michael J. Vieira Lazaroff)

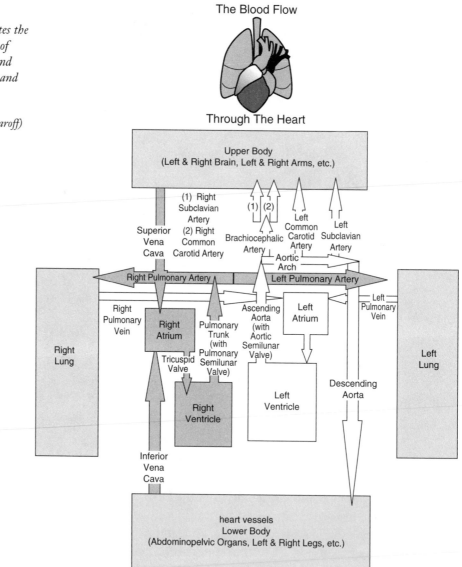

Remember, there is no particular place where all of this starts, given that the circuits are continuous. Let's start with the oxygenated blood in the arteries of the systemic circuit, leaving the left ventricle of the heart via the aorta. Immediately after leaving the top of the heart, all blood vessels enter and leave through the top of the heart, the aorta arches downward to send blood to the lower body. At the top of the arch there are three large branches that go to the upper body; in this way, the systemic circuit is divided in two.

Crash Cart

A common mistake is to define arteries as vessels carrying oxygenated blood, and veins as vessels carrying deoxygenated blood. Although this is *generally* true, there are two important exceptions, because of the true definitions of arteries carrying blood *away*, and veins carrying blood *to* the heart. The two exceptions, which make perfect sense, both involve the pulmonary circuit: The pulmonary arteries carry deoxygenated blood to the lungs *to be oxygenated*, and the pulmonary veins carry the newly oxygenated blood *away* from the heart!

After picking up and delivering various materials in the capillaries of the upper and lower body, becoming deoxygenated in the process, the veins drain into the largest veins in the body, the *superior vena cava* and the *inferior vena cava*. Anyone who works with quadrupedal animals should know that those same vessels are called the *anterior* and *posterior venae cavae* (plural for vena cava). The venae cavae drain into the upper and lower portions of the right atrium; since the right atrium is in the upper third of the heart, this inferior vena cava is still considered attached to the top of the heart.

As the right atrium contracts, the blood must pass through a valve between the atrium and the ventricle. This valve is called the tricuspid valve (for its three cusps or flaps), or the right atrioventricular (AV) valve. Once the blood is pumped out of the right ventricle, the right AV valve prevents backflow into the right atrium. The contraction of the right ventricle does pump the blood through another valve, the pulmonary semilunar valve (named for its half moon shape), and into the pulmonary trunk. Just as the aorta splits, so does the pulmonary trunk, but this time the blood splits into the left and right pulmonary artery, in order to go to both lungs. (To see what happens next, take a deep breath and read up on the respiratory system in Chapter 13.)

Flex Your Muscles

A good way to remember the difference between the two atrioventricular valves—the tricuspid (right AV valve) and the bicuspid (left AV valve)—is to think about the dissolved gases in the blood as it passes through those valves. The deoxygenated blood passing through the tricuspid valve contains CO_2, which contains three atoms (tri = three), and the oxygenated blood passing through the bicuspid valve contains O_2, which contains two atoms (bi = two). A pretty cool coincidence, considering the valves were named because of their *structure!*

Blood returning to the heart always returns from separate vessels, whereas blood leaving the heart always leaves from a single vessel and then splits to go in opposite directions. Having vessels in pairs makes sense, but single vessels leaving the heart? Why? Think about the shape of the heart. The cone shape of the apex gives a hint about the way the heart contracts. The contraction of the ventricles, which happens *simultaneously*, narrows the lumen of the ventricles, as well as shortening the length of the ventricles, which pumps the blood *up*! It is more efficient, in ensuring the equal flow to both lungs, for example, to have the blood leave one vessel, only to split later.

Oxygenated blood returns from the two lungs through the pulmonary veins, which attach to opposite sides of the left atrium. The rest of the trip is almost the same as on the right side: the left atrium pumps the blood through the left AV valve (or bicuspid valve) into the left ventricle, and the ventricle pumps the blood through the aortic semilunar valve into the aorta.

Just as the ventricular walls are thicker than the atrial walls (because of the difference in the distance the blood is pumped), the left ventricle, which has to pump to the entire body, has thicker walls than the right ventricle, which pumps blood only to the neighboring lungs. The thick left ventricular walls also provide a greater pressure on the left AV valve with each ventricular contraction. This valve, also called the *mitral* valve, can sometimes bulge into the left atrium, which is called *mitral valve prolapse*.

To help prevent such prolapses, there are fibrous, tendon-like cords called *chordae tendineae*. These connective tissue cords support the valve whenever the ventricles contract. Every time a ventricle contracts, there must be enough pressure in the contraction to exceed the pressure in the pulmonary trunk or the aorta, and thus push through the semilunar valves. This puts a great strain on the AV valves, so in addition to the chordae tendineae, there are small muscles attached to the bottom of the chordae tendineae, called papillary muscles, that contract whenever the ventricles contract.

So just one question: why are there no valves where the blood enters the atria? There are two reasons for this. The first is that the blood in veins returning to the heart is at extremely low pressure, so low that it could not easily push through the closed valves already in the veins. The other reason involves the weaker contraction of the atria. The atria contract when the ventricles are relaxed, which means that the lower pressure of the ventricles at that point will make it easier for the blood to flow in that direction than backward into the veins that are filled with blood.

The Heart's Own Blood

Why does the heart need its own circuit if blood is already going through it? The answer to that rests with our constant companion, surface area to volume ratio. Remember the issue of SA/V when I talked about capillaries? Well, the lumen of the chambers just doesn't have enough surface area to nourish the heart, especially with the thick ventricular walls, so there needs to be a separate blood flow to the heart, one that occurs on the capillary level. Without that, the heart would quickly die, and *so would you!*

Branching off opposite sides of the aorta, almost immediately after it leaves the left ventricle, are the left and right *coronary arteries*. These arteries pass between the atria and the ventricles (in case you ever have to find them in a practicum exam!), and I bet you can find the anterior and posterior interventricular branches. (They're between the ventricles, of course! Don't forget to keep using the names to figure things out!) This is the start of the coronary circuit, which nourishes the heart.

The return of blood to the heart from the heart—which should be thought of as a return of blood to the chambers of the heart from the tissues of the heart—starts by draining the tissues in the back of the heart at the *middle cardiac vein*, and at the front through the *great cardiac vein*. The blood finally empties into the right atrium from the *posterior coronary sinus*.

In a Heartbeat!

A heartbeat, that little twist of anatomical fate that keeps the vice president as a last-resort tiebreaker in the Senate, is a complicated thing. Despite all the emphasis given above to the left and right sides of the heart, the heart doesn't beat that way; the paired beat of your heart involves contraction from top to bottom. The first portion of the beat is the simultaneous contraction of the two atria pumping their contacts into the ventricles. The second part of the beat is the simultaneous contraction of the two ventricles pumping their blood into the pulmonary trunk and the aorta.

Remember those valves I told you about? When you listen to a heartbeat you are listening to the sounds of the valves slamming shut. Listening to the sounds of the heart, or any body sounds in fact, is called *auscultation*, and it is usually done with a stethoscope. If you close your eyes and imagine the sound of a heartbeat you can easily remember the rhythm of the beat since it's so distinctive—"lubb dupp, lubb dupp, lubb dupp," and so on. The first sound (S1 = "lubb") is the sound of the AV valves slamming shut during the contraction of the ventricle (ventricular *systole*). The second sound (s2 = "dupp") is the sound of the semilunar valves slamming shut when the ventricles relax (ventricular *diastole*).

The Conductor

Remember how atrial systole needs to happen during ventricular diastole (think about the rhythm of the beat)? Only then can the ventricles remain open to receive the blood from the atria. Now think about the last time you touched something *really hot*; didn't you pull your hand away incredibly fast? That rapid response time is due on part to the efficiency of nerve conduction (see Chapter 19) and to the speed of muscle cell stimulation (see Chapter 8).

Now think about the rhythm of the heartbeat. The two parts of a heartbeat come from the same initial impulse—which originates at the *sinoatrial* (SA) node, familiarly called the *pacemaker*. The length of the heart is rather short (about the size of your fist) when compared to the length of your arm, so why the delay between the first half of your heartbeat to the second half?

First off, there is a certain characteristic of cardiac cells that can be beneficial in the right hands, but fatal in the wrong hands (sinister music swells!). Cardiac cells have the capacity to be pacemakers, or *autorhythmic* (self-stimulating). The SA node is made of these autorhythmic cells; this means the heart, unlike skeletal muscle, does not need stimulation from nerves in order to contract. There is, nonetheless, a connection to the brain, but this connection, part of the autonomic nervous system (see Chapter 21), only serves to regulate the rate of the contraction (not your rhythm, but your tempo!).

 Medical Records _____

The existence of a pacemaker makes it possible for a person to survive a severed spinal cord, for the heart can continue to contract on its own. The autorhythmic capability of other cells can sometimes cause a problem, for if multiple parts of the heart start to try to contract on their own, the overall contraction of the heart will be too weak to effectively pump the blood. A *defibrillator* works by depolarizing (see Chapter 8) all the muscle cell membranes (sarcolemmas) so that when they repolarize, the cells can now contract in sync.

A heartbeat is the result of the *intrinsic conducting system* of the heart (see Figure 11.4). There are three basic parts of the conducting system: the SA node, the *AV node* (from its name, *atrioventricular node*, you should know its location—between the atria and the ventricle!), and conducting cells, which connect the two nodes, as well as sending the message to the ventricles. From the SA node's pacemaker cells (located at the top of the right ventricle, near the superior vena cava) the stimulus begins, and then spreads outward through the atrial walls via the internodal pathways (yet another name that makes sense) to the AV node. This takes about 50 milliseconds.

At the AV node something essential happens, for the narrower cells in the AV node, as well as the lower efficiency at sending messages from cell to cell, creates an essential delay (about 100 milliseconds) in the message being sent to the *AV bundle*, allowing for ventricular diastole to occur during ventricular systole. The AV bundle (or *bundle of His*) then sends the message on to the *right and left bundle branches*, which carry the message down through the ventricular septum, where they then travel up the outer ventricular walls via the *Purkinje fibers*, ultimately causing the ventricles to contract.

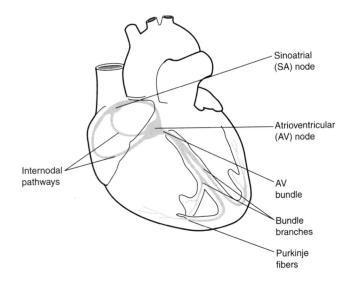

Figure 11.4

The intrinsic conducting system of the heart. Note: Although the atria contract downward, the pathway of the bundle branches and Purkinje fibers allows the ventricles to pump the blood upward.

(LifeART©1989–2001, Lippincott Williams & Wilkins)

Just prior to ventricular systole the papillary muscle contracts, being stimulated directly from the bundle branches, thus bracing the cusps of the valves prior to ventricular contraction. The message travels very fast along the Purkinje fibers, taking only about 75 milliseconds. The full time for the message to travel along the conducting system from the SA node to the end of the Purkinje fibers takes only 225 milliseconds, but the essential 100 milliseconds delay made the filling of the ventricles possible. Lastly, a relatively long *refractory period* (the period during which a second stimulus cannot be received) of 200 milliseconds ensures the spacing between successive heartbeats.

ECG: Measuring a Beat

The electrical conductivity that is an inherent part of the intrinsic conducting system of the heart allows the details of a heartbeat to be measured in great detail. This measurement, which is called an *electrocardiogram*, or *ECG* (it is sometimes called an EKG in order to separate it from the similar sounding *EEG* or *electroencephalogram;* see Chapter 20), is done by attaching numerous electric leads to the patient and measuring the voltage changes as a result of nerve impulses within the intrinsic conducting system.

There are three basic waves involved in an ECG tracing (see Figure 11.5): the P wave, the QRS complex, and the final T wave. The P wave is a gentle curve that indicates the depolarization of the atria. You may remember from Chapter 8 that depolarization is the start of the stimulus that initiates a muscle contraction, in this case the contraction of the atria. In order for the atria to contract, the cell membranes must repolarize prior to the next contraction.

Figure 11.5

The different portions of the ECG curve—PQRST—are used to illustrate atrial and ventricular depolarization and repolarization.

(LifeART©1989–2001, Lippincott Williams & Wilkins)

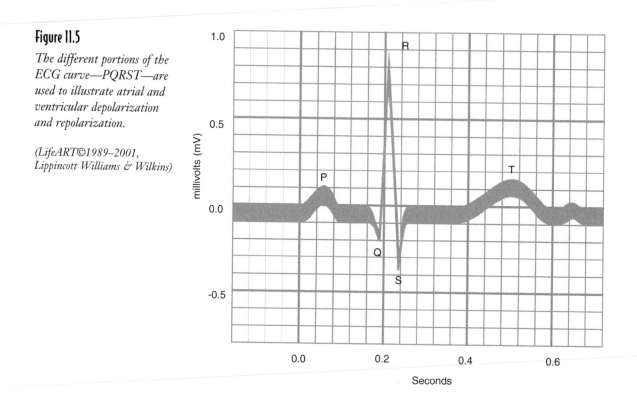

The repolarization of the atria is extremely small, and it is hidden by the sharp tracing of the depolarization of the much larger ventricles in the QRS complex. The sharp, angular nature of the curve explains the use of three letters, which are needed to map out the three parts of the inverted V shape. The last part of an ECG is the T wave, which looks similar to the P wave, but in this case illustrates ventricular repolarization.

The spacing of the intervals, not to mention the timing between successive PQRST cycles, says a lot about how the heart is functioning. A longer P-R interval could mean damage to either the conducting pathways or the AV node, or a long Q-T interval could indicate the presence of a dangerous congenital heart defect that could lead to sudden death.

The Cardiac Cycle

The *cardiac cycle* refers to the events in the heart not only during a heartbeat, but also during the full period between the start of one heartbeat and the start of the next beat. As you have seen earlier, in a cardiac version of "reading between the lines," diastole is just as important as systole. Numerous things are measured in a typical cardiac cycle (see Figure 11.6): (1) ECG, (2) heart sounds, (3) pressures in the atria, ventricles, pulmonary trunk, and aorta, and (4) left ventricular volume.

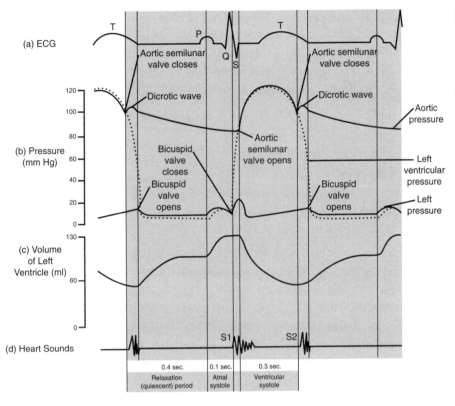

Figure 11.6

The phases of the cardiac cycle, showing the relationship between atrial and ventricular systole and diastole.

(a) ECG

(b) Pressure (mm Hg)

(c) Volume of Left Ventricle (ml)

(d) Heart Sounds

At the end of atrial systole the ventricles hold the maximum amount of blood. This maximum ventricular volume, a surprising small 130 milliliters for an average person at rest, is called the *EDV*, or *end-diastolic volume*. In the same sense, the smaller volume of ventricular systole, about 50 milliliters (the heart never fully empties), is called the *end-systolic volume (ESV)*.

So how much does your heart actually pump? Well, a single beat of the heart pumps blood out of the ventricles, and you have the before and after volumes already! The full ventricle—end-diastolic volume (EDV)—holds 130 milliliters, and an "empty" ventricle—end-systolic volume (ESV)— holds 50 milliliters, so the amount ejected, or *stroke volume (SV)*, is easy:

EDV–ESV = stroke volume (SV), or
130 ml– 50 ml = 80 ml

Flex Your Muscles

All this talk about systole and diastole will come in handy when you learn about blood pressure (see Chapter 12). The pressure in the arteries as a result of systole in the left ventricle is the *systolic pressure,* which is understandably higher than the *diastolic pressure* from left ventricular diastole.

That's not much per ventricle and per beat, but it adds up! Don't forget that sometimes the heart beats faster (*bpm = beats per minute*)! The *heart rate* (*HR*) times the stroke volume gives you your *cardiac output* (*CO*) for one minute: HR × SV = CO. You might remember the amount of blood you have from Chapter 10: females—4 to 5 liters, males—5 to 6 liters. Let's see how much of it you pump in a minute:

CO = (SV) 80 ml × (HR) 75 bpm = 6000 ml/min = 6 liters/min!

That's not much time for a round trip!

During ventricular systole, there is a period called *isovolumetric contraction*, during which the pressure in the ventricle rises while the valves remain closed. Have you ever tried to open a door when someone else was holding it closed on the other side? In order to open the door, you have to be able to push harder than the person on the other side, as you two play a kind of reverse tug-of-war.

Remember that the blood in the vessels is exerting a pressure on the vessel walls, not to mention the semilunar valves! The ventricles can pump the blood out only if the pressure that builds during the isovolumetric contraction exceeds that on the other side of the semilunar valves; at that point the aptly named *ventricular ejection* happens. For this reason, high blood pressure puts a strain on the left ventricle, for it must pump harder to force the blood through the aortic semilunar valve. To do this the ventricle increases in size; this increase, called *left ventricular hypertrophy*, ultimately weakens the heart and can lead to cardiac arrest.

Following ventricular systole, with the end-systolic volume, the ventricles start to relax, but the pressure in the ventricles exceeds that in the atria. The tug-of-war occurs at both ends of the ventricles! During this period the elasticity of the heart allows the ventricle to return to its original shape as it relaxes, in a period called *isovolumetric relaxation*.

Once again, when atrial systole exceeds the ventricular pressure, blood forces its way through the AV valves, and the cycle continues, on and on, for about 2,739,375,000 cycles, give or take several hundred thousand, depending on how long your heart continues to pump. Remember, after all, that ultimately everyone dies of heart failure (which makes cardiac arrest a worthless diagnosis of cause of death)! So keep on pumping!

The Least You Need to Know

- ◆ Due to a high need for oxygen, the four chambers of the heart allow the maximum separation of oxygenated and deoxygenated blood.

- ◆ Blood flows to and from the heart in three continuous circuits (artery to arteriole to capillary to venule to vein and back to the heart again): pulmonary (lungs), systemic (body), coronary (heart).

- ◆ The SA node (pacemaker) initiates the heartbeat, but the AV node is necessary to delay the contraction of the ventricles.

- ◆ The cardiac cycle is the alternation of contraction (systole) and relaxation (diastole) of the atria and ventricles (that is, when atria are in systole, the ventricles are in diastole).

Chapter 12

Shipping and Transport

In This Chapter

- ♦ Structure of arteries, veins, and capillaries
- ♦ Systemic circulation
- ♦ Fetal circulation
- ♦ Blood pressure
- ♦ Lymphatic circulation

We can't possibly look at all 12,000 miles (about 19,300 km) of blood vessels in a book so small, so I had better find an easy way for you to navigate your way around the vascular part of the cardiovascular system! In Chapter 11, you learned that in the course of about a minute, all of your blood could make one full circulation around the body.

Before you start telling everyone that your blood travels at 720,000 mph (1,158,000 kph), don't forget that none of the blood travels through *all* of the vessels in one minute! As you saw in Chapter 10, arteries split into arterioles, arterioles split into capillaries, capillaries merge into venules, and venules merge into veins. The *maximum* journey for any one blood cell is therefore *much* shorter, on the order of yards or meters.

In this chapter, you'll learn what makes the vessels different, not to mention some surprising quirks in both cardiovascular and lymphatic circulation. The unique life of a fetus, as you shall see requires some differences in the circulation, especially with that weird-looking umbilical cord. I'll also discuss what blood pressure is, and how to measure it.

Vessels To and Fro

As I discussed in the previous two chapters, arteries carry blood away from the heart, and veins carry blood to the heart, with capillaries doing the real work of exchange at the tissue level. The arterial and venous pathways are parallel most of the way, but that doesn't mean that arteries and veins are the same, nor are capillaries merely small arteries and veins. Their differences reflect both the pressure of the blood within the vessels, and the function of the vessels themselves.

Leaving and Returning Home

Most books are content to divide their illustrations into arteries and veins. Although they are opposite ends of the same circuit, you can best understand the nature of the blood vessels if you see them side by side, as shown in Figure 12.1.

Figure 12.1

This diagram shows the parallel pathways of the larger vessels in the systemic circulation.

(©2003 www.clipart.com)

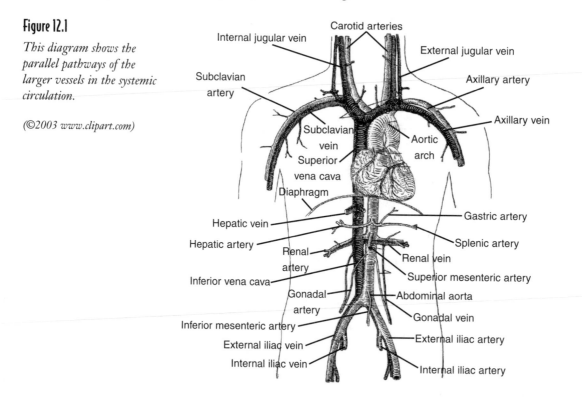

Several things about arteries and veins are identical. Both vessels have a thick outer layer called a *tunica externa*, with a layer of smooth muscle underneath called the *tunica media*. The hollow opening of each vessel is called its *lumen*, and the innermost layer, on the outer edge of the lumen is called the *tunica interna*, which is made of an epithelial layer (endothelium) and a basement membrane (see Chapter 4).

If you compare the structure of arteries and veins, two things immediately stand out. For one, the vessel walls are not the same; the arterial walls are much thicker than the venous walls, due in part to a larger smooth muscle layer. In addition, on either side of the muscle layer there are layers of elastic connective tissue called the internal and external elastic lamina. This layer is important due to the higher pressure the blood exerts against the arterial walls; this is, after all, what blood pressure is!

The other striking difference between arteries and veins is the presence of valves in the veins. Remember that the arteries are taking blood from the heart, whereas the veins are taking blood from the capillaries, so there is little pressure in the veins. Without these valves, the blood would easily pool in our legs. In addition, the contraction of our skeletal muscles in our limbs (*skeletal muscle pump*), as well as bellows-like pressure changes from the contraction and relaxation of our diaphragm (*respiratory pump*), help to pump the blood back to the heart.

Crash Cart

If you are ever in a wedding party, don't make one of the most common mistakes of posture, for you will easily embarrass yourself! Remember the importance of the skeletal muscles in returning the blood to the heart? Well, people who lock their knees, rather than using their muscles to support their lower legs, will have blood pool in their legs, eventually leading to a critical drop in blood pressure in the arteries to the brain. With enough time, and a big enough drop in oxygen to the brain, you will faint dead away, and, unless you hit your head on the way down, the only injury will be to your pride, as long as you are forgiven for screwing up the wedding!

Arteries, in turn, are divided into two basic types: *elastic* (*conducting*) and *muscular* (*distributing*). The elastic fibers in the conducting arteries temporarily expand and store some of the kinetic energy of the blood pumped with each heartbeat, and then transfers it back to the forward motion of the blood with each ventricular diastole (see Chapter 11) when the vessels recoil.

The muscular arteries are smaller, and more distant from the heart, and they are able to *vasoconstrict*, or contract the muscles in the vessel walls to narrow the lumen of the artery thus regulating the amount of blood to the tissues. This vasoconstriction is also carried out at the arteriole level. The relaxation of the smooth muscle around the arterioles leads to a dilation of the vessels, called *vasodilation*.

Arteries are probably what most people think about when they think about blood vessels. Arteries, after all, are the place where we feel our pulse, and where we measure our blood pressure. When people think of blood giving life—delivering oxygen, food, and water—they are usually thinking of arteries; it turns out, however, that things are more complicated than that. Arteries may deliver the goods, but they don't actually get off the truck until they get to the capillaries. What is even more surprising is that the highest amount of food, and the lowest amount of waste are not found in the arteries at all, but in certain specific veins (as I explain later in this chapter).

Pickup and Delivery

Capillaries may be small, but they do the important work! Without these unassuming little, and I mean *little*, vessels, none of our tissues would get oxygen, get rid of carbon dioxide and other wastes, receive nutrients, water, and so on. These capillaries connect the arterioles to the venules, but they branch extensively in the process, greatly increasing surface area to volume ratio (see Chapter 3), which is essential for all the absorption and filtration that capillaries do. (I discuss filtration and reabsorption in more detail in Chapter 15.) Diffusion (see Chapter 3) is a wonderful thing, but it needs *space* if we are going to be able to do enough diffusion for us to survive. Remember, we are all *glorified amoeba!*

There are a few things that make capillaries unique. For one thing, they are incredibly small. Their lumen is so small that the tiny erythrocytes (RBCs; see Chapter 10) must pass through single file! This also has the benefit of helping the pickup and release of oxygen from the RBCs, since they are so close to the endothelium. The endothelium is one cell layer thick, and since they are squamous cells, they are very thin. This thin layer helps the materials to travel through the cells. On the outside of the epithelium, as always, there is a basement membrane (see Chapter 19) for attachment to the outer tissues; that basement membrane is important when vessels get damaged, for it is involved in initiating the clotting process (see Chapter 10).

Capillaries also are involved in vasomotion, which is the movement of blood through the capillaries via a separate pumping action. It turns out that there is a special type of vessel that travels straight through a capillary bed from an arteriole to a venule, and this vessel is called a *metarteriole* (see Figure 12.2). This metarteriole has smooth muscles that pump the blood, and a smooth muscle precapillary sphincter in turn contracts and relaxes, thus regulating the blood into the true capillaries. The other end of the metarteriole, which lacks smooth muscle, is a low-resistance pathway called a *thoroughfare channel*, and it empties into the venule. It seems that there's a whole lot of pumpin' goin' on!

What I find cool is how the capillaries, as the end stage of the closed system, manage to transfer materials. The capillaries in our muscles, connective tissue, and lungs are called *continuous capillaries*, and they help out the flow of materials by having occasional gaps intercellular clefts—sounds like cheating to me! Other capillaries—in the intestinal villi, kidneys, ventricles of the brain, and in endocrine glands—are called fenestrated capillaries, which means they have large pores in the endothelial cell membrane—hey, maybe this system isn't so *closed* after all!

As if those weren't bad enough, there are other capillaries called *sinusoids*, which are found in such places as the liver, spleen, and adenohypophysis (anterior pituitary), that have larger spaces between endothelial cells, plus an incomplete or missing basement membrane— what a leaky pipe that is! All of these openings, however, are only big enough for plasma and phagocytic cells; RBCs are too big and inflexible to make it through. But perhaps the coolest way of sending materials out through the endothelium is a combination of endocytosis and exocytosis. Do you remember *pinocytosis* from Chapter 3? This miniscule "cell drinking" engulfs the plasma into little vesicles, immediately sends it to the outer edge of the epithelium, and dumps the contents into the interstitial fluid (see Chapter 10) by exocytosis. Pretty cool, huh?

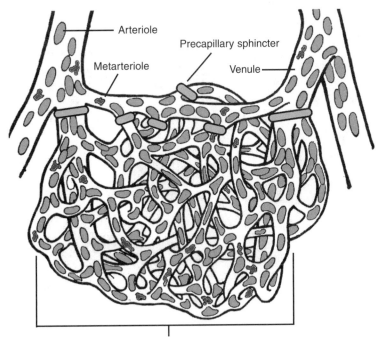

Figure 12.2

The profusion of small capillaries greatly increases SA/V, and the metarteriole regulates the blood flow into the capillaries.

(©2003 www.clipart.com)

Capillary bed (The greater surface area within the capillary bed makes it the site of pickup and delivery between the cardiovascular system and the tissues.)

Here, There, Everywhere

In Chapter 11, I discussed the three simultaneous circuits in our cardiovascular system: systemic, pulmonary, and coronary. Chapter 11 introduced you to the pulmonary and coronary circuits, and the pulmonary circuit will be explored further in Chapter 13, so in this chapter I concentrate on the systemic circuit. As you have seen from the blood flow in the pulmonary circuit, blood vessels tend to follow parallel tracks, to and fro. Later in this chapter I explore another parallel track—although not a complete circuit as it only goes one way—called the *lymphatic system*, which follows the pathway of the major arteries and veins.

Given the central location of so many structures, one thing you will find throughout the body are examples of *collateral circulation* (see Figure 12.3). Collateral circulation means that more than one artery feeds the capillary bed of an organ. When these arteries fuse it produces an *arterial anastomosis*; this is just another example of the body being clever, because a blockage on one side still means that the blood will get to the organ. As for the rest of the body, I will get you started with the major pathways here. Any good college textbook can get you the rest of the way. What follows will also help you to avoid some pitfalls.

Figure 12.3

Collateral circulation provides alternate routes to organs, which allows blood flow, even if a vessel is blocked.

(LifeART©1989–2001, Lippincott Williams & Wilkins)

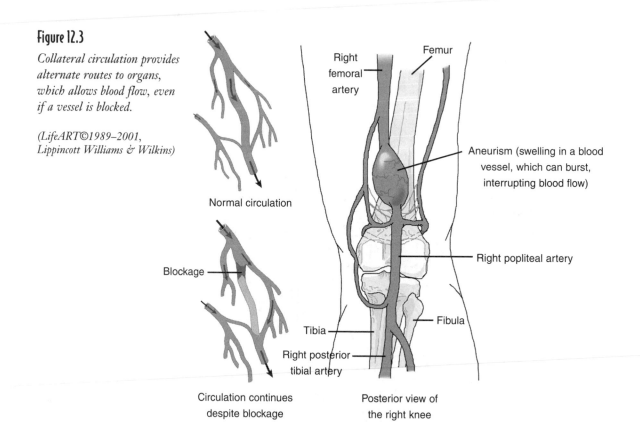

Normal circulation

Blockage

Circulation continues despite blockage

Right femoral artery

Femur

Aneurism (swelling in a blood vessel, which can burst, interrupting blood flow)

Right popliteal artery

Fibula

Tibia

Right posterior tibial artery

Posterior view of the right knee

Upper Body

When you look at the vessels of the upper body I want you to think about where the blood is coming from, and where it has to go. First of all, remember how blood leaves the heart in a single vessel? Well, since most of the body is below the heart, the aorta arches downward (in a part called the *aortic arch*, of course) almost immediately after it leaves the heart. At the top of the arch, however, the blood for the upper body leaves the aorta. The only problem is that there are only three vessels to go four places: the right arm, the right side of the head, the left arm, and the left side of the head. The names of the three branches provide a clue as to their destinations.

On each side of the neck there is a major artery—which you feel for when you take your pulse—called the *carotid artery* (left and right, of course). There are also branches that go to each arm; these branches are called *subclavian arteries*. (Guess what bone they go under? Yup! The clavicle!) Since the aortic arch is more on the left side of the body, the middle branch is the left carotid artery, and the left branch is the left subclavian artery (refer to Figure 12.1). The name of the right branch, which must go to both the right arm and the right side of the head is called the brachiocephalic (*brachio* = arm, *cephalic* = head) artery.

There are smaller branches from the subclavian that go up through foramina in the transverse processes of the cervicalvertebra, called the *vertebral arteries*. These vessels join in front of the brainstem (see Chapter 20) to become the *basilar artery*; the basilar later divides to become the *circle of Willis*, an example of collateral circulation around the pituitary gland.

Blood returning to the heart is mostly parallel to the arteries, with *subclavian veins, vertebral veins*, and (parallel to the carotids) *internal* and *external jugular veins*. The actual return to the superior vena cava, which is on the right side of the heart, is more symmetrical than the aorta, with both left and right *brachiocephalic veins*, both of which merge into the superior vena cava.

Abdominal Organs

Following the abdominal aorta down, the blood flow varies based upon the organs' locations. Many organs are large enough, or *long* enough, to receive blood from both sides of the aorta (you will find a similar situation in the orientation of the spinal nerves; see Chapter 21). In some cases there is more collateral circulation, such as the left and right gastric arteries (to the stomach), or the connecting branches of the inferior mesenteric (within the mesentery; see Chapter 13).

The blood flow to a few organs deserves special mention. The spleen, which is on the left side of the upper abdomen, next to the stomach, has the *splenic artery* and *splenic vein* that travel to and from the spleen, directly over the pancreas. As long as the blood travels over the pancreas, there are, of course, little *pancreatic arteries* and *pancreatic veins*.

The kidneys differ from many of the abdominal organs, simply because they are actually *behind* the abdominal wall, directly on either side of the abdominal aorta and inferior vena cava. The *renal arteries* and *renal veins* (*renal* means kidney) are thus kind of short, although the left and right vessels for each kidney will be slightly different, as the right kidney is closer to the vena cava than it is to aorta, and vice versa (refer to Figure 12.1). The pelvic blood flow is discussed in the next section. As for the return from the rest of the abdominal organs, well, that deserves its own section.

Going Their Separate Ways

The most dramatic point in the abdominal aorta is the abrupt division between the two *external iliac arteries* (and don't forget the parallel joining of the *external iliac veins* into the inferior vena cava). Abrupt, yes, but illogical, no, as each vessel ultimately goes to a separate leg (refer to Figure 12.1). Since it seems to make sense that they would divide in the pelvis, most people are surprised when they see how high up the split takes place: at about the level of the umbilicus (remember your belly button?). The reason for so high a split makes sense when you realize there is some work left to do, and the split makes it easier.

Remember the body cavities, from Chapter 1? Well, it's time to look at the "imaginary" pelvic cavity. As the aorta splits, it gives the external iliacs a chance to get out of the way of cramped pelvic organs. With so little space in the pelvis, well, there ain't nowhere for the organs to be but in the middle! This makes it easy to get blood to them; off the external iliac arteries are, you guessed it, *internal iliac arteries* (with the return trip taking *internal iliac veins*). Finally, the bulk of the blood goes down and up the legs via the femoral arteries and veins.

Hepatic Portal System

When students lay down the basic blood pathway in their end of the year project (a schematic diagram of all 11 body systems), they start with the arteries, and see the blood leading to the abdominal organs directly from the abdominal aorta. They then make a completely logical assumption and have all the blood return directly to the inferior vena cava, which runs parallel to the aorta. The only problem is that this is *wrong!*

Blood in vessels from the abdominal organs bypasses the inferior vena cava; the blood travels from capillary beds in those organs directly to capillary beds in the liver. Any vein that travels between two capillary beds is called a *portal vein*, and since this one goes to the liver—and hepatic refers to the liver—the name of this vessel is the hepatic portal vein (see Figure 12.4). This blood then travels through the liver and ultimately returns to the inferior vena cava via the hepatic vein. Since the blood in the hepatic portal vein is deoxygenated, it means that there needs to be a separate hepatic artery to bring the liver the needed oxygen.

Figure 12.4

The hepatic portal vein allows the liver to filter the blood prior to returning to the heart.

(LifeART©1989–2001, Lippincott Williams & Wilkins)

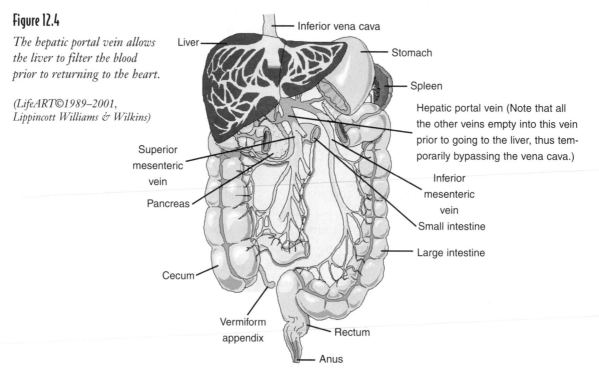

Hepatic-Portal Circulation

Think of where the blood is coming from: spleen, pancreas, stomach, small intestine, and large intestine. The blood from the small intestine, in particular, has a very special quality to it; this blood, carried in the superior and inferior mesenteric veins, has a higher level of food in it than anywhere else in the body! That blood will ultimately mix with blood in other veins, blood that is very low in food, before it is ultimately pumped out of the heart. Blood from the left ventricle may be very rich in oxygen, but it is not that high in food.

Medical Records _____

Just as blood from the heart is not as rich in food as you might expect, that same blood has a lot more waste than you might expect! After all, blood going to nourish the tissues also goes to the kidneys to be filtered of wastes! The "clean" blood from the kidneys ultimately joins up with waste-filled blood before it gets pumped out of the heart again. This is all because in the systemic circuit, venous blood always mixes with blood from multiple places before it reaches the heart, but arterial blood is all the same as it branches out from one source!

When the blood travels to the liver directly from the abdominal organs, the liver immediately goes to work (see Chapter 14 for a summary of the liver's functions). The gastrointestinal tract absorbs a lot of toxins, which the liver does its best to break down before sending it to the rest of the body. Excess food absorbed is turned into fat for storage by the liver. Insulin (see Chapters 1 and 18) and glucagon released by the pancreas travel to the liver first, where the liver either starts making glycogen (with insulin) or breaking down stored glycogen (with glucagon).

The Baby's Lifeline

Babies have a very different life than ours. Think about it. They really don't eat, or drink, or even breathe. A growing baby needs an extraordinary amount of energy, not to mention raw materials for growth. All that material must come from its mother, in the ultimate pampering. You know … womb service. Blood must reach the placenta so that material exchange can occur, removing wastes and CO_2, taking in food and O_2. You might assume that the fetal blood that is the most deoxygenated would go to the placenta, but the body, once again, has some surprises in store.

Although much of the circulation looks the same, it must differ in terms of both the respiratory and digestive systems (see Figure 12.5). Rather than sending blood from the inferior vena cava to the placenta, the blood is *arterial!* There are two umbilical arteries that branch off of the two internal iliac arteries, and extend onward to the placenta. The developing urinary bladder nests between these umbilical arteries.

Figure 12.5

Fetal circulation. Note the two umbilical arteries, the umbilical vein, the foramen ovale, and the ductus arteriosus.

(LifeART©1989–2001, Lippincott Williams & Wilkins)

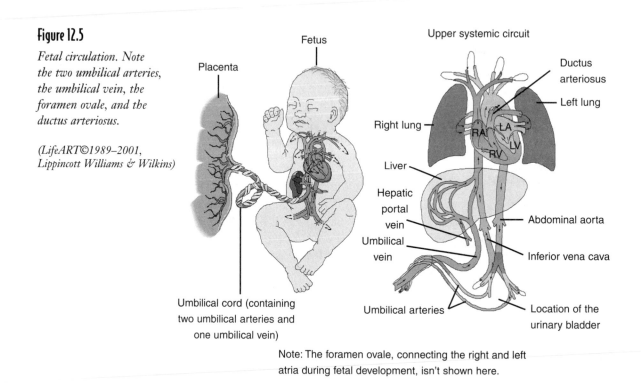

Umbilical cord (containing two umbilical arteries and one umbilical vein)

Note: The foramen ovale, connecting the right and left atria during fetal development, isn't shown here.

The trip back along the single umbilical vein—which, like the pulmonary vein in an adult, carries oxygenated blood—once again does not go directly to the heart. Like the adult, the blood with the most nutrients goes straight to the liver (in the fetus it drains into the ductus venosus, which connects to veins inside the liver), and for all the same reasons. These three vessels, however, are not the only differences. There are two openings in the heart, which disappear after the baby is born.

One opening, called the *foramen ovale*, connects the right and left atria. This allows blood from the placenta, by going from the right atrium to the left, to bypass the pulmonary circuit and go straight to the systemic circuit. This makes sense, considering that the lungs are not involved in oxygenation yet. The other opening is a connection between the pulmonary trunk and the aorta called the *ductus arteriosus*; its function is the same as the foramen ovale. These connections close after the baby is born, for the mixture of oxygenated and deoxygenated blood would ultimately be a handicap (and a somewhat reptilian one at that!).

Oh, the Pressure!

Blood pressure is an amazing thing. It is simply the pressure exerted by the blood on the vessel walls. It doesn't sound like much, but without it, we would never be able to pump oxygenated blood against gravity to the brain. You have probably experienced a miniature form of this if you stand up too quickly. Simple inertia—a body in motion stays in motion, and a body at rest stays at rest, unless acted upon by a force—will keep the blood in place, despite the sudden elevation of your head. The subsequent loss of blood pressure is what makes you feel dizzy.

Systolic and Diastolic

First of all, even though every vessel experiences pressure against its walls, when you have your blood pressure taken, the measurement is of your *arterial* blood pressure. Even though the arteries are deeper than the veins, the pressure is still strong enough to stretch the arterial walls with each heartbeat, and you can feel that as a pulse. Remember that even though the heartbeat is a double beat (the sound based on the slamming shut of the valves; see Chapter 11), the blood flow out of the heart is a single pulse out of both ventricles. That is why your pulse is a single beat.

The high and low pressures make perfect sense when you think about the heart (see Chapter 11). The left ventricle pumps the blood out to the systemic circuit. The higher number, or the *systolic pressure*, is a measurement of the pressure when the left ventricle contracts, or is in systole. (Makes sense, no?) When the left ventricle relaxes, or is in diastole, there is still a pressure being exerted on the vessel walls; this is the smaller number, or the *diastolic pressure*.

Medical Records

For some reason, some people are terrified of having their blood pressure taken. This fear response (from the sympathetic nervous system; see Chapter 20) raises the patient's blood pressure. This situational high blood pressure has been affectionately called *white-coat hypertension*.

Normal blood pressure varies from childhood to adulthood. High blood pressure, or *hypertension*, is when one's systolic pressure exceeds 139, and diastolic exceeds 89. The risks of hypertension include heart attack (see Chapter 11) and stroke. The risks (which are less severe, in general, than those for hypertension) of hypotension, or low blood pressure, include dizziness or *syncope* (fainting).

Measuring off the Cuff

It turns out that measuring blood pressure is an extremely simple concept. By compressing an artery with a cuff attached to a meter, the blood pressure also exerts pressure against the cuff. If the pressure of the cuff is higher than the systolic pressure, all blood flow stops. This is the same principle behind using a tourniquet to stop blood flow, whether it be to stop bleeding, or to take a blood sample. On the other hand, if the pressure of the cuff is lower than the diastolic pressure, the blood flows freely through the vessel. Despite the simplicity of the concept, taking a blood pressure reading is a definite skill; some may even call it an art.

The artery being compressed when taking blood pressure is the brachial artery, and the true name of a blood pressure meter is a sphygmomanometer. First, raising the pressure of the cuff must stop the blood flow. As the pressure on the cuff is released, a sound can be detected with a stethoscope as the pressure of the blood first breaks through the cuff; this sound, a bit of a thudding with each pulse (if there is a needle on the meter, you will see it jump with each pulse), is the systolic pressure.

Medical Records

The second sound is so quiet that it cannot be heard if there is too much ambient noise. At the scene of a car accident, for example, a paramedic can palpate, or feel the pulse return at the systolic pressure. This type of reading is called a "BP palp."

In the same way, the sound will slowly get quieter, becoming a *swishing* sound, until it finally stops when the pressure of the cuff is lower than the diastolic pressure. These sounds are called *sounds of Korotkoff*; the name is a Russian one, and due to differences between the English alphabet and the Cyrillic alphabet, the name is sometimes spelled *Korotkov*, or *Korotkow*.

Drainage System

As you saw in Chapters 3 and 10, as well as earlier in this chapter, the plasma in the blood vessels leaves the capillaries, becomes interstitial fluid, and then is drained into lymphatic capillaries, where it is called lymph; lymph is ultimately drained back into the bloodstream via the subclavian veins. This constant movement of blood *out* of the capillaries has two potential problems: a drop in blood pressure as a result of the loss of blood volume, and a buildup of fluid in the peripheral tissues, known as *edema*. Yet, the role of the lymphatic system is far more complex than just drainage, as we shall see.

Lymphatics and Lymph Nodes

Two things differentiate lymphatic vessels from blood vessels: structure and direction. Although there are a surprising number of openings in capillary walls, the gaps in the endothelium of lymphatics are especially large. Unlike cardiovascular capillaries, which are continuous with arterioles and venules, lymphatics have blunt ends in the tissues. Those blunt ends illustrate perhaps the most important difference, which is that fluid in the lymphatics travels in only one direction: back to the heart.

Please understand that more than just fluid is being drained. Chemicals released into the interstitial fluid enter both types of capillaries. These chemicals include CO_2 (to the lungs), wastes (to the kidneys), hormones (target organs all over the body), and so on. In this way, the lymphatic system is another delivery person. Some of those deliveries, such as bacteria or cancer cells, are not really wanted!

Have you ever stepped on a tack? AAAARRRGGHH!! A puncture wound—heck, even a little cut—introduces bacteria where it normally wouldn't be. The very outside layer of our skin is dry and dead (epithelial tissue is avascular, without blood nearby, the topmost layer will die; see Chapter 4), which acts as a barrier to infection (see Chapter 16). Like

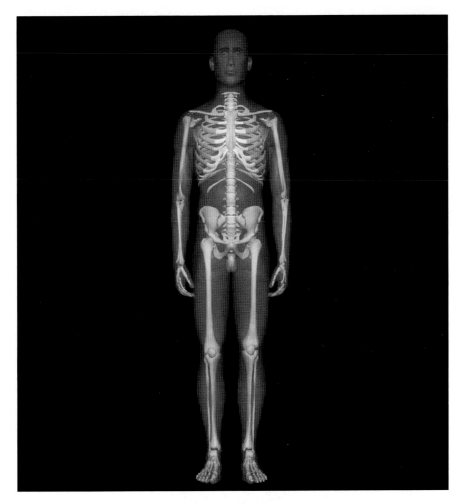

Skeletal System

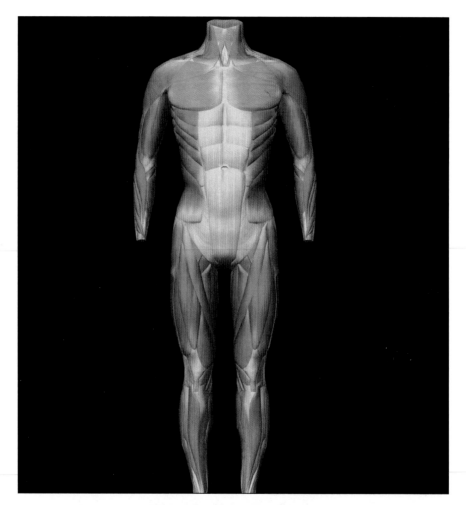

Muscular System—Front

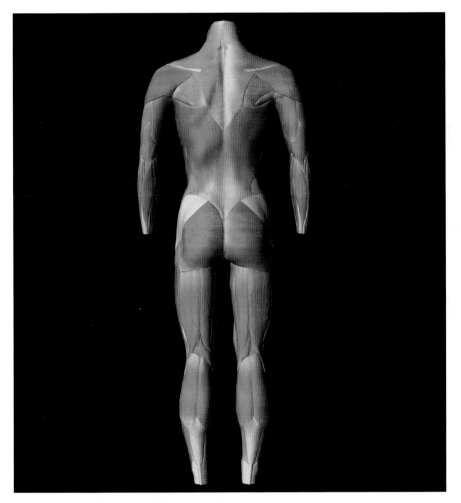

Muscular System—Back

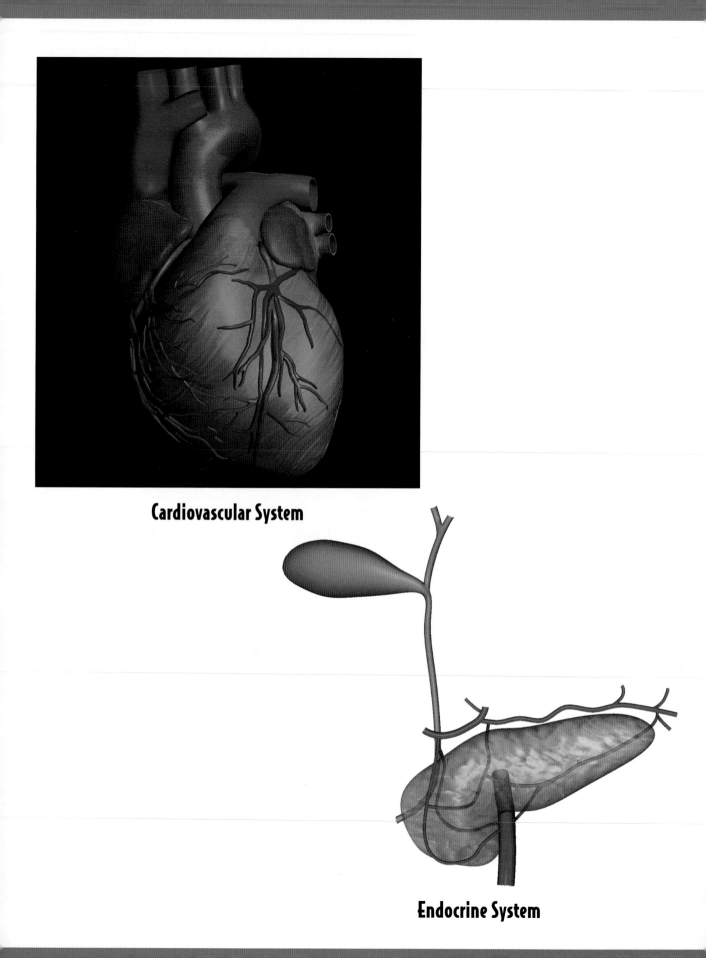

Cardiovascular System

Endocrine System

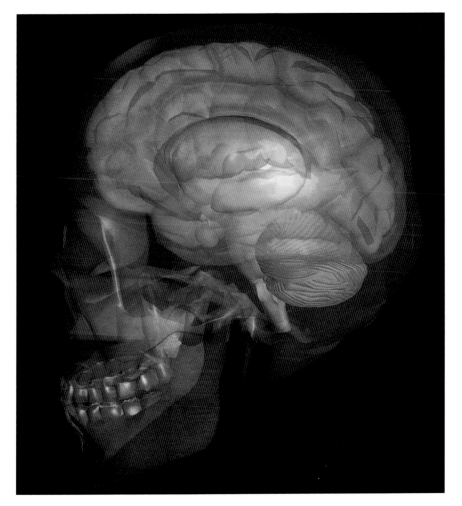

Nervous System

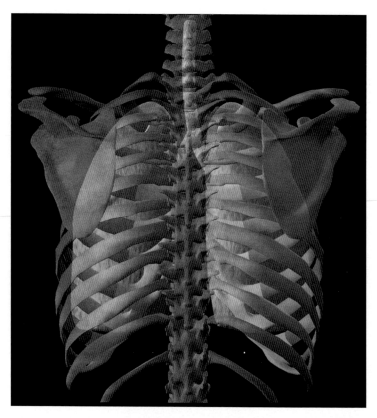

Respiratory System

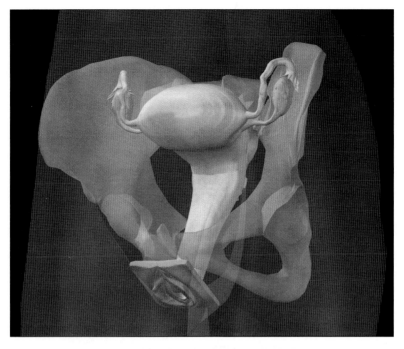

Reproductive System—Female

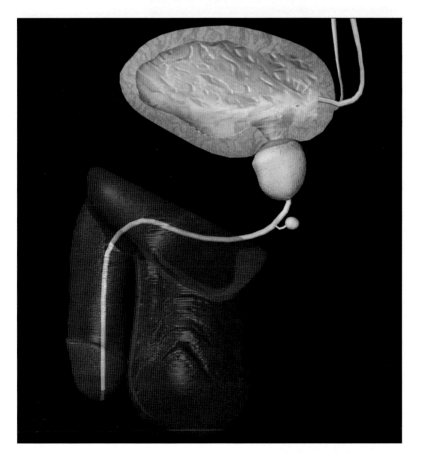

Reproductive System—Male

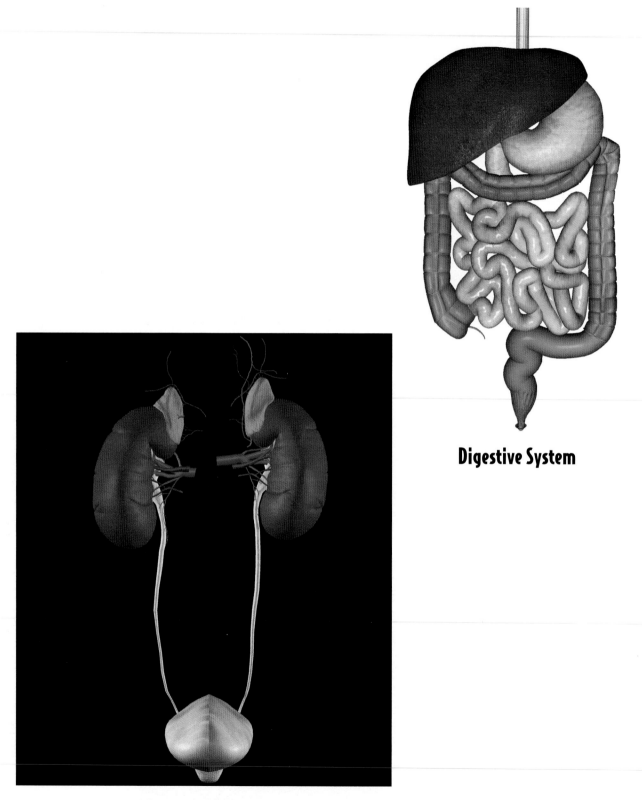

Digestive System

Excretory System

any other living thing, bacteria need food, water, and warmth—sounds like a good description of the inside of your body.

Bacterial infections run the risk of infecting the whole body. The real risk is the introduction of bacteria into your bloodstream, which makes your blood *septic*, a condition known as *septicemia*; if untreated, this will kill you. The large gaps in the lymphatic endothelium are an easy route for bacteria. Since the lymph will be dumped into the bloodstream eventually, there had better be a way of getting rid of the bacteria first!

Luckily, there is! Lymph must first travel through lymph nodes (see Figure 12.6), and it is here that the main fight against infection happens. The vessels entering the nodes, called *afferent vessels*, and those leaving, called *efferent vessels*, have valves (see Chapter 11) to prevent backflow; indeed, these valves are present in all lymphatic vessels, as the lymph is at extremely low pressure, just like the veins, and the contents are pumped back the same way.

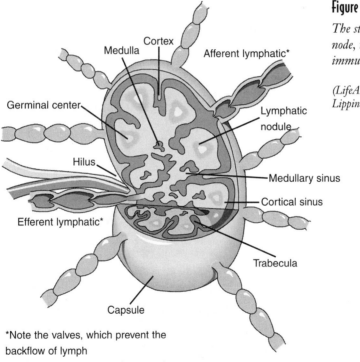

Figure 12.6

The structure of a lymph node, the major site of immune activity in the body.

(LifeART©1989–2001, Lippincott Williams & Wilkins)

*Note the valves, which prevent the backflow of lymph

The Big Picture

One insidious way the body connects one system with another is due to the lymphatic system. Cancer cells (see Chapter 3) have a habit of breaking away from tumors and traveling along lymphatics. They will divide when they reach the nodes, and they can ultimately spread to the entire body. Each of these new tumors, distant from the primary tumor, is called a *metastasis*. This is the main argument for early detection, for cancer is more likely to be cured if it can be treated before the malignant cells find their way on the lymphatic highway.

Given how they save our lives every day, it's amazing how small lymph nodes are (1–25 mm or up to an inch). There is an outer connective tissue *capsule* with walls called *trabeculae* that extend inside the node. In between these walls are pyramid-shaped areas, the wider portion of which is the cortex, and the narrower part is the medulla. Within each pyramid is a *lymphatic nodule* or *follicle* with those antibody factories, the B lymphocytes, inside, and T lymphocytes and macrophages on the outer edge. Together, these cells work to wipe out the infection before it spreads (see Chapter 17 for more details).

Cisterna Chyli, Thoracic Duct, Return

Lymph nodes are both superficial and deep. Beyond that, and in addition to nodes scattered throughout the body, nodes can be found in larger groups that follow the lymphatics: inguinal nodes (in the groin), iliac nodes (along the external iliacs—remember the idea of parallel pathways?), intestinal nodes, axillary nodes (in the armpits), and cervical nodes. Lymphatics from the legs, the intestinal mesentery (see Chapter 14), and other abdominal regions, all collect in a dilated vessel beneath the diaphragm called the *cisterna chyli* (see Figure 12.7).

Figure 12.7

The cisterna chyli and the thoracic duct are part of the return to the cardio-vascular system.

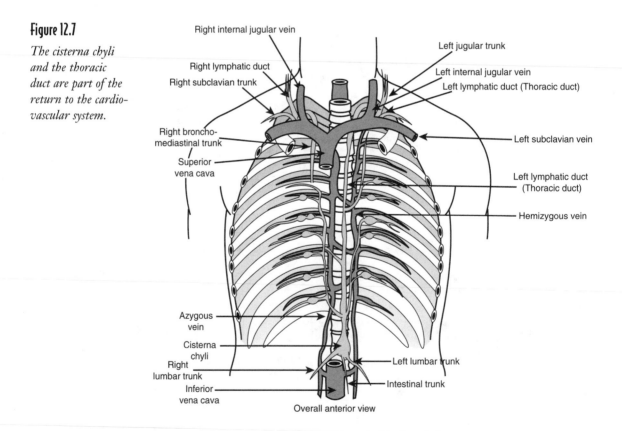

Right internal jugular vein

Left jugular trunk

Right lymphatic duct

Left internal jugular vein

Right subclavian trunk

Left lymphatic duct (Thoracic duct)

Right broncho-mediastinal trunk

Left subclavian vein

Superior vena cava

Left lymphatic duct (Thoracic duct)

Hemizygous vein

Azygous vein

Cisterna chyli

Left lumbar trunk

Right lumbar trunk

Intestinal trunk

Inferior vena cava

Overall anterior view

Flex Your Muscles _____

Don't assume that everything is split down the middle. The lymphatic system is one that doesn't follow a simple bilateral plan. The entire lower part of the body and the left arm and left side of the head and chest are drained into the left subclavian vein, but only the right arm and the right side of the head and chest drain into the right sub-clavian vein. That's not exactly even, but the job gets done.

From the cisterna chyli there is a large duct that travels next to the spinal column, up the left side of the thoracic cavity. Name it. Oh, come on, you know it has to be called the *thoracic duct!* This duct is also called the *left lymphatic duct* because there is a much smaller, much shorter duct on the right hand side (the *right lymphatic duct*). These ducts merge with ducts from the arms (the left and right subclavian trunks) and the head (the left and right jugular trunks), and dump into each of the subclavian veins. And with that, the fluid, and all its contents (minus bacteria and viruses, thanks to those wonderful nodes) make their way back to the bloodstream.

The Least You Need to Know

- Arteries and veins differ in their structure, due mainly to differences in blood pressure.

- Capillaries are the location for the transfer of materials, and they use different types of gaps and transport to get the job done.

- Arterial and venous blood flow is mostly parallel, and bilateral arteries sometimes lead to collateral circulation.

- Blood pressure measures left ventricular systole and diastole by stopping the blood flow through an artery, and measuring the pressure on a cuff.

- The parallel lymphatic system drains fluids back to the heart, but not until pathogens are removed in the lymph nodes.

Part 4

What Goes In Must Come Out

Death on the installment plan. Not a very cheery way of looking at life, I grant you, but it holds a kernel of truth. That's not to say we shouldn't feel joy in each other's embrace, and take wonder in the small details of our lives. Still, without a *constant* concern for getting and using a few essential materials, not to mention getting rid of others, we would all die. Each breath, each bite of food, each trip to bathroom, keeps us alive.

The three body systems in this section—respiratory, digestive, and excretory—are our lifelines to the outside world. We need to move matter in and out, be it gases (respiratory), food and water (digestive), or wastes (excretory), and parts of all three systems are in constant action to do just that. If you have ever held your breath too long, waited too long between meals, or crossed your legs, hoping to get to a bathroom soooon, you have felt the power of that need.

So relax, take a few deep breaths, have a snack, and start your journey through these amazing systems. Just make sure you go before we leave!

Chapter 13

Waiting to Exhale

In This Chapter

◆ The functions of the respiratory system

◆ Organs of the conducting and respiratory regions

◆ The larynx and voice production

◆ Gas pressure laws and breathing

◆ Pulmonary volumes and capacities

◆ Control of breathing by the brain

It's interesting how many expressions we use that refer to the act of breathing: breathtaking, breathing life into something, and so on. Have you ever been swimming in the deep end and decided to swim down to get something shiny off the bottom of the pool, only to realize that you have been underwater just a bit too long, and that it's an awfully long way to the top! In moments like that, it becomes obvious how important breathing is! Despite that, breathing is so automatic that it usually escapes notice.

We all think about lungs when we think of breathing, but that is only the half of it! There is much more going on, and many safeguards to keep you from, well, dying! Of all the basic needs—food, water, air, and waste removal—lack of air, or the 21+ percent that contains oxygen anyway, is the fastest way to kick the bucket. Within minutes the lack of oxygen causes brain damage, and, ultimately, death. With all this urgency, there are many players involved, whose roles give you everything from life itself, to a voice all your own to express your thoughts.

Function Junction

A lot of people think the function of the respiratory system is just breathing. Nothing could be further from the truth. There are, in fact, six basic functions:

♦ **Exchanging gas** between the air and the bloodstream, and also providing a surface area for that to happen (the respiratory surface).

♦ **Moving air** in and out of the lungs.

♦ **Warming, humidifying, and filtering air** before it reaches the respiratory surface.

♦ **Protecting the respiratory system** from infection by pathogens.

♦ **Detecting air molecules** and sending the message to the brain via the olfactory nerve—in a word: smelling.

♦ **Producing sound** for speech, singing, and all that jazz …

Lungs, Bronchi, Trachea, Nose

I know, I know, breathing always makes you think about lungs, but how does the air in the lungs get there? Despite the primary function of gas exchange, a significant portion of the respiratory system is only for transport, but the body takes advantage of this extra, non-respiratory space, and puts it to good use. You can see the respiratory organs in Figure 13.1.

Figure 13.1

The organs of the upper and lower respiratory system.

(LifeART©1989–2001, Lippincott Williams & Wilkins)

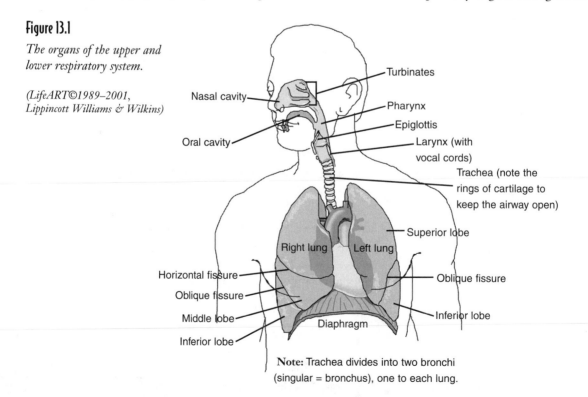

Note: Trachea divides into two bronchi (singular = bronchus), one to each lung.

Conducting Portion

A typical breath (or tidal volume) is about 500 ml, but of that portion only about 350 ml actually are involved in gas exchange (O_2 in, and CO_2 out); that region is called the *respiratory portion*. That leaves a volume of 150 ml that the air must pass through in order to get to the *respiratory portion*; that smaller volume is also called the *conducting portion*. This region consists of the nasal cavity, mouth, pharynx, trachea, bronchi, and bronchioles. This anatomical dead space limits the size of snorkel tubes; a snorkel tube merely extends the anatomical dead space, and if the tube were too long, the tidal volume would keep circulating air within this larger dead space, and you would not get the oxygen you need (this also explains why old diving suits needed pumps to continually force air down to the divers below).

Imagine seeing the sun shining through the window. What do you see in the air? Have you ever noticed the light reflecting off all the dust floating in the air? Well that dust—or rather dead skin cells, plant matter, hair, and dust mite feces—is in every breath you take! Gross! That's not all! Have you ever noticed areas that haven't been dusted in a while? That gray coating on the furniture has the potential of forming inside your lungs!

This is where the glory of the conducting portion comes in. Just as the conductor of a train has many tasks, far beyond just collecting tickets, the conducting portion is also a master at multitasking. For one thing, the conducting portion of the respiratory system, particularly the nose, is the major gateway for infection, so the body has to be prepared with defenses. Despite the conducting portion being very big on the division of labor, there is, nonetheless, quite a bit of overlap. And so, let's go on down.

Nasal Cavity

Once again, a very useful way to learn what a part of the body does is to observe the effect of its absence. Even though most breathing occurs through the nose, the mouth is also useful, especially when you are out of breath. If you have ever slept with such a badly stuffed-up nose that you had to breathe through your mouth all night, you probably woke up with the effects of having no nose: Your throat was probably very sore.

Why? First of all, let's look at the passageway when you breathe through the nose (see Figure 13.2). After the opening vestibule (nostril or nare) the air passes through the passage and down the nasopharynx, which is the portion of the pharynx behind the nose (the oropharynx is the portion, as you would imagine, behind the mouth).

Inside your nose you have a series of structures called *turbinates*. Whenever you breathe, the air is forced over the turbinates. Every turbinate is covered with a mucous membrane. This mucus serves two functions. One function is to trap a lot of the dust that you see floating in the sunlight from your window. This function continues all the way down to the lungs. This is very helpful, because a lot of bacteria and viruses are carried along with that dust. Given that the mucus will dry over time, much of it gets caught in the nose hair, which are called *vibrissae* (which incidentally is the same name given to the whiskers of a cat, too).

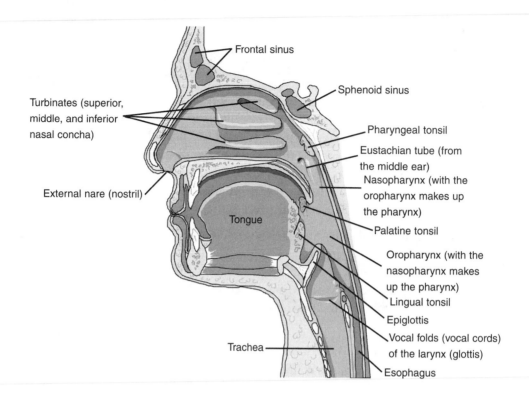

Frontal sinus

Sphenoid sinus

Turbinates (superior, middle, and inferior nasal concha)

Pharyngeal tonsil

Eustachian tube (from the middle ear)

Nasopharynx (with the oropharynx makes up the pharynx)

External nare (nostril)

Tongue

Palatine tonsil

Oropharynx (with the nasopharynx makes up the pharynx)

Lingual tonsil

Epiglottis

Vocal folds (vocal cords) of the larynx (glottis)

Trachea

Esophagus

Figure 13.2

The nasal cavity helps to warm, filter, and humidify the air before it reaches the pharynx.

(LifeART©1989–2001, Lippincott Williams & Wilkins)

As you can see from Figure 13.3, turbinates are surprisingly curly. This shape is directly related to the second function of the mucous membrane on the turbinates. That shape creates a great deal of turbulence in the air whenever you breathe; the turbulence forces air over the mucous membrane, both warming and humidifying the air. Breathing through your mouth all night prevents the air from moistening, which results in your sore throat the morning after.

The turbinates make another good use of that turbulence whenever you pass a beautiful flower garden, or savor the aroma of an upcoming meal. At the very top of the nasal cavity is the olfactory nerve. Now close your eyes, and imagine smelling your food to determine what special ingredient is a part of the recipe. Did you imagine yourself breathing in more deeply? By breathing in more deeply you create more turbulence in the nasal cavity, which forces the air to the very top of the nasal cavity exactly where the olfactory receptors are found. Funny how so many things make sense when you think about them!

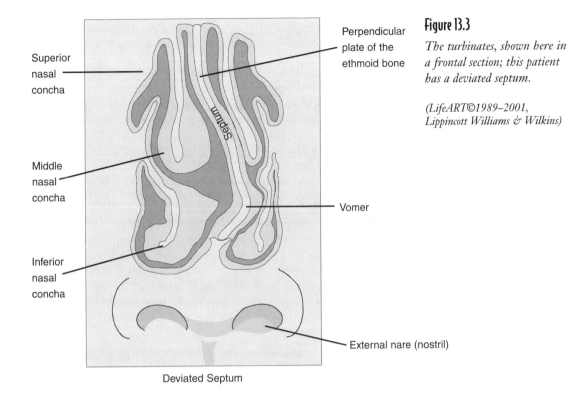

Superior
nasal
concha

Middle
nasal
concha

Inferior
nasal
concha

Septum

Perpendicular
plate of the
ethmoid bone

Vomer

External nare (nostril)

Deviated Septum

Figure 13.3

The turbinates, shown here in a frontal section; this patient has a deviated septum.

(LifeART©1989–2001, Lippincott Williams & Wilkins)

Pharynx

At the back of the nose, we start our journey down the pharynx. The first stop is in the nasopharynx, followed by the oropharynx, which is right behind the mouth, the hypopharynx, and then the laryngopharynx, which is the last stop. Just as some infections are only a finger's length away (a cautionary note to nosepickers of all ages), some infections not only make it past the nose, but they make it all the way down into the lungs.

The Big Picture

Remember, thinking about the body one system at a time is always a mistake. Even in the most traditional textbooks you will find diagrams of the respiratory system with a heart in view. That, quite simply, illustrates the fact that the lungs are exchanging gases with the cardiovascular system. But such details as the epiglottis at the top of the larynx illustrate this system's connection with the digestive system, and the tonsils within the pharynx illustrate the connection with the lymphatic system. If you learn to look for it, you will see that every body system is connected with every other one!

As an extra measure of protection, the pharynx is lined with several tonsils—the adenoids in the nasopharynx, the palatine tonsils at the back of the mouth, and the lingual tonsils at the back of the tongue—that also help in the fight against infection. This is crucial, because bacterial infection inside the brain can be fatal. The brain is limited in terms of fighting off infection, and so the tonsils, which are a part of the lymphatic system (refer again to Figure 13.2), fight the infection before it can ever get to the brain!

One interesting thing about the tonsils is their shape. Rather than having a distinct, smooth outer edge, the tonsils have a series of open pits called crypts. These crypts are designed for catching bacteria and viruses as they pass through the pharynx. It's as if the tonsils are saying, "Go ahead … make my day!"

Medical Records

A child's body is exposed to a remarkable range of pathogens. This exposure helps the body's immune system develop, especially given the memory capabilities of the body's B cells (so named because they are made in the bone marrow). Undue concern with infection, think of the ever-growing passion for antibacterial soaps, has led to parents over-protecting their children. Without proper exposure to foreign antigens, the body's immune system fails to develop properly, leading to an increase in the incidence of asthma and fatal food allergies; these problems are rare among people raised near large mammals, with the ever-present manure. The irony of this is that the best way to develop protection from ill-nesses is to roll around in cow dung as a child!

The other cool structure in the throat is the epiglottis (see Figure 13.4), because eating is a potentially life-threatening situation! This is because the trachea needs to be held open to allow an almost constant airway (except when swallowing). This is done by the cartilaginous rings around the trachea, which you can feel if you gently rub your fingers down the center of your throat. If food travels down through the glottis (larynx) to the trachea, the airway can get blocked and the person can choke to death. Only by coughing, or, in extreme cases the Heimlich maneuver, can you remove food that has gone down, well, the wrong tube. The open tube of the trachea makes the existence of the epiglottis *crucial*.

Figure 13.4

The flap-like epiglottis is the only protection between the ever-open trachea and the food you swallow.

(LifeART©1989–2001, Lippincott Williams & Wilkins)

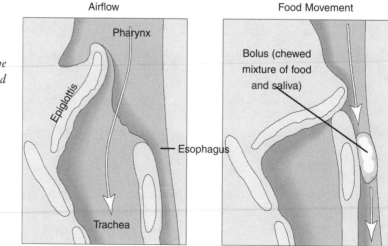

Flex Your Muscles

Memorizing the definition of the word epiglottis is nowhere near as effective as analyzing the word: *epi-* means above, and the *glottis* is the larynx (voice box), which is at the top of the trachea. As such, knowing the location of the flap-like epiglottis, as implied by the name, reveals its function: to block the entrance to the glottis when swallowing, and thus forcing the food down the esophagus.

Giving Voice

You might remember from Chapter 6 a cool little bone called the *hyoid*. The hyoid bone is found at the very top of the larynx; its function is to hold the opening to the airway open. In front of the larynx, below the hyoid bone, is a thick area of cartilage called the thyroid cartilage. As you can imagine, right in front of that you can find the thyroid gland.

Sound is based on a few mechanically simple properties. In a nutshell, sound is based on the vibration of air molecules. That vibration is caused when air is forced past the vocal folds, or vocal cords, which are flaps inside the larynx. The movement of air causes the vocal folds to vibrate. The tighter the folds, the higher frequency the vibrations and thus the higher pitch to the sound; by relaxing the vocal folds, the frequency of the vibrations and the pitch of the sound will be lower.

Think again about the thyroid cartilage (or Adam's apple). Sexual dimorphism in humans is the idea that males and females are of a different size and body shape. One area of that size difference is the size of the thyroid cartilage. Because the thyroid cartilage is right in front of the larynx, it makes sense that the larynx would be a different size as well. Just as the cello produces a deeper sound then the smaller violin, a larger larynx gives men a deeper voice.

The larynx alone, however, is not responsible for the difference in sound between one person's voice and another. Have you ever played acoustic or electric guitar? Is there a difference in the volume and quality of the sound of an electric guitar when it's plugged in and when it isn't? Why is an acoustic guitar able to produce a louder sound without electricity? Remember, the body of an electric guitar is solid, whereas that of an acoustic guitar is hollow and has a circular opening. The hollow portion of the acoustic guitar acts as a resonating chamber for the sound; the nature of that chamber alters not only the volume but also the quality of the sound.

Think back again to Chapter 6 and the skeleton. Remember learning about the sinuses? Those hollow chambers in the frontal, ethmoid, sphenoid, and maxillary bones act as resonating chambers for the voice. Since the shape of the sinuses is partly genetic, it makes sense that you sound a little like your parents. Since those sinuses are also lined with mucous membranes, it also makes sense that when you have a cold or bad allergy the extra mucus in the sinuses will change the quality of your voice, just as the champagne dulls the clink of the glasses during a toast.

More Than Just a Pipe

Once again, you die rather quickly from the lack of oxygen. Oxygen enables you to get 18 times more energy from the breakdown of glucose than doing so without oxygen. Food is the source of energy, and without oxygen, you can't get enough energy; so when you suffocate, you are in effect starving your cells to death! Since your brain requires more energy at rest than the rest of your body, it is the first organ to die. Obviously the exchange of gases is critical, and so your body emphasizes this point by making the trachea stay open through a series of rings of tracheal cartilage.

Remember when I talked about the dust floating in the air that gets trapped by the mucus in your nose? Only so much of that dust gets trapped in the nose, and if you breathe through your mouth, little or none gets trapped. Luckily the trachea is one long mucus factory. Remember from the tissues chapter, mucous is produced by the Golgi apparatus, which are in very large quantity in goblet cells.

I always liked the name goblet cells, particularly knowing the contents of mucus. Mucus is made of water, proteins, and polysaccharides; that's almost a balanced meal! You could probably survive on a bucket of mucus a day! But I digress …. These tracheal goblet cells are found in between groups of ciliated columnar epithelial cells. After the dust gets caught in the mucus, these tiny balls of mucus and dust are propelled upward, allowing them to ultimately travel down the esophagus (see Figure 13.5).

Figure 13.5

Dust in the air you breathe is trapped by mucus, and is then moved upward toward the pharynx by cilia on the epithelial cells of the mucous membrane.

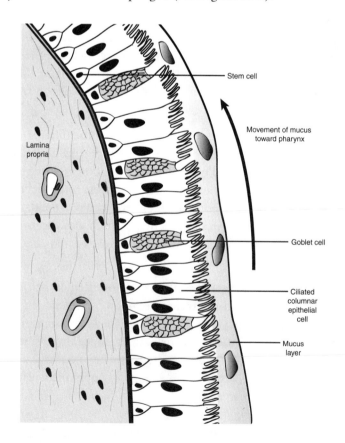

Stem cell

Movement of mucus toward pharynx

Lamina propria

Goblet cell

Ciliated columnar epithelial cell

Mucus layer

A Different Kind of Tree

Our old friend surface area is coming into play again! Our bodies are so large that there is no way for the skin to absorb enough oxygen to keep you alive. The first place for increasing the surface area, the bronchi, also serves another function: to divide the airflow to the two lungs. Beyond that, the airway in each lung continues to divide and divide, greatly increasing the surface area. This ever-increasing division of the airway is known as the bronchial tree.

Throughout all this, because of the need to keep the respiratory surface moist (a necessity for gas exchange), the bronchial walls contain goblet cells, which perform the same role as in the trachea. These goblet cells go into overdrive whenever you get a bacterial infection in your lungs. This inflammation of the bronchial tubes, complete with the excess mucus production, is called bronchitis.

Respiratory Portion

The lungs would be useless without the heart. Remember, never think about systems in isolation! The respiratory surface is entirely focused on exchange of gases with the cardiovascular system. For good measure, there are even some white blood cells called alveolar macrophages, hanging around the site of gas exchange (the alveoli) to clean up any foreign material! As I explored in Chapter 11, the four chambers of the heart are dedicated to keeping deoxygenated blood from the systemic circuit (in the right atrium and ventricle) separate from the oxygenated blood from the pulmonary circuit (in the left atrium and ventricle).

The smallest portion of the respiratory system is the alveolus. There are so many alveoli that the total surface area is equal to the size of a doubles tennis court (see Figure 13.6). Of course, we could also just have evolved as a large, flat pancake, but it would be awfully hard to go dancing! Every alveolus is surrounded by capillaries, and it is this junction where the actual gas exchange occurs. Two things make this possible in every alveolus. One is the moist respiratory surface; gases cannot diffuse through a membrane unless there is water present. The other is directly related to the structure of the capillary and the alveolus. The walls of each of these structures are only one flat cell thick; these incredibly thin walls allow the gases to move freely across the divide. These walls also allow water from the blood to be exhaled, as anyone has seen walking on a cold winter's morning. In a similar sense, any alcohol a person imbibes will travel through the bloodstream, only to diffuse through those thin walls to be detected on the breath!

Figure 13.6

The alveoli are the location of the gas exchange between the air and the bloodstream.

(LifeART©1989–2001, Lippincott Williams & Wilkins)

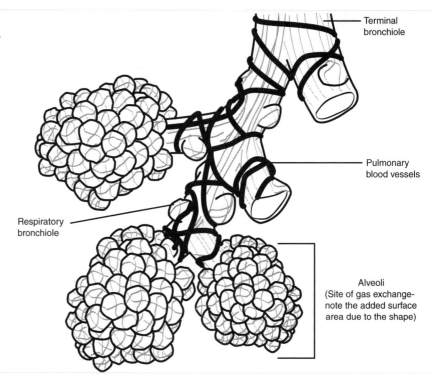

Terminal bronchiole

Pulmonary blood vessels

Respiratory bronchiole

Alveoli
(Site of gas exchange-note the added surface area due to the shape)

Gas Laws and Breathing

Every chemistry student learns three basic gas laws: Charles's law, Boyle's law, and Dalton's law. In terms of respiration, Charles's law is the least applicable since body temperature rarely changes by much. Charles's law states the given constant pressure as the temperature of the gas increases so does the pressure. Boyle's and Dalton's laws, however, very much apply.

Flex Your Muscles

A great illustration of this is to take an empty 2-liter soda bottle with the cap off and squeeze it. It doesn't put up much resistance. Now put the cap on and try to squeeze it again. At this point, with the enclosed container, the gas molecules inside are affected by the change in volume, increase the pressure, and push back against the muscles of your hand.

Contents Under Pressure

Boyle's law refers to a simple inverse relationship between volume and pressure. As the volume of a container increases, the pressure of the gas within the container decreases. Conversely, a decrease in the size of a container will increase the pressure of the gas inside. In terms of containers, think about the thoracic cavity. The thoracic cavity is enclosed by the rib cage and by the diaphragm. Although the inside of the lungs is directly open to the outside environment, the area outside the lungs is not. As you will see, this is very important in your ability to breathe.

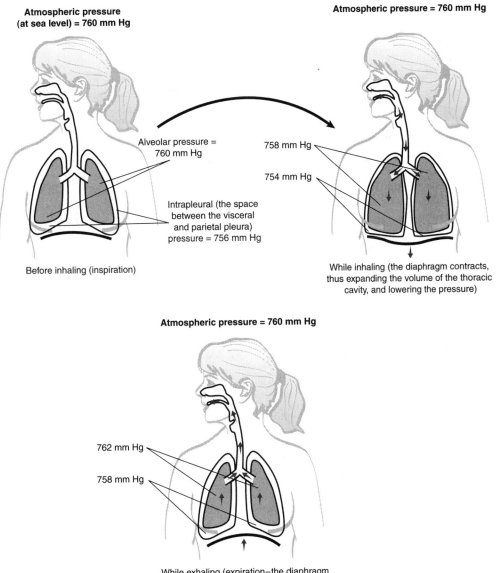

**Atmospheric pressure
(at sea level) = 760 mm Hg**

Alveolar pressure =
760 mm Hg

Intrapleural (the space
between the visceral
and parietal pleura)
pressure = 756 mm Hg

Before inhaling (inspiration)

Atmospheric pressure = 760 mm Hg

758 mm Hg

754 mm Hg

While inhaling (the diaphragm contracts,
thus expanding the volume of the thoracic
cavity, and lowering the pressure)

Atmospheric pressure = 760 mm Hg

762 mm Hg

758 mm Hg

While exhaling (expiration–the diaphragm
relaxes, thus decreasing the volume of the
thoracic cavity, and raising the pressure)

Figure 13.7

The control of gas pressure (through Boyle's law) is controlled by the contraction of the diaphragm and the rib cage.

Don't Stop Bellowing

Many muscles are involved in breathing. Some, such as the internal and external intercostal muscles, control the movement of the rib cage; others control the movement of the abdomen. The most important muscle, however, in breathing is the diaphragm. The diaphragm is a dome-shaped muscle, and like every muscle, it moves by contracting. The dome shape is important because when the diaphragm contracts, the curve of the dome becomes flatter and shallow. As such, when the diaphragm contracts, the thoracic cavity gets bigger (see Figure 13.7).

The increase in the volume of the thoracic cavity means a slight drop in pressure. At this point the pressure of the outside air is slightly higher than the pressure inside the thoracic cavity. Since pressure moves, just like diffusion, from high to low, the outside air rushes in and fills the lungs. The dome shape kicks in again when the diaphragm relaxes, because relaxation causes the diaphragm to resume its dome shape in a process known as elastic rebound. This reduces the volume of the thoracic cavity, increases its pressure, and forces the air out. The change in pressure need not be large, however, for in normal breathing the pressure inside the thoracic cavity changes only slightly: air pressure at sea level = 760 mm Hg (mercury), normal inhalation = 759 mm Hg, normal exhalation = 761 mm Hg!

Crash Cart

One of the most common mistakes students make is to refer to "sucking in air," or "sucking on a straw." When people think of suction, they think that the air is being pulled into the straw, or into the vacuum cleaner. Nothing could be further from the truth! The *lower* pressure inside the *mouth* allows the *higher* pressure in the *air* to push down on the surface of the fluid you are drinking, thus pushing the fluid up the straw. There is no "sucking force" in science, also known as "Nothing Sucks in Science!"

Dalton's Law and Partial Pressure

Dalton's law states that each of the gases in a gas solution (such as air) exerts its own pressure based upon its concentration in the solution (see Figure 13.8). The air you breathe is made predominantly of two gases: nitrogen (78.6 percent) and oxygen (20.9 percent). The rest, only 0.5 percent, is mostly water, although in the summer on the East Coast it sure feels like more! Surprisingly, carbon dioxide is only 0.04 percent of the air!

Using "P" or "p" for partial pressure, air follows this formula for partial pressure:

$$pN_2 + pO_2 + pH_2O + pCO_2 = 760 \text{ mm Hg}$$

with the percentages from above:

$$597 \text{ mm} + 159 \text{ mm} + 3.7 \text{ mm} + 0.3 \text{ mm} = 760 \text{ mm Hg}$$

Dalton's Law

Figure 13.8

Each of the different gases in the air exerts a different partial pressure (Dalton's law).

(©Michael J. Vieira Lazaroff)

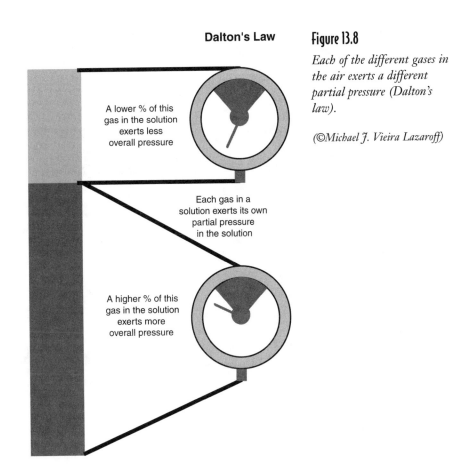

A lower % of this gas in the solution exerts less overall pressure

Each gas in a solution exerts its own partial pressure in the solution

A higher % of this gas in the solution exerts more overall pressure

In the alveoli, the carbon dioxide reaches up to 5.2 percent. With a partial pressure of 40 mm Hg, this is 1,000 times higher than the pressure in the air! As you can imagine, you have no problem getting rid of carbon dioxide! On the other hand, the levels of oxygen drop as low as 13.2 percent, or a partial pressure of 100 mm Hg. Clearly you don't use all the oxygen with every breath—good news for people trapped with a limited supply of oxygen!

Volumes and Volumes

To truly understand breathing, it is necessary to look at the pulmonary volumes and capacities (see Figure 13.9). The in and out of normal breathing takes place within the tidal volume, whose name makes perfect sense since tides also go in and out. That alone might be enough, except for the fact that you sometimes need to either take in, or release, extra air.

A simple cough demonstrates this very easily. If you have ever inhaled enough dust to irritate your lungs, the incredible urge to cough takes over. If, however, you only had a tidal volume, you would have no hope of ever being able to get rid of the irritation. The expiratory reserve volume allows you to force the dust out by applying far more pressure behind the dust than there is in front of it.

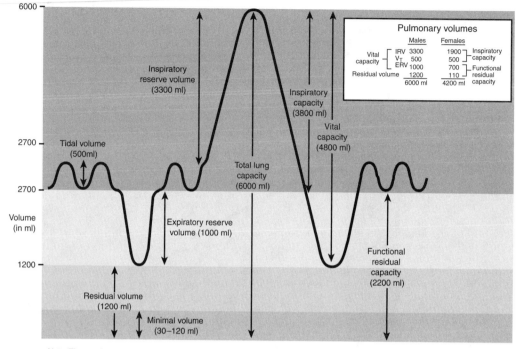

Note: The numbers on this graph reflect those of a typical adult male; the numbers for adult females are, on average, lower.

Figure 13.9

Your normal breathing is your tidal volume, but reserve volumes necessary to your survival increase your total lung capacity.

The residual volume is well known if you have ever had the wind knocked out of you. The uncomfortable sensation of struggling to take your first breath afterward is due to the fall forcing out the residual volume. The residual nature of that volume prevents you from having to inflate a fully deflated lung with every breath; this is just like the way a balloon is easier to blow up after it has a little air in it.

Medical Records

Imagine having to re-inflate your lungs with *every breath!* Premature babies have this very problem if they are born before the eighth month. If you have ever tried to float a paper clip on fresh water, you have seen the effect of surface tension. Adding dish soap breaks the surface tension, causing the paper clip to sink. The surface tension of the water in the lungs has the effect of collapsing the roughly spherical alveoli. Surfactant, a protein that starts being produced in the ninth month, breaks the surface tension, allowing the alveoli to stay open.

Swimming would not be possible without inspiratory reserve volume, which allows you to inhale well beyond your tidal volume. This enables you to hold your breath by taking in extra air. This also explains why the first desire when holding your breath is to release the breath you had been holding! This brings up an important point. We have always been told of the primary importance of getting oxygen; equally important, however, is the need to get rid of carbon dioxide. As you will see, the buildup of carbon dioxide is crucial to the regulation of your breathing.

Controlling Breathing

Have you ever thought about how often you breathe? Adults breathe about 12 to 18 times a minute, while children tend to breathe 18 to 20 times a minute. If you imagine breathing in 500 ml of air on a normal tidal volume, then, at 12 times a minute—12 × 500 ml = 6000 ml = 6 liters—you are breathing 6 liters of air (three 2-liter bottles) every minute. Of the 500 ml in each breath, only 350 ml actually reach the alveoli. As a result, the alveolar ventilation is less: 12 × 350 ml = 4200 ml = 4.2 liters a minute. That might not be enough to meet your gas exchange needs, so your brain must have a way to change the rate you breathe.

To understand how the brain controls breathing, you need to remember a little about pH, or the powers of hydrogen. As I mentioned earlier, O_2 is carried mainly in the hemoglobin of red blood cells (RBCs), and CO_2 is carried mainly in blood plasma, which is mostly water. The interaction of the CO_2 and the water can be seen in the reversible reaction below:

$$CO_2 + H_2O <\longrightarrow> H_2CO_3 <\longrightarrow> H^+ + HCO_{3-}$$

carbonic acid

Whenever cells use up the O_2 in cellular respiration, and thus release CO_2, your blood pH changes. With the addition of the H^+ ions, the blood becomes acidic.

Blood pH must remain within a narrow range; if the pH strays too far from neutral (pH 7), enzymes will fail to function. There is a word for that condition: death! As such, o by monitoring the brain must carefully monitor the CO_2 levels, and it does spH. The brain does so in the respiratory centers of the medulla oblongata, where the brain and the brain stem meet.

The Least You Need to Know

- The respiratory system is divided into two portions: a conducting portion where air is filtered, warmed, and moistened, and a respiratory portion where gas exchange occurs.

- The epiglottis provides protection from choking by blocking the entrance to the trachea when food or liquids are swallowed.

- The movement of air is accomplished via changes in gas pressure as a result of the diaphragm and ribcage changing the volume of the thoracic cavity.

- Pulmonary volumes and capacities illustrate the extra capabilities of the lungs to take in or expel air in larger quantities than normal breathing.

- Gas exchange happens between the thin-walled alveoli and the equally thin-walled capillaries in the lungs.

- Breathing is controlled by changes in pH as a result of the buildup of CO_2 in spinal fluid carried to the medulla oblongata.

Chapter 14

It Dices, Slices, Chops ...

In This Chapter

- The organs of the gastrointestinal tract
- Mechanical versus chemical digestion
- Enzymes, substrates, and the organs involved
- Nutrition and usage of food molecules

We are biological machines who need a constant supply of materials and fuel in order for us to function. In addition to being a fuel source (as in the case of carbohydrates), food contains vital raw materials, in the form of small monomers and chains of polymers (see Chapter 2) that we use for repair, growth, defense, and so on. Simply put, if we stop eating we will ultimately die. The need for fuel and materials is so strong that our bodies will start to break down what we have already built (such as our fat reserves and even our muscles) in order to survive.

Despite the intimate relationship we all have with food, far too many people are ill informed on the basics of the digestive system. This is a shame, because this system of tubes, transport, breakdown, and ultimately distribution runs the gamut from the wonderful to the disgusting. The organs, our protagonists, run the risk of destroying themselves because they are made of the same basic materials as the food we eat, and they have thus evolved many clever ways of protecting themselves. Given that we are what we eat, we will also follow the many twisted pathways from the moment our food touches our lips, to the point where it reaches our hips (whether it stays for a lifetime or not!).

Function Junction

The glory of the digestive system exists, in part, because of the many different functions it carries out. Eating alone is not enough. As you have seen before, the structure of the different organs will vary according to their functions. The functions of the digestive system are …

- **Ingestion.** Ingestion is the act of eating. Food is fuel, and we have to get it inside somehow!

- **Mechanical digestion.** Much of what we eat is rather large by cellular standards, and it needs to be broken down into smaller pieces (without actually changing the molecules) to increase the surface area available for the enzymes. Mechanical digestion consists of tearing, chewing, and the churning effects of peristalsis.

- **Peristalsis.** Once we swallow, the ride begins! Food needs to travel about 30 feet (9 meters) before all is said and done; luckily muscle tone in our organs shortens that length by about half. Peristalsis is the rhythmic contractions of our digestive organs that propel the food.

- **Chemical digestion.** As the food travels along the digestive tract, various enzymes are released, which chemically change the complex polymers into the basic monomers that our bodies will ultimately use.

- **Secretion.** It is important to note that digestive enzymes can digest your own body! The organs thus need to protect themselves by secreting various fluids.

- **Absorption.** The monomers and water in the gastrointestinal tract need to be transported to the body via the bloodstream, but they first need to be absorbed into the blood.

- **Storage and toxin processing.** All of the blood from the abdominal digestive organs passes through the liver, which stores some sugar (as glycogen) for emergencies, converts other sugars into fat, and breaks down toxic chemicals before too much gets to the body.

- **Excretion and egestion.** Egestion is the release of wastes (feces) we do not use, via the anus; this release is also called defecation (to "de-feces" oneself).

Stops Along the Way

In order to understand the digestive system, it is important to think about the different organs and the varied roles they play (see Figure 14.1). First let's get to know their names. In terms of the organs of the digestive tract itself, if it's easier, think "**M**y **P**ink **E**lephant **St**ill **Sm**ells **L**ike **R**otten **A**pples." Whaaaat? Simple. **M**outh, **P**harynx, **E**sophagus, **St**omach, **Sm**all Intestine, **L**arge Intestine, **R**ectum, **A**nus. These amazing organs, despite the wonderfully gross things they do, still need help. The accessory organs (given their importance, it's a bit odd to call them that)—salivary glands, pancreas, liver, and gallbladder—will be discussed in more detail later.

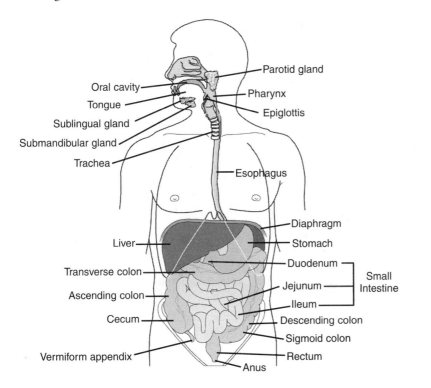

Labels on the figure:
- Parotid gland
- Oral cavity
- Pharynx
- Tongue
- Epiglottis
- Sublingual gland
- Submandibular gland
- Trachea
- Esophagus
- Diaphragm
- Liver
- Stomach
- Duodenum
- Transverse colon
- Small Intestine
- Jejunum
- Ascending colon
- Ileum
- Cecum
- Descending colon
- Sigmoid colon
- Vermiform appendix
- Rectum
- Anus

Figure 14.1

This view shows the overall layout of the major digestive organs, although some are hidden from view.

(LifeART©1989–2001, Lippincott Williams & Wilkins)

What They Have in Common

Despite their differences, it's a good idea to find what all these organs have in common. The organization of our digestive tract in our basic body plan hasn't really advanced much beyond that of the earthworm (phylum Annelida); as such, it is hundreds of millions of years old. Our basic body plan is that of a tube within a tube. The inner tube is, of course, your digestive tract, but the outer tube is your larger hollow body cavity, which is divided into the thoracic and abdominopelvic cavities (see Chapter 1).

Medical Records

Given our evolutionary past as organisms that absorbed food through our outer layer, the lining of our mouth to the lining of our intestines is all continuous with the outside of the body; this makes the *inside* of your *stomach* technically the outside of your *body*! The outside of our body protects itself by making the outer layer dead tissue (see Chapter 16), but given the need to do everything from producing enzymes to absorbing food, the inside of our digestive tract needs to be alive. For this reason, we risk bacterial invasion from both ends, and a normal, healthy person has bacteria in both the mouth and the large intestine.

Given that every organ is made of many tissues, it is important to look at the histology, from the inside out, of the layers of the typical digestive organ. All these layers are illustrated in Figure 14.2. After all, as with people, they are more alike than they are different.

Figure 14.2

This view of the small intestine illustrates all the major layers of every organ of the digestive tract.

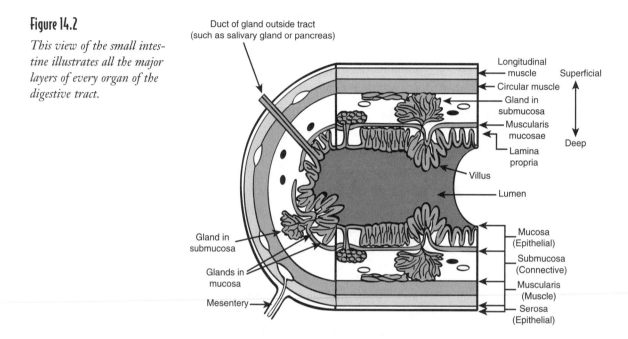

The innermost layer is called the *mucosa*. The mucosa, which is, as you would imagine, a mucous membrane (see Chapter 3), varies most from organ to organ among all the layers. Just outside the mucosa is the *lumen*, or hollow cavity, of the organ. This lumen is where all the mechanical and chemical digestion happens. This type of digestion is called *extracellular digestion* (meaning outside the cell).

The cells of the mucosa produce enzymes that are released by exocytosis (see Chapter 3), and the food is digested outside the cells, which is literally outside the body. None of the food actually enters the body until it is absorbed by the mucosa. This layer is composed of epithelial tissue (see Chapter 3), which is avascular (without blood vessels). It is this layer that secretes the enzymes, as well as the chemical protection from those enzymes.

The *submucosa*, which is as the name implies under the *mucosa*, is, like all layers beneath epithelial layers, made of connective tissue. This is where blood vessels and *lymphatics* (see Chapter 12) can be found; the food absorbed by the mucosa is then absorbed into these vessels. Many of the glands of the mucosa also extend into this layer.

Peristalsis, the movement of food along the digestive tract, is accomplished in the next layer, the *muscularis externa*. This layer consists of two separate layers of smooth muscle (see Chapter 3), a longitudinal layer, and a transverse, or circular, layer; peristalsis consists of alternating contractions of the two layers. The immense amount of churning in the stomach requires the existence of a third layer of muscle, which is an oblique layer.

The *serosa*, which is the outermost layer, is a combination of another connective tissue layer and an outermost epithelial layer. In addition to being a simple outer boundary, this layer performs a crucial function. As the *muscularis externa* contracts, the organ itself will actually move! Given the tight quarters in the abdominal cavity, these organs run the risk of slowly wearing each other down due to the friction they create. The serous membrane (see Chapter 3) prevents the resulting *scrape*, by releasing a simple, tear-like secretion to lubricate the organs.

The Big Picture

Within each of the connective tissue layers are simple connections to other body systems. The capillaries connect the organs to the *cardiovascular system*. Hormones of the *endocrine system* also travel through these capillaries. Drainage and uptake of lipids are accomplished through vessels of the *lymphatic system*, known as *lymphatics*. Nerves from, of course, the *nervous system* can also be found here, otherwise our brain would have no idea what is happening down there!

Lastly, in the layers of connective tissue found in the *submucosa* and the *serosa* are a number of nerves. These nerves are of three types: sensory (ever had an upset stomach?), sympathetic (to slow down or inhibit muscle contraction), and parasympathetic (to increase muscle contraction). The action of the last set of fibers will also indirectly (through the contraction of the *muscularis externa*) stimulate the sensory neurons, as we have all felt when we do more than *hear* our stomach growl.

You've Gotta Start Somewhere

With so much attention we pay to the food we eat, it is rather ironic that out of the 16 to 24 hours it takes for food to complete the journey through our alimentary canal (gastrointestinal tract), food is only in our mouths for about 10 to 20 seconds before we swallow! The mouth contains several specialized structures that aid in its basic function of mechanical digestion. I will discuss the tongue later when I explain the senses in Chapter 21.

Teeth, of course, deserve special mention. We have come a long way from the rows of almost identical teeth found in the shark. Our teeth come in several very specific shapes that vary according to their function. Numbering of teeth is traditionally done in quadrants:

◆ **Incisors** are thin, sharp teeth that are best for cutting. We have two per quadrant (total = 8).

◆ **Canines** are conical teeth that aid in tearing. We have one per quadrant (total = 4).

◆ **Premolars** are used in grinding food. We have two per quadrant (total = 8).

◆ **Molars** are larger and are also used in grinding, but these are the only teeth that don't descend in infancy. We have three per quadrant, and the third molar is called a *wisdom tooth* (total = 12).

The first three groups of teeth appear in two sets, the first of which, as memories of the tooth fairy will attest, fall out. These "baby teeth" are also called milk teeth or deciduous teeth. I've always loved the latter name because trees that lose their leaves are called deciduous trees; I'm just glad our teeth don't turn brown before they fall out!

Flex Your Muscles

If you ever get a chance to study comparative anatomy or comparative zoology, you will get a special bite out of your study of teeth and their role in different feeding strategies. Some of my favorites: the continuously growing incisors of rodents, the sharp molars of carnivores, and the broad, flat molars of herbivores. Indeed, the rather remarkable lack of specialization in our teeth is a reflection of both our status as omnivores and the evolution of tool use.

It's a Long Way Down

The mixture of masticated (chewed) food and saliva we swallow is called a *bolus* (sort of a big spit ball). The role of the pharynx, and especially the epiglottis, is discussed in Chapter 13. Quite an adventure will ensue, but the first leg of the journey is rather uneventful. The esophagus is less than a foot long (some 25 cm). Not being in constant use as with the trachea, it lacks the trachea's cartilaginous rings, opening mainly when we swallow. The esophagus is mainly a tube, somewhat lengthened to allow room for the thoracic organs; it doesn't even have a strong sphincter muscle at each end.

The muscular entrance, called the *upper esophageal sphincter*, is called such merely because it mimics the effect of a sphincter muscle. The exit, or the lower esophageal sphincter, is also called the *cardiac sphincter*, given its proximity to the heart. That name, however, sometimes causes some confusion, so it is increasingly known as the *gastroesophageal sphincter*. By putting the prefix *gastro-* before the esophagus, this helps to emphasize that the primary function of this "sphincter" is to prevent backflow, similar to, though less efficient than, the valves in the heart, in veins, and in lymphatic capillaries. When that backflow does happen, the poorly protected lining of the esophagus is digested by gastric juices, causing the sensation known as *heartburn* (there is no actual connection to the heart, except for the pain being felt *behind* the heart).

Protein Buster

Whenever people think about digestion, the first organ that comes to mind is the stomach. In fact, the stomach does very little digestion at all (liquids reside there for only about

$1^1/_2$ to $2^1/_2$ hours, and solids anywhere from 3 to 4 hours). When asked to point to their stomachs, most people would probably point to the area of the umbilicus ("belly button"), and the idea of touching the bottom of the ribcage on the left side probably wouldn't occur to them. In many ways, the stomach is a surprising organ.

Medical Records

One of the most difficult things about the digestive system is the fact that many disorders will manifest similar systems. Heartburn or ulcers have some of the same symptoms as stomach cancer. It is often advisable to have an endoscopy when symptoms are in question. Stomach cancer, like most cancers, runs in families, so if you have a family history of it, you should be prepared to consider reducing the amount of salt in your diet. Stomach cancer is one of the few cancers to decline in the twentieth century, but for a surprising reason ... the invention of refrigeration (salt was no longer needed to preserve meats)!

The expression "a full stomach" makes a lot of sense when you consider the many large folds in the lining of the gastric mucosa, known as *rugae*. These rugae (singular, ruga) allow the lumen of the stomach to expand after eating a large meal; you might be surprised to learn that the average adult's stomach can hold anywhere from 1 to $1^1/_2$ liters of food! Is it any wonder that some people's clothes feel tighter after a big meal?

The contents of the stomach—a mixture of food, saliva, and gastric juices—are called chyme. It may surprise you to know that your stomach gets started before you even take your first bite! This first phase of gastric activity is called the cephalic phase (*cephalic* = head). In this phase the gastric secretions actually start as a result of seeing, smelling, tasting, or even thinking about food!

The Big Picture

The way to a man's heart (or a woman's for that matter) may not be through the stomach, but the way to the stomach *is* through the brain! The vagus nerve (N-X), a part of the parasympathetic branch of the autonomic nervous system, will increase the activity of the digestive system when the body is in a rest and repose state (as opposed to fight or flight state). If that brilliant Russian scientist, Pavlov, had been able to measure the gastric activity of his dogs, he would have found the bell triggering more than just salivation!

The second phase, aptly named the gastric phase, doesn't start until the food actually enters the stomach. At this stage, the change in pH—brought about by the cephalic phase—the presence of your chewed-up meal (especially the presence of protein, the stomach's forte), and the stretching of the stomach lining by that food, get the gastric phase into full gear. Gastrin, a hormone produced by the G-cells at the very bottom of the gastric pit, will increase gastric secretion—the irony is that, as a hormone, it must make a full circuit through the blood to get to the cells that are right next door! Stretch receptors also send nervous signals that ultimately lead to greater gastric juice.

> ### Crash Cart
>
> It used to be believed that the best thing to do when experiencing the pain of heartburn or ulcer was to drink milk. In a way this was logical; given the acidity of the gastric juice (pH 2) and the more neutral nature of milk (pH 6.4 to 6.8), the combination of the two does, if not produce a neutralization reaction, at least dilute the stomach acid and thus raise the pH. Any relief, however, is temporary, because ultimately the protein in the milk will increase the gastric secretions, making the pain even worse!

The last phase, which starts up to four hours after the start of the gastric phase, is quite sensibly named the *intestinal phase*, is the one in which, as you can imagine, the chyme is sent on its merry way into the first part of the small intestine, the duodenum, through the pyloric sphincter (in this case, a *real* sphincter). The lion's share of the digestion happens in the small intestine, so it is important that the digestion happens efficiently. If too much food enters the small intestine at once, the enzymes will not be able to digest it all effectively.

This is regulated by something called the *enterogastric reflex*, in which a slowing of gastric contractions follows the "squirting" of chyme into the duodenum. This prevents too much chyme from entering the duodenum at once, and also enables the duodenum to prepare, through the production of extra buffers, to counteract the low pH of the gastric chyme. If this regulation doesn't function properly, a duodenal ulcer is the result!

> ### Flex Your Muscles
>
> In remembering the three gastric phases, concentrate on the other areas of the body involved. The *cephalic phase* is triggered by nerve impulses from the brain when you *think* about food. The *gastric phase* concentrates on the actions of the stomach itself once the food actually enters the stomach. Lastly, the *intestinal phase* is, quite simply, the phase where the contents of the stomach actually enter the duodenum of the small intestine through the pyloric sphincter.

Okay, you get the idea that the stomach releases digestive enzymes to digest proteins, but aren't there proteins on the outside of every cell, including those that line the stomach? So how does the stomach keep from digesting itself after pumping out so much stomach acid and enzymes? For one thing, it doesn't always! Sometimes the stomach starts to digest its own lining, basically turning itself into a cannibal! That, in a nutshell, is the basis of an ulcer—I know some people get hungry from time to time, but really! Still, most people never develop ulcers, so how does the stomach keep from eating itself?

Medical Records _____

To understand the three types of ulcers, refer to the layers of the typical digestive organ. Since extracellular digestion occurs in the lumen of the stomach, either a lack of mucus or an excess of HCl can cause extracellular digestion—digestion of the mucosa itself. The simplest ulcer is just this, and is a reddening of the gastric mucosa. Since the epithelium of the mucosa is avascular, there would be no bleeding. A bleeding ulcer involves digestion of the submucosa's vascular connective tissue (gastric juice makes blood turn black). If the bleeding isn't enough to stop you, and the digestion continues through the muscularis and the serosa, the perforation of the stomach wall gives this type of ulcer its name: a perforated ulcer.

For one thing, the gastric mucosa (which lines the lumen) makes good on its name and churns out large quantities of mucus. Since mucus is made of water, protein, and polysaccharides, the water helps to dilute the acid, thus raising (buffering) the pH of the gastric juice closest to the mucosa; the enzyme pepsin gets to work on the mucous proteins, rather than the mucosal proteins! That is all well and good, but what about when the HCl and the pepsin are first released? What protects the cells then?

Medical Records _____

HCl acid does more than activate the pepsin. Proteins are so complex that their digestion is a long process. The HCl acid breaks the hydrogen bonds that make the secondary, tertiary, and quaternary structures of the proteins, making it easier to break the peptide bonds in the primary structure. The HCl acid also breaks down plant cell walls (although we cannot actually digest the cellulose in them), not to mention breaking down the connective tissue in meat. The HCl acid also helps to kill bacteria in food. This, however, is not foolproof, because certain bacteria can make it past the stomach's defenses, not to mention the recent discovery of Helicobacter pylori, the species of bacteria whose daily activity is the cause of many gastric ulcers!

In a tricky maneuver, the stomach takes advantage of the shape of the gastric mucosa, and a little chemical trickery, to get away with it (see Figure 14.3)! The gastric mucosa is filled with pits called gastric pits. At the base of these pits are chief cells, which release pepsinogen, an inactive form of the enzyme pepsin. Being inactive, the protein poses no problem for the mucosa. At the neck of these gastric pits are parietal cells (remember that word from Chapter 1), which release HCl acid. That alone would be a problem, but for the cell's clever habit of releasing H^+ ions and Cl^- ions separately, thus protecting itself from the HCl acid! Sure, fine, but how does the pepsinogen turn into pepsin? By mixing with the HCl acid! Those two don't mix until they meet in the neck of the gastric pit, and the enzyme doesn't become active right away, thus buying enough time to wait for the chemicals to reach the lumen! This is just one more example of how true understanding of the anatomy of the body (in this case, of the gastric pits) is only understood when one also looks at the physiology!

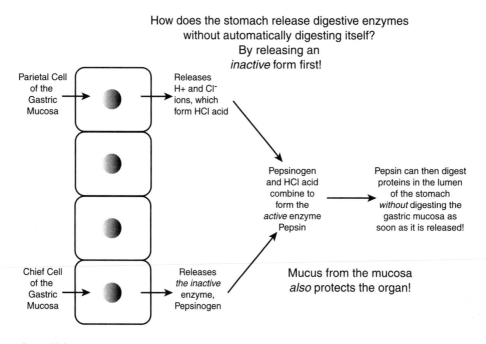

Figure 14.3

Why enzymes are released in an inactive form.

(©Michael J. Vieira Lazaroff)

The greatest risk when the stomach starts digesting itself is that such an ulcer can actually digest so deep a hole in the stomach wall that an opening is made into the abdominal cavity; this is called a perforated ulcer. Luckily, the body has an ace up its sleeve in a somewhat disgusting sheet of connective tissue called the greater omentum (see Figure 14.4). This sheet literally hangs down in front of the small intestine, acting in part as a place for depositing excess fat, but more importantly, the greater omentum can adhere to any perforations in the stomach wall, thus preventing the contents from emptying into the abdominal cavity. There is also a *lesser omentum*, but it just connects the liver to the stomach.

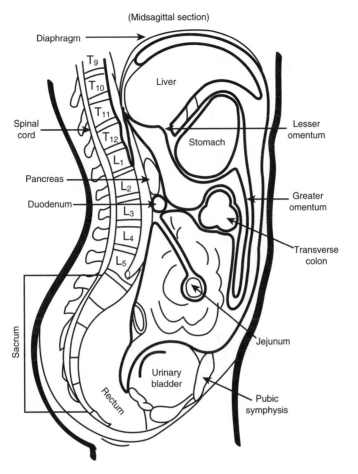

(Midsagittal section)

Diaphragm

T₉
T₁₀
T₁₁
T₁₂
L₁
L₂
L₃
L₄
L₅

Spinal cord

Liver

Stomach

Pancreas

Duodenum

Sacrum

Rectum

Urinary bladder

Lesser omentum

Greater omentum

Transverse colon

Jejunum

Pubic symphysis

Figure 14.4

This sagittal section shows the organs within the lesser omentum and the greater omentum, as well as the intestinal mesentery.

www.clipart.com

Jack-of-All-Trades

The next step, and by far the most eventful one, is the three- to five-hour ride through the three parts of the small intestine (in order: duodenum, jejunum, and ileum—the last length is closest to the pelvis, which should make its location easy to remember, being so close to the ilium). The action of enzymes is intense here, as most of the digestion happens here. Some of the enzymes come from the epithelium of the small intestine itself, while the rest come from the pancreas, with bile (not an enzyme, but an emulsifier), alone, coming from the liver, via the gallbladder.

The enzymes, their origin, location of action, and the action of the enzymes themselves, are all outlined in a table further on. A primary function that has yet to be outlined is the absorption of nutrients. Since absorption occurs through the cell membrane surfaces along the lumen of the mucosa, it is crucial that that surface area be maximized. That is accomplished in part through folds in the lining called *plicae*, each lined with thousands of finger-like villi that make it feel like velvet. On a microscopic scale, the edge of each epithelial cell along each villus contains even smaller projections called microvilli (see Figure 14.5). The sum total of all these surfaces makes the surface area of the small intestine the size of a doubles tennis court!

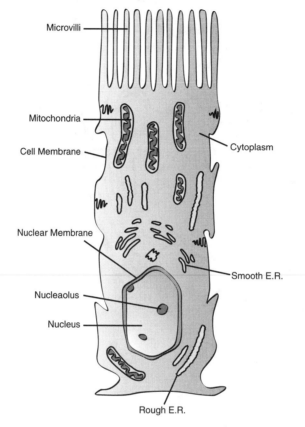

Figure 14.5

The cells of the villi increase the surface area for absorption through a "brush border" of microvilli.

Once absorbed through the microvilli, the nutrients are then absorbed into the vessels in each villus. Lipids are absorbed into the lacteal, which then empties into the cisterna chyli before ultimately emptying into the blood at the left subclavian vein. All other monomers are absorbed into the capillaries and then sent directly into the liver via the hepatic portal vein, as discussed in Chapter 13. This allows excess nutrients, as well as toxins, to be processed directly by the liver.

There are some rather somber-sounding intestinal crypts at the base of the villi. There are stem cells within that constantly produce more epithelial cells to replace the intestinal mucosa. The epithelial cells last only a few days because of the constant wear and tear on the small intestine. The gastroenteric reflex stimulates peristalsis of the small intestine, whereas the gastroileal reflex causes the ileocecal valve to relax, thus stimulating movement into the large intestine, our next stop.

Trash Compactor

The *large intestine* (also known as the *colon*) is primarily concerned with the absorption of water, which is the most important step in the making of feces. The parts of the large intestine can be seen in Figure 14.6. Remember, quite a large amount of mucus is secreted into the digestive tract, both to ease peristalsis, as in the esophagus, and to protect you from digesting yourself, as in the stomach.

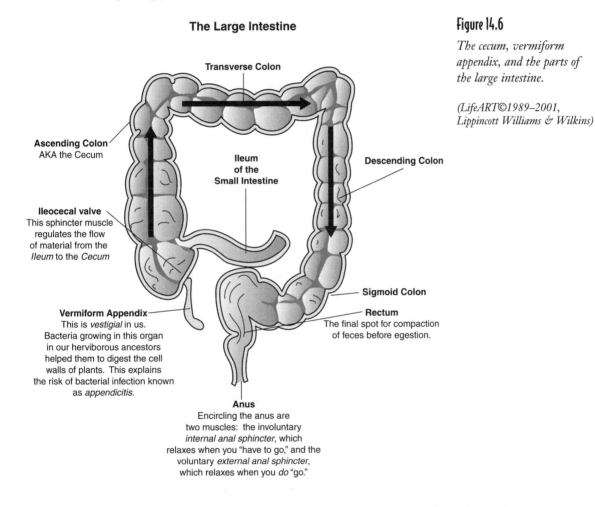

The Large Intestine

Transverse Colon

Ascending Colon
AKA the Cecum

Ileum of the Small Intestine

Descending Colon

Ileocecal valve
This sphincter muscle regulates the flow of material from the *Ileum* to the *Cecum*

Vermiform Appendix
This is *vestigial* in us. Bacteria growing in this organ in our herviborous ancestors helped them to digest the cell walls of plants. This explains the risk of bacterial infection known as *appendicitis*.

Sigmoid Colon

Rectum
The final spot for compaction of feces before egestion.

Anus
Encircling the anus are two muscles: the involuntary *internal anal sphincter*, which relaxes when you "have to go," and the voluntary *external anal sphincter*, which relaxes when you *do* "go."

Figure 14.6

The cecum, vermiform appendix, and the parts of the large intestine.

(LifeART©1989–2001, Lippincott Williams & Wilkins)

If we didn't have a way of recovering some of that fluid, or at least the water portion, we would quickly dehydrate (see Figure 14.7). To give you some perspective, about 1,200 to 1,500 ml (remember, a 2-liter bottle has 2,000 ml!) enter the *cecum* (or start of the colon) through the sphincter known as the *Ileocecal valve*, but only about 150 to 200 ml of feces are expelled every day.

Figure 14.7

The movement of fluid in and out of the digestive tract every day.

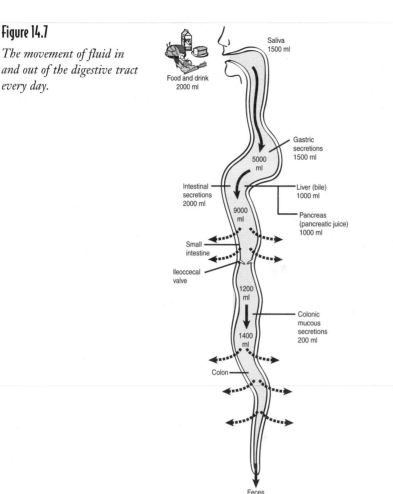

This task, however, does present the potential of some problems if the pace of the movement through the colon is not normal. The amount of time food spends in the large intestine varies from 8 to 15 hours. While it's there, a colon's gotta do what a colon's gotta do. So if the passage is too slow, and too much water is absorbed, the dehydration of the feces causes constipation. On the other hand, if the movement (pun intended, unfortunately) is too rapid, and too little water is absorbed, the person suffers diarrhea. The loss of so much water through diarrhea can result in systemic dehydration.

In addition to absorbing water, the large intestine also absorbs other substances, such as vitamin K (which the liver uses to make clotting factors), biotin (for many reactions), vitamin B5 (for making steroids and neurotransmitters), urobilinogen, bile salts, and toxins (all three to be excreted by the kidneys). Pigments from the breakdown of urobilinogen are what make feces brown!

Flex Your Muscles

Once again, make the words work for you! Can you figure out what the "coli" in E. coli refers to? If you thought about its home, the colon, you are right!

A little space needs to be made here for our companions, the bacteria known as intestinal flora. The most famous of these is Escherichia coli (E. coli to its friends). People these days have a misguided idea of bacteria. The common misconception is that all bacteria are bad; nothing could be further from the truth. Apart from the ecological impact of bacteria, especially in terms of the nitrogen cycle in soil, bacteria are necessary for our survival!

About half of our daily vitamin K needs come from intestinal bacteria and is absorbed through the walls of the colon. These bacteria also break down undigested peptides and plant matter in the feces, thus producing a medley of gases, including hydrogen sulfide (H_2S), known as *flatus* (which is why farting is known in kinder circles as flatulence). So, next time you, shall we say, "break wind," it's not your fault; blame it on bacteria!

The End of the Tunnel

The final resting place for the compacted feces is in the last 6 inches (15 cm) known as the *rectum*. One interesting thing to note about the rectum is the pair of sphincter muscles that surround the anal opening. Without knowing about their existence, we have all experienced the effects of each of these sphincters. As I stated in Chapter 9, given that all muscles contract, a circular sphincter muscle will close the aperture when contracted, and the aperture will open when the muscle relaxes.

That being said, let me introduce you to the *internal and external anal sphincters*. Perhaps what distinguishes them is their nervous control. The internal muscle is involuntary, but the external is under our voluntary control. Imagine now that it has been a long time since you have defecated, and the feces is starting to pile up and exert pressure on the stretch receptors along the rectal wall. This will trigger the relaxation of the *internal anal sphincter*, which will cause a great deal of pressure on the *external anal sphincter*. At this point the only thing standing between you and extreme social embarrassment is the contraction of that voluntary external sphincter! In essence, when the internal sphincter relaxes, you "have to go," and when the external sphincter relaxes, you *"do go"*!

Running Along the Tract

Oh, sure! It's easy to say that the food travels *along* the gastrointestinal tract, but how does it move? Through a cool process known as *peristalsis*. Remember the two layers of muscle in the muscularis layer of every digestive organ? The longitudinal layer shortens the length of the organ, while the circular layer constricts the lumen, and thus pushes the food along the tract, rather like pushing a present through a Christmas stocking. The alternation of these contractions propels the food forward.

In addition, the alternating contractions also help to mix the food in a process known as *segmentation*. Contraction at various points pinches the small intestine at different points, making it look a lot like a string of sausages—interesting, because the gastrointestinal tract of cows, pigs, sheep, and goats are still used today as a sausage casing! These segments, through repeated contractions, churn the food, allowing it to mix well with the various enzymes at work.

Accessorize with Accessory Organs!

These organs are all well and good, but they cannot function alone. They need help, and that help comes from the accessory organs. Some of you may take exception to having them called *accessory*, when their function seems so crucial. You couldn't survive, for instance, without a pancreas. Their accessory function merely implies that the food never passes through these organs.

Primary Organ	*Its Accessory Organ(s)*
Mouth	Salivary glands
Small intestine	Liver, gallbladder, and pancreas

Our Little Spit Factories

The mouth may be the start of it all, but without the salivary glands, it wouldn't be a very fun ride! Have you ever chewed dry crackers when your mouth was already dry? Even worse, have you ever *swallowed* dry crackers when you had a dry, scratchy sore throat? SCRAAAAAPE!! Luckily, that is not a very common experience due to the wonders of spit, or should I say, saliva.

Saliva, from any of our three main salivary glands (parotid, sublingual, and submandibular), first and foremost provides lubrication for the chewed food, so that when it is swallowed, the spit and food combo (called a *bolus*) can slide easily along the esophagus. The other two functions are to start the digestion of starches by using salivary amylase (more about that later), and to raise the pH of the mouth. Saliva is slightly basic (pH up to 7.5). This serves two functions: to neutralize the acidic secretions of bacteria in the mouth (the scourge of dentists everywhere), and, under certain circumstances, to neutralize the effect of stomach acid after vomiting.

The Big Picture

Secretion of saliva is controlled by both parasympathetic nerve pathways and conscious thought. Placing food in your mouth will trigger receptors on the three of the cranial nerves (N VII, N IX, N X), but the mere thought of food can cause salivation. Thinking about the vinegar taste (an acid) in a pickle is enough to cause a lot of people to salivate, once again, to lower the pH!

Not Just Good with Onions

The liver keeps very busy, as a quick glance at the following list shows. To start off, the fats we eat are often in clumps, especially given the way fats are stored in our bodies along connective tissue. These globs of fat have a very small surface area to volume ratio (see Chapter 3). Bile functions as an emulsifier, breaking these globs into smaller globules (just the way dish soap breaks up the fats on our dishes). These globules, with their smaller volume, have a much larger surface area to volume ratio, and it is on the surface area of the food that the enzymes do their work. More surface area means faster digestion, just the way a spoonful of sugar will dissolve faster than a sugar cube.

The bile is produced in the hepatocytes in each liver lobule. The bile is then collected in little tubes called *bile canaliculi*, and then collected in slightly larger bile ducts. These little bile ducts then leave the liver as the hepatic duct. The hepatic duct branches both to the gallbladder as the cystic duct (which comes from the gallbladder's other name, the *cholecyst*), and to the duodenum of the small intestine as the common bile duct. The common bile duct joins up with the pancreatic duct, to form the *hepatopancreatic ampulla*. The gallbladder stores the bile (up to 70 ml) until a hormonal signal from the duodenum (cholecystokinin or CCK) signals its release.

Medical Records

While in the gallbladder, water is absorbed from the bile. Similar to the large intestine, the longer the bile remains, the more concentrated it becomes. During pregnancy, the elevated release of progesterone slows the contraction of smooth muscle, including in the gallbladder walls. The bile salts can ultimately crystallize, producing choleliths (gallstones). The release of cholecystokinin (CCK) can force the gallstones into the much smaller common bile duct, causing severe pain. Given the continuous, slow release of bile from the liver, one treatment is the removal of the gallbladder in a cholecystectomy.

As I mentioned in my discussion of blood vessels, blood drained from the abdominal viscera (excluding the kidneys and the pelvic organs) is drained into the liver through the hepatic portal vein. An understanding of this blood flow helps you to understand some of the many functions of the liver, which include the following:

- **Carbohydrate metabolism** Excess glucose is converted into glycogen via glycogenesis, and, when needed, the glycogen is broken down via glycogenolysis.

- **Lipid metabolism** The amount of lipids in the bloodstream is regulated through both release and uptake in the liver. Since lipids are absorbed into lacteals and then returned to the blood only later, the lipid levels are regulated after they reach the liver via the hepatic artery.

- **Amino acid metabolism** The liver removes excess amino acids, for either protein synthesis or energy use.

- **Waste removal** Whenever amino acids are used for energy, through either breakdown or conversion, the amino group must be removed in a process called deanimation. These amino groups ultimately become the urea that is excreted by the kidneys.

◆ **Vitamin and mineral storage** Iron and fat-soluble vitamins are stored in the liver, to be released only when your diet doesn't have enough.

◆ **Drug inactivation** Drugs you ingest have a limited duration of effect because the liver inactivates the drug over time. One side effect of this is that the liver only has so much capacity to inactivate drugs. Dosage limits are crucial here, for the strain on the liver of exceeding a drug's recommended dosage can be fatal, as in the case of a gifted former anatomy student of mine who accidentally destroyed her liver by taking too much over-the-counter acetaminophen for pain over a short period of time; the only reason she is alive today is through the gift of life— she had a liver transplant.

Pancreatic Panacea?

The pancreas is my favorite organ, if only because mine doesn't work so well (see Figure 14.8). Beyond the endocrine role as a result of the release of insulin and glucagon (which, ironically, control the products of digestion), the pancreas is also an accessory digestive organ. Many of the enzymes used in digestion are produced here. A listing of them can be found in the following section.

Figure 14.8

The layout of the pancreas, liver, and gallbladder.

(LifeART©1989–2001, Lippincott Williams & Wilkins)

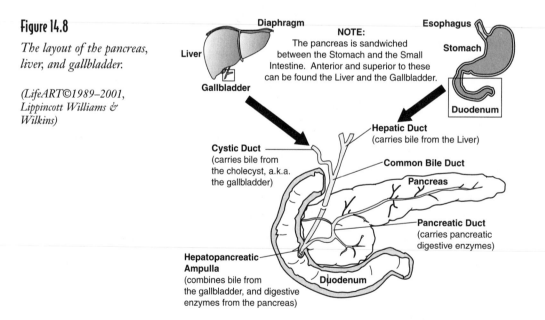

In order to prevent wasteful production of enzymes, the pancreas needs to know what to produce, and when. This is due to the hormonal control of pancreatic secretions. These secretions are exocrine in nature, traveling to the small intestine through the pancreatic duct, which joins up with the common bile duct to form the hepatopancreatic ampulla.

Given that these pancreatic juices go to the duodenum, it only makes sense that the duodenum should be sending the message to make the enzymes. This chemical message is the duodenal hormone cholecystokinin, which causes the release of pancreatic enzymes. In a similar fashion, the duodenum's release of the hormone secretin causes the pancreas to release a buffer solution (pH 7.5 to 8.8) to counteract the acidity of the chyme from the stomach.

Tear 'Em Up!

Remember monomers and polymers from Chapter 2? Well, what follows is a table of the basic enzymes produced by each of the major organs of the body, the location where that enzyme is used, the polymers they break down, and the monomers that are produced.

Digestive Enzymes—Their Production, Use, and Effects

Organ	Enzyme Produced	Enzyme Used	Substrate	Products	Organic Molecule
Mouth	Salivary amylase	Mouth and stomach	Starch	Maltose	Carbo-hydrates
Mouth	Lingual lipase	Mouth and stomach	Fats	Monoglyceride 2 fatty acids	Lipids
Stomach	Pepsin	Stomach	Proteins	Polypeptides	Protein
Stomach	Gastric lipase	Stomach	Fats	Monoglyceride 2 fatty acids	Lipids
Liver	Bile	Gall-bladder	Fat globs	Globules (More SA/V)	Lipids
Pancreas	Pancreatic amylase	Small int.	Starch	Maltose	Carbo-hydrates
Pancreas	Trypsin	Small int.	Poly-peptides	Peptides	Proteins
Pancreas	Chymo-trypsin	Small int.	Poly-peptides	Peptides	Proteins
Pancreas	Carboxy-peptidase	Small int.	Peptides	Terminal amino acids	Proteins
Pancreas	Cholest. esterase	Small int.	Cholest. esters	Cholesterol and fatty acid	Lipids
Pancreas	Pancreatic lipase	Small int.	Fats	Monoglyceride 2 fatty acids	Lipids
Pancreas	Deoxyribo-nuclease	Small int.	DNA	DNA nucleotides	Nucleic acids

continues

Digestive Enzymes—Their Production, Use, and Effects   (continued)

Organ	Enzyme Produced	Enzyme Used	Substrate	Products	Organic Molecule
Pancreas	Ribo-nuclease	Small int.	RNA	RNA nucleotides	Nucleic acids
Small int.	Maltase	Small int.	Maltose	2 Glucoses	Carbo-hydrates
Small int.	Sucrase	Small int.	Sucrose	Glucose, fructose	Carbo-hydrates
Small int.	Lactase	Small int.	Lactose	Glucose, galactose	Carbo-hydrates
Small int.	Amino-peptidase	Small int.	Peptides	Amino acids	Proteins
Small int.	Dipep-tidase	Small int.	Dipeptides	Amino acids	Proteins
Small int.	Nucleo-tidase	Small int.	Nucleotides (Phosphate, Sugar, and Base)	Separates Nitrogenous Base and Sugar	Nucleic acids
Small int.	Phosphat-ase	Small int.	Nucleo-tides	Removes Phosphate	Nucleic acids

You might remember from Chapter 3 that enzymes are basically large proteins with an active site where the substrate, or molecule to be digested, fits. The way an enzyme functions is by changing its shape, and, in the case of digestive enzymes, separating polymers into either monomers or smaller polymers. In the first case, sucrase breaks down sucrose, or table sugar, into glucose and fructose, which are both monomers. Larger polymers, such as starch, are digested in a multistep process; they need to first be broken down by either salivary or pancreatic amylase into small polymers (disaccharides) of maltose, and then the maltose is broken down by what else, maltase, into two molecules of glucose. This is yet another example of the division of labor.

What We Do with Our Food

As I said earlier, the monomers that are ultimately produced through all these processes are absorbed by the small intestine's villi into the bloodstream and the lymphatic capillaries. Once they get to the cells of the body, these molecules have multiple functions. Some of them are used by the body for immediate energy, such as glucose, while some are stored as energy in the form of polymers. Many are used to build other polymers.

The following table of molecule use demonstrates the idea of the *absorptive state* and *postabsorptive state*. Whenever we are in the process of absorbing nutrients into the bloodstream, we have a constant supply of easy energy. For about 12 hours out of every day we are in the postabsorptive state, however, and we must rely entirely on the body's reserves. Blood glucose levels are critical here. Since we don't need all the glucose we eat as soon as we eat it, by reducing blood glucose in the absorptive state, we have freed the glucose up to be converted into other molecules that will remain in reserve until the postabsorptive state.

The Big Picture
We get so much energy from carbohydrates, it makes you wonder how carnivores get their energy, especially if their prey has little or no fat. The answer is that the amino acids can be broken down for energy use. The only problem is that amino acids, unlike carbohydrates, contain nitrogen. The result of their breakdown is the buildup of nitrogenous (containing nitrogen) wastes. These wastes are toxic, so they are released from the cells into the interstitial fluid, drained by the lymphatic system into the blood, and then filtered from the blood by the kidneys, where the wastes are released from the body as urine.

Molecule Use

Type of Organic Molecule	Uses in the Body
Carbohydrates	Quick energy, energy storage
Lipids	Energy storage, production of lipoproteins
Proteins	Quick energy, production of proteins (hormones, structure, and so on)
Nucleic acids	Quick energy (RNA), protein synthesis, replication

All this talk of polymers illustrates the idea that much of what you do with your food is tantamount to recycling! You break down the polymers of others, and then use the resulting monomers to build your own polymers. Much of your protein, fat, and even DNA, are merely reshuffled monomers of others. When you eat meat, the muscles of animals, you take those protein polymers and break them down into amino acids that you then use when, for example, you build up your own muscles. In a nutshell, you *are* what you eat!

The Least You Need to Know

- Digestion is a multistep process. The first step is mechanical digestion, which increases the surface area of the food upon which the enzymes work. Chemical digestion, in turn, may require several steps.

- Our bodies use extracellular digestion, which is the breakdown, outside the cells, of polymers into monomers, which are then small enough to be absorbed into the cells.

♦ Since the cells are made of the same substances that are being digested, every organ must protect itself through the production of mucus, as well as the continual replacement of the mucous membrane.

♦ Once chemical digestion is completed, the organs must not only absorb the food, but also the water, in the organs for distribution to the body via the lymphatic and cardiovascular systems.

Chapter 15

Cleaning Out the Works

In This Chapter

- The structure of the kidney, ureters, bladder, and urethra
- The structure of nephrons
- Filtration, secretion, and absorption
- Micturition (urination)

When many people look at the cardiovascular system they often tend to accentuate the positive contents in the blood—food, water, and oxygen—and rarely focus on the negative. True, I did talk about CO_2 and the lungs, but what about other wastes, particularly nitrogenous wastes from the breakdown of proteins. This chapter rectifies that error and concentrates on eliminating the negative.

Have you ever looked at the alcohol content of wine? It is always around 12 percent because at that concentration the alcohol (a toxic waste) kills the yeast that makes it. If we did not have our own sewage system, we would poison ourselves! This condition, a toxic level of urea (a toxic, nitrogenous waste) in the blood, is called *uremia*.

Not all of the waste is removed by the stars of this chapter, the kidneys, but the kidneys are members of the only body system concerned mainly with waste removal. All the other organs that also act as waste removers are covered elsewhere, and they are members of other systems: the lungs (remove CO_2, heat, and H_2O), skin (removes heat, H_2O, CO_2, salts, and a little urea), and the gastrointestinal tract (removes solid wastes, H_2O, CO_2, salts, and heat). This chapter focuses on the kidneys.

Function Junction

Compared to some other body systems, the excretory (or urinary) system has a fairly short list of functions. All the functions, however, are related to the composition of the blood, which undergoes some dramatic changes before it makes its way out through the renal vein. The kidneys, in fact, receive anywhere from 20 to 25 percent of the blood pumped out of the heart every minute (about 1200 ml of the cardiac output). The excretory system …

♦ **Regulates blood volume and composition** by removing wastes (and thus making urine) and excess water, and by excreting specific wastes (such as H+ ions—lowering pH).

♦ **Regulates blood pressure** through release of renin, and the action of the nephron.

♦ **Aids the metabolism** by making new glucose (*gluconeogenesis*), stimulating production of RBC, and converting D_3 made by the skin into the hormone *calcitrol*.

Waterworks

As I mentioned in Chapter 12, the kidneys like to stay in the background, so much so that they are actually hidden behind the abdominal wall! Figure 15.1 shows the left kidney a little higher than the right kidney, which makes sense with that honkin' big liver in the way! Their familiar shape is like a large, uh, kidney bean that has been punched in on one side. That punched-in portion, like that of the lung, is called the *hilus*, and it is the location for the renal artery and renal vein, as well as the *ureters*. Ureters carry the waste-filled urine made by the kidneys down to the next stop, the urinary bladder.

More primitive organisms simply release wastes constantly, but we have a urinary bladder to store it until there is a socially acceptable opportunity to get rid of it, through the urethra, that little tube from the bladder to the outside world.

Medical Records

Since not all of our evolutionary ancestors were concerned with decency in terms of urination, why bother storing the urine in the first place? There are two reasons why this trick, for which we can thank the bladder, evolved. The first reason makes a lot of sense right away; these are wastes, after all, so do you really want them in your den? Wouldn't it be healthier to be able to leave it farther away? The second reason makes sense if you have ever had a male dog; dogs don't just pee, they mark their territory! To do that, you need to be able to pee on command. I just wish with my dog Milo that he would pee on *my* command, and not *his!*

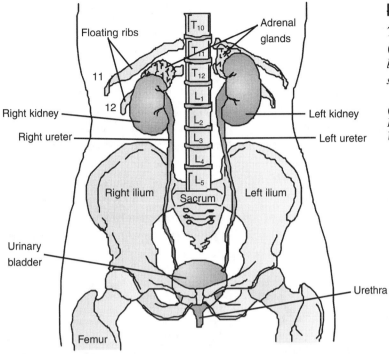

Figure 15.1

The organs of the excretory (urinary) system dump the blood's wastes into the outside world.

(LifeART©1989–2001, Lippincott Williams & Wilkins)

Not Just a Bean ...

The kidneys don't quite look how most people expect them to look, having seen so many clean-looking diagrams (such as Figure 15.1). This is because diagrams usually show only the innermost of the three outer layers, the *renal capsule*. This layer is smooth, fibrous, and even transparent, and it prevents infections from reaching the kidney; this layer is also contiguous with the hilus.

The next two layers are more important in terms of protection from trauma. The kidneys, after all, are very close to the dorsal surface of the body, and they can be easily damaged, which is why the kidney punch is illegal in boxing. The middle, fatty layer is, and rightly so, called the *adipose capsule*, and the outermost layer of dense irregular connective tissue is called the *renal fascia*. In addition to protection, together these two layers anchor the kidney in place.

The inside of the kidney looks a little like the inside of a lymph node, complete with the pyramid-shaped structures, this time called *renal pyramids*. In between

Medical Records

One danger of being too thin—you know, the media-driven craze to look like a stick insect—is *nephroptosis*, or floating kidney. If the adipose capsule is too thin the kidney may drop out of place, making a kink in the ureter, blocking the flow of urine. In addition to pain, this can also damage the kidney—whoever said all fat is bad?

the pyramids are columns of tissue called, of course, *renal columns*. Also similar to the lymph nodes is an outer cortex (the *renal cortex*), followed by an inner medulla (the *renal medulla*).

The inside of the kidney is where the filtering of wastes happens, but where do wastes go after that? Think about the pipes in your home. You probably have two or more sinks, at least one toilet, and one bathtub. Where does all that water, and everything else go? Smaller pipes merge with larger pipes, until the largest pipe makes its odoriferous way to the sewer or septic tank. The kidney works the same way! The *collecting system* starts with small collecting ducts that empty the urine into a larger duct, which empties into a minor calyx, which empties into a major calyx, which empties into the renal pelvis, which finally dumps its contents into the ureter (see Figure 15.2).

Flex Your Muscles

Don't fall into the trap of thinking that the cortex is the core. You should, however, think of the medulla as the middle, for that's what it is!

Figure 15.2

The anatomy of that urine-spitting machine, the kidney.

(© 2003 www.clipart.com)

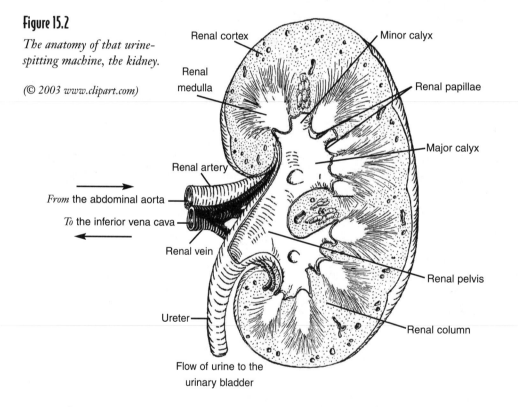

Renal cortex

Renal medulla

Minor calyx

Renal papillae

Major calyx

Renal artery

From the abdominal aorta

To the inferior vena cava

Renal vein

Ureter

Renal pelvis

Renal column

Flow of urine to the urinary bladder

I Need to Get Out!

The rest of the trip is rather uneventful. After collecting in the renal pelvis, the urine makes a long trip down the ureters, which are hiding behind the posterior peritoneum, or posterior abdominal wall (refer to Figure 15.1). The two ureters attach separately to opposite sides of the dorsal surface of the urinary bladder. Females, on average, have a smaller bladder capacity than do males due to the position of the uterus curving directly behind and above the bladder.

Medical Records

Now imagine that the uterus is occupied. During pregnancy, the size and weight of the uterus increase—don't forget to add the weight of the amniotic fluid, too! Some of that mass will extend forward, causing a pregnant woman to rearrange her footing to account for the different center of gravity, and some will be supported by the flared ilia of the female pelvis (see Chapter 7), but the rest of the mass compresses the urinary bladder. Keep that in mind the next time a pregnant woman is behind you in line for the bathroom!

From the bladder it's a short trip outside through the urethra. Men have a longer urethra, due to its position running down the length of the penis. One of the implications of this is the greater frequency of bladder infections in women than in men. As to the exit from the urethra, it is similar to the anus (see Chapter 15) in that there are two sphincters: the involuntary *internal urethral sphincter* (which relaxes when you *have to* pee), and the voluntary *external urethral sphincter* (which relaxes when you *do* pee). Once again, babies have yet to develop voluntary control over that external sphincter. If you have a child, chances are you've been peed on!

Where It All Happens

Are you a coffee addict? I have to get up very early because the first class I teach is at 7:30 A.M. My biggest complaint is when coffee tastes too weak; brown water doth not a good cup of coffee make! I remember asking someone behind the counter at some coffee house if the coffee was strong. His answer showed no understanding of the science behind coffee, which, surprisingly enough, has an awful lot to do with nephrons!

The young man behind the counter said that the beans were strong, and that was the only thing that mattered. While it's true that some beans have a stronger flavor than others, simple logic tells you that strong coffee has to have more dissolved solutes. The simplest way is to make the coffee with more coffee beans! In the same way, stronger, darker urine, which is produced when the body has to preserve water, has less water and thus more solutes.

Another way to make the coffee stronger, in other words to get more solutes into the water, is to grind the beans smaller. The smaller the grind, the greater the surface area to volume ratio of the beans, and the greater the ratio, the more solutes get into the coffee. In the same way, the proliferation of capillaries in the development of the kidney—in particular the cluster of about 50 fenestrated (see Chapter 12) capillaries in the *glomeruli* (singular, *glomerulus*), where the solutes first leave the capillaries—leads to a greater SA/V in the kidney. The higher the SA/V ratio in the kidneys, the stronger the urine you'll make!

If you are a real coffee aficionado, then you have probably discovered espresso, that rich, strong coffee. So what's the secret to making espresso so strong? It's more than just a finer grind (although espresso grind is extremely fine). The secret is the use of steam rather than hot water. What's the difference? The steam is at higher pressure, which increases the strength of the filtration process.

In the nephron (see Figure 15.3), the blood enters the glomerulus through an afferent (entering) arteriole, flows through the fenestrated capillaries of the glomerulus, and then exits through a smaller efferent (exiting) arteriole. Just like putting your thumb over the end of a flowing garden hose, the smaller size of the efferent arteriole increases the pressure. The higher the pressure, the stronger the filtrate (the material that is filtered out), or, in this case, the urine. That means that we have two little espresso makers in our body! Just don't start drinking your urine!

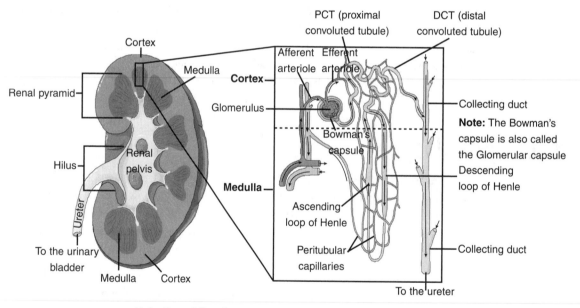

Note: If you follow the arrows in the blood vessels and the tubules you will see that they go in opposite directions. This is the basis for the countercurrent mechanism.

Figure 15.3

The structure of a nephron, the part of the kidney where the urine is actually made.

(LifeART©1989–2001, Lippincott Williams & Wilkins)

Smooth muscle cells that wrap around the afferent arteriole and the distal convoluted tubule (which we will meet shortly) are used to help regulate blood sugar. This odd ring of muscle has a rather scary name, hidden behind an innocuous abbreviation: the JGA, or juxtaglomerular apparatus (the thing near the glomerulus, or words to that effect). Blood pressure is also regulated through a hormone called renin, which is released by the cells of the JGA.

Renin converts a plasma protein from the liver called angiotensinogen into angiotensin I, which in turn is turned into angiotensin II in the lungs. Angiotensin II raises blood pressure, especially in the glomerulus (the start of the nephron); it also stimulates the release of aldosterone from the adrenal glands (see Chapter 18) and the subsequent reabsorption of Na^+ ions (which helps to reabsorb the water—see the countercurrent mechanism). Stimulating thirst also helps raise blood volume, as does angiotensin II's habit of stimulating the release of ADH, or antidiuretic hormone, which prevents water loss in the urine. All in all, that's a pretty busy schedule of hormonal activities!

> ### The Big Picture
>
> This connection between the kidneys and blood pressure in the cardiovascular system has ramifications for blood pressure medication. Higher blood pressure, as indicated in the discussion of renin and angiotensin II, leads to less urination. In order to lower blood pressure, one type of medication, called a diuretic, relaxes the JGA and causes a release of more water in the urine. This is why diuretics make you pee!

Next, I'd like you to look at the structure of the nephron. There are two basic parts, the *renal corpuscle*, where the fluid is filtered (made of the Bowman's capsule and the glomerulus), and the *renal tubule*, actually a series of tubules, where the filtered fluid travels. These tubules are, in order, the *proximal convoluted tubule* or *PCT* (named both for its proximity to the glomerular capsule and for its twisted shape), the *loop of Henle* (with its descending limb and its ascending limbs), the *distal convoluted tubule* or *DCT*, and finally the collecting duct (which we mentioned earlier, the start of the collecting system). Each portion of the renal capsule, as we shall see, differs in its structure, as well as its function in the urine-making process.

Filtration

Filtration is the process of selectively removing, or filtering, materials out of a solution, and it happens in the glomerulus, producing a protein-free fluid, *filtrate*, that is somewhat akin to blood plasma (see Figure 15.4). The only problem is that blood has an awful lot of things we *don't* want put in the urine, not to mention things we do want to release! Now that it is in the filtrate, we have to get the good stuff back, and the job of getting it back will be discussed in the next section.

Figure 15.4

Filtration in the glomerulus.

(©Michael J. Vieira Lazaroff)

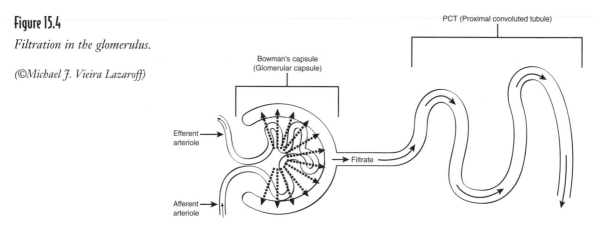

Note the movement of filtrate *out* of the glomerular capillaries and into the PCT

I already mentioned the afferent and efferent arterioles, on the other end of the glomerulus, and the high pressure produced that helps with the filtration, but what about the rest of the renal corpuscle? The glomerulus lies in a capsule known as the *glomerular capsule* or the *Bowman's capsule*. This capsule is made of simple squamous epithelium (see Chapter 4), which makes up the parietal epithelium (see Chapter 1). The visceral epithelium consists of cells called *podocytes*, which have numerous small "feet" called *pedicels*. These pedicels, and the basement membrane (called the *lamina densa*), help to keep the proteins from entering the filtrate.

An amazing amount of filtrate is produced every minute. The *glomerular filtration rate* (GFR) is about 125 ml a minute, or about 10 percent of the volume of the blood entering the kidney. One of the reasons this is possible is the enormous filtration surface of 6 m^2 per kidney. Once the filtrate leaves the capillaries, the coils and loops of the renal tubule, and the capillaries around them, called *peritubular capillaries*, change the concentration of the filtrate, through reabsorption and secretion.

Reabsorption

It's a bit ironic that once the material has been filtered out, so much of it is reabsorbed! Some of this makes sense when you look at the amounts. Remember that the GFR is 125 ml/minute; that adds up to 180 liters of filtrate a day. If we didn't reabsorb some of that material we would run out rather fast! Since we average about 5 liters of blood, and 55 percent of that is plasma, we would be *out* of fluid in 22 minutes! It turns out that we reabsorb between 178 to 179 liters a day. That leaves 1 to 2 liters to be urinated away every day. Think of that the next time you open a 2-liter soda bottle; for some reason the word humbling comes to mind!

Much of this movement of material starts right away in that twisted tube right off the glomerulus called the *proximal convoluted tubule* or *PCT* for short (see Figure 15.5). The inside of this tubule, like the cells on the outside of the villi of the small intestine (see

Chapter 14), has a brush border of microvilli, which of course increases surface area for the reabsorption; the major difference here is the cuboidal shape of the cells (see Chapter 4), unlike the columnar cell of the villi. I am talking here about reabsorption into the blood, specifically into what are called *peritubular* (around the tubes) *capillaries*.

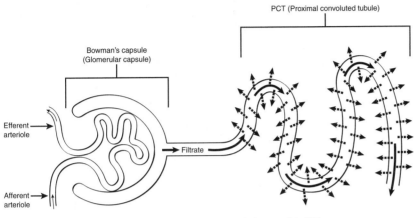

PCT (Proximal convoluted tubule)

Bowman's capsule
(Glomerular capsule)

Efferent arteriole

Filtrate

Afferent arteriole

Note the reabsorption of water and solutes *out* of the PCT and into the peritubular capillaries (not shown).

Figure 15.5

Reabsorption in the PCT and descending loop of Henle.

(©Michael J. Vieira Lazaroff)

You might remember from Chapter 3 that the movement of materials in and out of vessels occurs via both diffusion (which, as passive transport, requires no energy) and active transport (which, going against the concentration gradient, requires energy). For diffusion to occur there needs to be a difference in the concentration of material (solute or solvent) on both sides of the membranes. In addition, cells cannot perform active transport to reabsorb water, and so the only way to get the water back into the capillaries is to create osmotic pressure by greatly increasing the concentration of the blood (that is, making it hypertonic; see Chapter 3). The easiest way to do this is to have a *countercurrent mechanism.*

A countercurrent mechanism means that the flow of the filtrate through the tubes passes in the opposite direction from the flow of the blood, despite both starting at the glomerulus in the Bowman's capsule: The filtrate goes directly from the glomerulus to the PCT and loop of Henle, and then to the DCT, whereas the capillaries go from the glomerulus, skip over to the DCT, and then move backward to the loop of Henle and the PCT.

In this way, increasing the concentration of Na^+, Cl^-, K^+, and HCO_3^- in the capillaries (through reabsorption and secretion from the DCT and the ascending limb of the loop of Henle, see the following table) will thus make the plasma hypertonic (see Chapter 3). When the blood makes its way to the ascending limb of the loop of Henle and the PCT, the water will be returned, along the concentration gradient, to the capillaries. On the peritubular side, simply by removing the water in the descending limb, the ascending limb will have a higher concentration of ions than the plasma, thus making its movement into the capillaries possible. It may take a little effort to see it, but in the end it's worth the effort, because beneath the complexity of the details, the actual countercurrent mechanism is elegant in its simplicity!

Secretion

In addition to reabsorption of materials from the filtrate into the blood, other materials from the plasma are being placed into the filtrate in a process known as *secretion* (see Figure 15.6). Secretion plays a major role in the balance of ions in the blood (see the following table). Excess K^+, for instance, can cause cardiac arrest, so the level can be lowered by secreting it into the collecting duct (in the presence of aldosterone).

Figure 15.6

Reabsorption in the DCT and ascending loop of Henle.

(©Michael J. Vieira Lazaroff)

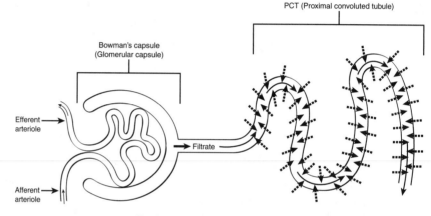

Note the secretion of solutes *out* of the peritubular capillaries (not shown) and into the PCT.

Reabsorption and Secretion in the Nephron

Location	Reabsorption	Secretion
Glomerulus (Filtration *only* of H_2O, glucose, amino acids, Na^+, Cl^-, HCO_3^-, K^+, urea, uric acid, creatinine, etc.)	*None*	*None*
Proximal convoluted tubule	Na^+, Cl^-, HCO_3^-, K^+, H_2O, glucose, urea, amino acids	H^+, NH_4^+, creatinine
Descending limb, loop of Henle	H_2O	urea
Ascending limb, loop of Henle	Na^+, Cl^-, K^+	urea
Distal convoluted tubule	Na^+, Cl^-, HCO_3^-, H_2O, glucose	*None*
Collecting duct	Na^+, Cl^-, HCO_3^-, urea, H_2O*	H^+, K^+

*Water is reabsorbed here as a result of ADH (anti-diuretic hormone—see Chapter 18).

Normal blood pH is slightly basic, falling within the narrow range of 7.35 to 7.45 (7 is neutral; see Chapter 3). When blood gets too acidic, H^+ ions are secreted into the PCT and the collecting duct, thus raising the pH to normal. In addition, the blood pH can be raised by reabsorbing HCO_3^- (which binds with the H^+ ions) from the PCT, DCT, and the collecting duct. As you can see, the use of reabsorption and secretion plays an important role in homeostasis.

When You Gotta Go ...

Okay, so you've made the urine. Now what? You can't keep it forever! So how does your bladder let you know that it's full? After all, you don't exactly dribble it out! Similar to the uterine walls, there are stretch receptors in the walls of the urinary bladder. Although the capacity of the urinary bladder is around 750 ml (about the size of a bottle of wine), the stretch receptors start to respond when the volume reaches somewhere between 200 and 400 ml.

Sensory neurons send a message to the brain that makes us consciously perceive a full bladder, and a desire to *urinate*, or *void* one's bladder, an act that also carries the high-falutin' name *micturition*. In addition to the brain's conscious control, there is a reflex action (that evolved first) located in the sacral region of the spinal cord (obviously quite close to the bladder). This reflex is called the *micturition reflex*.

The reflex starts with parasympathetic nerve impulses (see Chapter 21) that cause the urinary bladder walls to contract, and the internal urethral sphincter to relax. This, of course, puts a great deal of stress on the walls of the urethra, particularly the external urethral sphincter. Once again, as in the case of the anus, the conscious contraction of our external urethral sphincter is the only thing standing between us and embarrassment!

The Least You Need to Know

- The kidneys not only remove wastes, but also regulate the concentration of water and solutes in the blood.

- Blood pressure regulation is controlled in part by the action of the kidney.

- The movement of materials in the nephrons of the kidney is the result of three processes: filtration in the glomerulus, and reabsorption and secretion in the tubules.

- Urination (micturition) is the result of both reflexive and conscious neural activity.

Part 5

Holding the Fort

Our body is always at risk of being overrun by outsiders. Our skin is literally covered with bacteria. The air we breathe, and the dust it carries, has hitch-hikers. Bacteria, viruses, fungi, not to mention the occasional flatworm or roundworm, are all anxious to get a piece of the action.

Think about it. The bacteria on our skin need food, moisture, and warmth. Sounds a lot like the inside of our bodies, and that's the point. Although certain pathogens, disease-causing organisms, are specific to certain tissues, in general the inside of the body is like unclaimed ore during a gold rush. Bacteria that live innocuously in your gut, will kill you if it gets into your blood.

Before you go feeling sorry for yourself, know that we have a few tricks up our sleeves, both literally and figuratively. The skin under your cuff keeps out various and sundry gate-crashers, but even if a few get through the weakest of your defenses, your mucous membranes, your lymphatic system's immune response is always there to save your life! Find out how in the next two chapters!

Chapter 16

Your Birthday Suit

In This Chapter

- ◆ Functions of the skin
- ◆ Epidermal and dermal structure
- ◆ Glands
- ◆ Hair and nails

When we find ourselves physically attracted to someone, we are attracted to her or his skin. True, the way the skin covers the framework of bones and muscles is also a factor, but what we see first is skin. I remember my own high school anatomy and physiology teacher, Mr. Otis McCain, making a comment about lovers walking along the beach, holding hands, "Dead skin against dead skin." Our first attraction, that joyous celebration of life, is ironically based on the sight of dead tissue! Underneath that dead layer, however, the skin is very much alive.

We all grew up with the wrong idea about organs. When people think about organs, they usually think about the heart, or the lungs, or the stomach. Those are all fine, but as we've seen, every bone in our skeleton is an organ, as is every skeletal muscle. We are used to thinking that we have maybe a few dozen organs, when in reality the number is closer to a thousand! When students first dissect an animal, they are usually surprised at how large the liver is, the largest internal organ. Yet most people take no notice of the largest organ of all, our skin.

Covering, and I mean *covering*, 22 square feet (or 2 square meters), and weighing in at 10 to 11 pounds (or 4.5 to 5 kilograms)—these measurements vary according to height and weight and so on—this overlooked organ has a wide range of tasks, not the least of which is keeping us from dying by massive infection! That this organ is *overlooked* is ironic, because it is the first organ we see when we *look ourselves over!*

Function Junction

For an organ "on the outside," it sure has its fingers in a lot of pots. An expression like "beauty is only skin deep," though a noble sentiment, gives a false impression of the ultimate importance of the skin to our survival. Far more than decoration, our skin's importance to the body is deeper than it appears on the surface:

- **Protection**   The outer dead layer protects delicate living tissues underneath from abrasion and infection.

- **Excretion**   As I mentioned in Chapter 15, the skin is involved in the secretion of salts, wastes, and water.

- **Sensation**   Nerve endings in the skin are involved in the perception of touch, pain, pressure, and temperature.

- **Regulation**   The maintenance of blood flow throughout the skin, and the excretion of water in sweat, help to regulate body temperature.

- **Storage**   The skin, in particular the subcutaneous fat layer, helps to store nutrients.

- **Synthesis**   Vitamin D_3, a steroid that is converted into the hormone calcitrol by the liver, is produced in the skin.

On the Outside

The outer epithelial layer of skin is called the *epidermis*. As epithelial tissue it is avascular (see Chapter 4), but it is also very thick. With so many layers, the uppermost layer is so distant from moisture, nutrients, and so on, that it dies. As I mentioned earlier, that dead tissue is essential in terms of protection. There are, however, many other living epidermal layers that are very much alive, as you shall see (see Figure 16.1).

The Replacements

The epidermis is made of stratified squamous epithelial tissue (see Chapter 4). The scale-like shape is important in terms of abrasion; any scrape, rather than tearing into delicate cells, is more likely to take away (slough off) entire cells. We lose an estimated 30 to 40 thousand skin cells every minute, or anywhere from 40 to 60 million cells a day. That's an incredible toll. If we didn't replace them all the time, we would quickly run out and start bleeding from our newly exposed blood vessels.

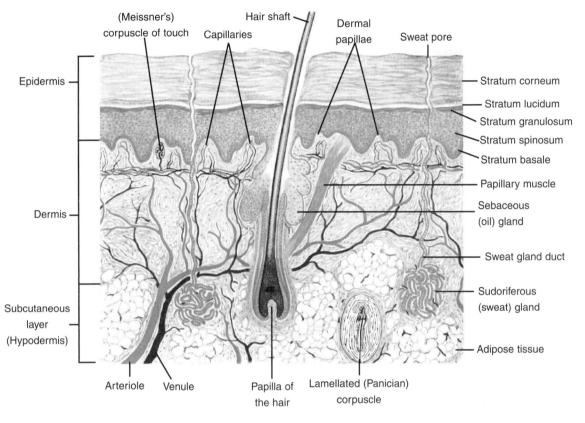

Figure 16.1

The multiple layers of the skin, each with its own role to play.

(©2003 www.clipart.com)

The lower levels of the epidermis need to be alive, and nourished by the blood, to provide those replacements. The blood is provided by the lower *dermis*, which is made of connective tissue. Despite much of our outer skin appearing essentially flat, the underside of the epidermis is anything but. A series of peaks and valleys in the underlying dermis, looking very much like sheets of foam packing material, form what are called *dermisdermal papillae* ("nipples"). The layer of epidermis immediately above the papillae is called, of course, the *papillary layer*. This arrangement increases the surface area of the papillary layer, and is the most efficient way to provide blood to the epidermis, as well as providing a placement for nerve endings.

In addition to the dermal layer, there are other larger ridges in the dermis, directly underneath the epidermal ridges (see Figure 16.2). Now look at your fingertips. Those are examples of familiar ridges, but look closely at the palms of your hands (or the soles of your feet, if you are limber enough!). All such ridges are important in terms of increasing the surface area, and thus the friction, of those areas of the skin that need friction (for grabbing or walking). The combination of those ridges, and secretions from the sweat and oil glands in our skin, is what produces fingerprints.

Figure 16.2

The arrangement of dermal papillae allows for the placement of blood vessels and nerve endings beneath the epidermis.

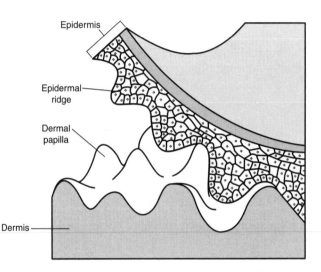

Medical Records

Fingerprints are genetically determined, but as I discussed in Chapter 3, the expression of the genes (remember the operon?) depends upon the environment. Those genes are set on their path rather early, and so our fingerprints don't really change over time; they just get larger. On a side note (a side note in a sidebar?), the action of operons means that even identical twins have different fingerprints!

So what about the layers of the epidermis? First of all, 9 out of every 10 skin cells are *keratin*-producing cells called *keratinocytes*. Keratin, which is a protein, keeps the skin waterproof, in conjunction with oil glands in the dermis. The uppermost layer of the skin is called the *stratum corneum*, which is made entirely of dead cells. This layer varies in depth, not just from the dermal papillae, but also according to the location on the body. Areas of the skin that don't receive a lot of abrasion, such as your forehead or eyelid, will have a very thin stratum corneum, but the palms of your hands and the soles of your feet need more protection, so the stratum corneum will be thicker there.

Down below the 25 to 30 layers of the stratum corneum is a layer that is usually only found in the thick-skinned areas mentioned above: the *stratum lucidum*, named for its almost transparent appearance. These three to five layers produce a precursor to keratin, called *keratohyalin*. This substance is also made in the next layer, called the *stratum granulosum*, named because of the appearance of dark granules of keratohyalin. The nuclei in the cells of this layer start to break down, causing the cell to die.

All this death is actually a good thing, for this is what provides the first line of defense against infection, but the constant loss of skin cells means that there is a constant need to replace them. The replacements are made in the lowest layer, called the *stratum basale*, because it is at the base of the epidermis, and the *stratum germinativum*, because of the presence of stem cells constantly undergoing mitosis to produce the replacements. Some of the stem cells also migrate into the dermis to produce the oil and sweat glands, which are, after all, in the dermis, despite being made of epithelial tissue.

One type of cell in the stratum basale deserves special mention: the *melanocyte*. This cell gets its name from its production of melanin, the brown pigment that gives skin much of its color (see the next section). These bizarre-looking cells have long cytoplasmic extensions that extend up into the next layer called the *stratum spinosum*, named because of the cytoskeleton filaments that extend into the desmosomes (see Chapter 3), which hold the cells tightly together. The melanin makes its way into the cells of this layer through phagocytosis.

Medical Records

In Chapter 2, I mentioned the waterproofing done by oils in the skin. Understand that your skin is never *really* waterproof. If you have ever spent too much time in a bathtub or a pool, you have noticed your fingertips and toes getting all wrinkled. This is one of those "why is that?" questions, because people think that wrinkled means dry. But what if you gained, say, 100 pounds overnight? Ever notice how people show the extra weight by making extra chins? In the same way, too much time in the tub washes off the oil, causing you to take in more water, thus swelling your skin and making folds!

The Shallowness of Skin Color

I am a big fan of jazz and its history. The history of jazz in America is basically a history of being African American, from the nineteenth century to the present. One cannot truly understand jazz without understanding racism in America; it is especially interesting when you realize that musicians such as Louis Armstrong, Coleman Hawkins, and Miles Davis were accepted almost as if they were royalty in Europe. Racism is one of the real blights on American history, from slavery, to Japanese internment camps in World War II, to modern-day fears of innocent Arab Americans.

Racism is odious in itself, but it is especially scary when one considers how *little* of the skin is actually involved in color! To put things in perspective, I want you to remember that we are all one color, brown; the only thing that really differs is the *shade*. First off, the concept of race is a questionable one, as no reliable genetic markers have been found as to race. The concept of three races can be traced to the history of European colonization: south to Africa, and east to Asia. This is all ironic, as we all came, as a species, from Africa; we are *all* Africans!

Using the false concept of three races, we can see the differences are subtle, at best. The brown color we all share is *melanin* produced by *melanocytes* at the lowest level of the epidermis, the *stratum germinativum* (or *stratum basale*). The melanin provides protection against UV light. The skin uses UV light to produce vitamin D, but only about one hour of exposure a week is necessary to produce what we need. In certain areas of the world, particularly near the equator, the exposure to sunlight far exceeds what is necessary; too much exposure increases the risk of mutation and skin cancer (see Chapter 3). The melanin absorbs the UV light, thus protecting the DNA. You can see this under the microscope, because the majority of the melanin actually collects directly above the nucleus.

It thus makes sense that when humans evolved in Africa, the intense UV light around the equator necessitated a great deal of melanin, which is the source of the darker color in people of African ancestry. As people migrated to other regions of the world, particularly areas far north of the equator (remember, the majority of the world's land mass is north of the equator) the need for UV light to produce vitamin D exceeded the need for protection from UV. The low levels of melanin in people of European ancestry make the skin fairly transparent, which means that most of the pinkish hue comes from the hemoglobin in the capillaries of the dermis.

Medical Records

Of course, melanin exists in all populations, and the levels can vary dramatically within populations. Levels can also change over the course of a person's lifetime, or even over a summer. In response to a larger amount of UV, the skin churns out more melanin, and, voilà, a tan! In the opposite sense, the recluse or shut in, as in the case of the elderly, will be very pale from lack of melanin, and may be very low in vitamin D as a result. The most extreme case, of course, is the albino, who lacks the ability to make melanin, and looks the pinkest of all!

People of Asian ancestry may have a slight (once again the hue varies enormously) yellowish tinge. Carotene, which is needed to produce vision-related pigments, is found in the dermis, as well as the highest level of the epidermis, the *stratum corneum*. One look at Figure 16.1 shows how *little* of the skin is actually involved in skin-color difference. After all, we all have hemoglobin, we all have melanin, the only differences are the presence of absence of carotene, and the *amount* of melanin. Given that people with the darkest skin only show that in the lowermost level of the epidermis, it seems like an awfully small amount upon which to base centuries of violence and hatred, especially in what is nominally called the land of the free!

I've Got You Under My Skin

One of the great things about skin is its elasticity. It sounds a little gross, but think about it. Our skin may be bendable and stretchable, but the sign of its elasticity is its ability to resume its original shape. Did your mom ever tell you that if you continued to make that face that it would stay that way? I know what you thought, "Coooool!" No such luck. If skin wasn't elastic, it would stay pinched long after your Aunt Zina has gone home.

Medical Records

If you see more wrinkles, crow's feet, and baggy skin as you get older, there's a good reason for that. As people age, the elastic fibers in the dermis of their skin start to lose their elasticity, and collagen fibers start to stiffen and break. Replacing them gets harder as the number of cells that make those fibers (fibroblasts) declines. The result? The skin loses its elasticity, and becomes a little looser, so to speak. By cutting the skin and stretching it over the bone, in other words, a facelift, a person can achieve the *appearance* of youth, but the real youth is in your mind!

Remember that elasticity is possible in connective tissue due to the matrix, and the wide spacing of the cells, which prevents damage to them under stress. Running throughout that matrix, however, are many structures. As I mentioned earlier, the epidermis is avascular, so all the blood vessels are in the dermis. In addition you will find lymphatic capillaries, glands, and nerves.

Medical Records

Have you ever gotten a paper cut? It always amazes me how much something so harmless hurts! Apart from the pain being way out of proportion, the lack of blood from a typical paper cut illustrates the lack of blood vessels in the epidermis. A deeper cut will go into the dermis, reach the blood vessels below, and cause capillary bleeding. Longer, more profuse bleeding means that the arterioles and venules of the lower dermis have been damaged.

One of the important aspects of the dermis relates to the wide separation of the cells. Think about the last time you got a paper cut. Now try to remember which finger it was on. If you can't remember, within a few days you won't be able to find any evidence of it! The tightly packed epidermis does that easily, with the cells reproducing until they touch each other (see cell junctions in Chapter 3). The dermis, however, cannot easily repair a deep cut.

In a deep cut, the first phase of repair is an *inflammatory phase*, in which blood loss causes a clot to form all the way up to the surface (see Chapter 10), which helps to hold the sides of the cut together. The migratory phase follows, getting its name from the migration of epithelial cells growing underneath the clot. *Fibroblasts* in the dermis make scar tissue, which at this point is called *granulation tissue*. The third phase is the *proliferative phase*, in which more epidermal cells are made, as well as more collagen fibers, not to mention regrowth of blood vessels.

These new blood vessels are fragile, not to mention shallow, until more epidermal cells are made. That's why it is best to wait until the scab dries and sloughs off on its own after the *maturation phase*, or the renewed bleeding that will result will just start the whole process over again. In the second and third phases the poor growth of scar tissue cannot keep pace with the epidermal growth, causing a noticeable mark, usually some form of indentation, known as a scar. Thus a scar is less a sign of character than a sign of poor dermal growth!

Sensation

The dermal papillae have more than just capillaries, they also contain sensory nerve endings. These general senses—as opposed to special senses such as vision, hearing, taste, and smell—include pressure, temperature, and pain. At the very bottom of the epidermis is a cell called a *Merkel cell* (technically an epidermal cell), which forms a *tactile*, or *Merkel*, *disk*, which is used in the perception of touch. In addition there are dermal touch receptors in the papillae called *Meissner's corpuscles*, or *corpuscles of touch*.

The bottom of the dermis is called the *reticular layer* because of its dense, irregular connective tissue, filled with collagen fibers. This layer also contains fat deposits, hair follicles, nerves, and blood vessels. This layer gives skin much of its elasticity, but too much stretching, from weight gain, edema (see Chapter 10), or pregnancy can result in *striae*, also known as *stretch marks*. Still, better to have stretch marks than *tears* in the skin! This layer also contains pressure receptors called *lamellated*, or *Pacinian corpuscles*.

Medical Records

Ironically, third-degree burns, even with the extra swelling, are not as painful as severe second-degree burns. This makes little sense until you think about it: with the nerve endings destroyed, messages for pain are not sent. Pain, yes, but not as severe.

As far as temperature is concerned, cold is perceived at the bottom of the dermis, and at the *subcutaneous* (*sub* = under, *cutaneous* = skin) *layer*, made of superficial fascia (see Chapter 8), which connects to muscles, which are surrounded by deep fascia. This makes sense, as this is the layer that needs to start constricting the blood flow in order to prevent loss of heat. In the same way, the high risk of damage from burns means that the heat receptors need to be higher up, in the middle and top layers of the dermis, thus speeding up reaction time!

One of the results of such sensation is the extreme level of pain one can feel with severe burns. We have all experienced first-degree burns at one point or another. They hurt much more than you would expect since your skin just looks a little red; the damage is limited to the epidermis (see Figure 16.3). Second-degree burns, on the other hand, look and feel much worse, adding a raised blister to the mix. The damage in this case is more severe, having extended into the dermis. Third-degree burns involve not only greater depth, but also actual destruction of the tissues.

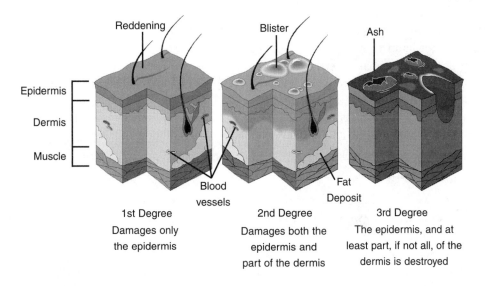

Figure 16.3

The three degrees of burns vary in terms of depth, but unlike first- and second-degree burns, third-degree burns actually involve destruction of tissues.

(LifeART©1989–2001, Lippincott Williams & Wilkins)

Treatment for third-degree burns thus involves the use of skin grafts, in which excess skin from one part of the body is used to replace the dead skin of the burn. Given the many important functions of skin, treatments of severe burns vary according to the amount of skin damaged. The extent of the third-degree burns is extremely important because a small burn is minor, but burns over a larger percentage of the body can be life-threatening, or even fatal. As such, a quick way to measure the extent of a burn, and thus the danger the patient is in, is important; the "rule of nines" does just that by dividing sections of the body into percentages, which are based on multiples of 9 (18 percent for the front of the body, 9 percent for each leg, and so on).

The Big Picture

Skin, after all, is more than just covering, and so its loss from third-degree burns means a loss of its functions as well. This means that extensive third-degree burns have systemic ramifications. The destruction of blood vessels means the loss of water and plasma, causing shock. The water loss will decrease the production of urine. Temperature regulation, from reduced blood flow, is extremely difficult. Perhaps the greatest impact is the lack of protection from infection. This puts the immune response into overdrive, which can tax the body's resources, making the body more susceptible to opportunistic infections.

Glands Galore

Oil glands, or *sebaceous glands*, are found attached to the side of a hair follicle, which explains what happens to your hair after you haven't washed it for a few days. In addition to preventing epidermal water loss, the oil, or *sebum*, produced in these glands prevents brittle, dry hair, and also stops the growth of some bacteria. As anyone who has gone through puberty knows, however, some bacteria like the oil, causing pimples or deep boils, sometimes called *whiteheads* because of the color of the pus, which is made of white blood cells (more about them in Chapter 17) and bacteria. Blackheads, which also form from bacterial growth, turn black not because of dirt, but rather from oxidation of oil and the depositing of melanin.

The other evil-sounding gland is called the *sudoriferous*, or *sweat gland*, which produces, well, sweat. There are two different kinds of sweat gland: *eccrine* and *apocrine*. Eccrine glands, which are by far the most common, release the watery sweat through pores in the epidermis, which is interesting given that it is made in the subcutaneous layer. Apocrine glands, on the other hand, produce a thicker, more viscous and pungent sweat. These glands—found in the armpits, pubic area, and on the areola (the pigmented area) of each breast—release their sweat into the hair follicles. Given the growth of hair in those areas during puberty, it makes sense that the glands start to function, not to mention stink, during puberty.

The only other gland left is called the *ceruminous gland*, which is found in the ear canal, or *external auditory meatus* (see Chapter 21), and makes ear wax. These glands open both directly out of the epidermis and into oil-gland ducts. Their main function is to prevent foreign bodies (especially the kind with six or eight legs!) from getting too far down the ear canal and damaging the ear drum, or *tympanic membrane*.

Red in the Face

The blood vessels in the dermis have more to do than just feeding and watering the tissues. The hydrogen bonds between water molecules (see Chapter 2) make the plasma in blood a very efficient means of carrying heat generated by working muscles to all parts of the body. The blood flow to the skin is one of the primary means of losing the excess heat produced in our muscles. Remember the metarterioles in the capillary beds? The opening and closing of the precapillary sphincters along the metarteriole (in other words, part of vasodilation and vasoconstriction) is one of the main means of regulating body temperature.

Think about the last time you were exercising regularly. Typically you would have appeared red in the face; as your vessels dilated, your body heat radiated away from the skin. The combination of sweat and increased blood flow is very efficient, for the sweat absorbs the heat from the blood, making it easier to evaporate. Thus the evaporation of the sweat takes the heat from your body. Perhaps the only exercise that is an exception is swimming; even though swimming is an excellent cardiovascular workout, the water in the pool will absorb your body heat, so there is no need to sweat!

Hair and Nails

Hair and nails are called *accessory structures*, but they do have important functions. Nails, for instance, make it much easier to grasp objects because of the added support to the fingertip. They also help to protect the small, fragile, distal phalanges (the bones of the fingertips and toes). Hair also has its functions, which include protection from heat loss and UV light (particularly on the head), as well as keeping bugs, and other foreign particles, out of your eyes, ears, and nose. Hair grows continuously, since you lose about 70 to 100 hairs off your head every day.

Nails are hard because of the layers of tightly packed keratinized cells. Similar to hair, there is a hidden *nail root*, and a *nail body*. The epidermis under the nail, especially near the root, is responsible for building more nail and pushing it forward at the rate of 1 mm a week. Ahead of the root, at the proximal end of the nail body, is the external *cuticle*, or *eponychium*, which is a layer of epithelium that helps attach the nail body to the skin behind it. Just in front of the cuticle is a whitish semicircle called the *lunula*, named because it looks like a little moon; its pale color is a reflection of a thick stratum basale, which blocks out the color from the blood vessels beneath.

We inherited our hair from nonhuman ancestors who had a greater need for insulation, hence the existence of the *erector pili muscle* attached to the hair. Goose bumps, which we get when we are cold, raise the hair and thus increase insulation. As to why we get goose bumps when we are frightened, that may be similar to a cat's raising its fur, thus making us look bigger and more threatening. An interesting note, not only do females and males have the same

Flex Your Muscles

You lose approximately 70 percent of your body's heat from your head, given the incredible profusion of blood to your head (and its not just for the brain). As such, long hair makes more sense in the winter than in the summer, given its ability to insulate. Think about that 70 percent the next time you go out in the winter without a hat!

number of hairs (although they may differ in size), but also humans have the same number of hairs as a chimpanzee!

The structure of a hair obviously includes the *shaft* (above the surface of the skin) and the *root* (below), but there is more to it than that. The base of the root (refer to Figure 16.1) forms a wide bulb, with an indentation at the base, the *papilla*, for capillaries. The shaft may be dead, but the root is very much alive and needs nourishment for the hair to continue to grow. Surrounding the actual hair is a double layer of cells called the *hair* follicle, made up of *external* and *internal root sheaths*, which are epithelial tissue that is continuous with the stratum basale of the epidermis.

Since most people know that the hair shaft is dead, they may make the mistake in assuming that it is not made of cells, when in fact it is basically long columns of dead cells, which, like those in the epidermis, are keratinized. A cross section of a hair shows three layers: the outer *cuticle*, the *cortex*, and the inner *medulla*. The cuticle is made of squamous cells arranged like shingles, but not like a house, with the edges pointing up. The medulla has, in addition to pigment granules, air spaces. The intervening cortex, which is the largest part of the hair contains long cells that will be filled with pigment granules in dark hair, but white hair will contain mostly air in these dead cells.

At the hair papilla are melanocytes, providing the hair with its color, melanin. Blond and red hair also contains melanin, but the melanin produced contains iron and more sulfur than brown hair. In order to make melanin, melanocytes use an enzyme called *tyrosinase*, which is produced less and less as people age, producing graying hair, or in the absence of all melanin, white hair. The age at which this happens varies dramatically. I am 40 years old, and my hair is almost completely gray/white (refer to Figure 7.7). I got my first gray hair when I was five, but I kid my students that it is their doing (don't tell them what I just said)!

The Least You Need to Know

- The skin is our body's first line of defense, and as such needs a quick means of self-repair.

- The cells in the epidermis need to be constantly replaced due to regular abrasion.

- Skin's elasticity is due to fibers in the connective tissue of the dermis.

- The oils in skin prevent water loss.

- Skin is important in temperature regulation due to blood flow so close to the surface.

Chapter 17

Keeping the Invaders Away

In This Chapter

- ◆ Nonspecific resistance
- ◆ Specific resistance
- ◆ Innate versus acquired immunity
- ◆ Antibodies

We are constantly at risk of infection from *pathogens*, bacterial, viral, and even fungal agents of disease. As far as protection from infection goes, the integumentary system is a bit brutish. Its form of protection is rather like building a 12-foot wall around a town. That might keep out most people whom you want to keep out, but chances are some people will get through. The immunity provided by the lymphatic system is much more sophisticated. Imagine a high-tech competitive software company. Due to the threat of piracy, such companies may invest in sophisticated biometrics, or body measurement, techniques to determine every person's identity (such as fingerprint, facial, and retinal scans). In other words, this level of protection is specific, rather than general.

You might remember my introduction to the lymphatic system in Chapter 12. The clever connection of drainage and immunity, thanks to those workhorses, the lymph nodes, managed to kill two birds, and countless pathogens, with one stone. In that chapter, however, I didn't cover the actual means by which the invaders are vanquished. This chapter explores the cells involved in protecting the body and how they do just that.

One of the first tasks is to be able to recognize self versus non-self. In other words, the body needs to know to attack only foreign cells or viruses. All cells and viruses display surface proteins and glycoproteins. The ability to recognize the combination of antigens specific to yourself, the *major histocompatibility complex* or *MHC*, is essential to this whole process. (MHC antigens are also called human leukocyte-associated antigens, since they were first found on the leukocytes.) Class I MHC antigens (or MHC-1) are found in all cells but the red blood cells (RBCs); Class II (MHC) antigens (MHC-2) are posted on the membrane of cells exposed to a pathogen. Despite how specific the MHC is and how sensitive the body is in responding to invasion, the entire process would quickly overwhelm the body were it not for some basic protective measures to keep pathogens out!

Keeping Out No One in Particular

Although I might have roughed up the skin a bit above (in reputation only), I should be fair and point out that other parts of the body offer general defensive measures, known as *nonspecific resistance* to disease. Think about it. Here you have this dead outer layer that you wear like armor, but you are in constant threat of invasion nonetheless. To stay alive you can't simply close yourself off. As an open system (one in which both matter and energy can travel in and out), you have to allow nutrients, fluid, and gases access to your living tissues, not to mention allowing light to reach your eyes. Each of these essential pathways is a vulnerability in the defense of your body.

Mucous membranes are especially helpful here, as part of a general mechanical protection from infection. Think about all that you are breathing in besides nitrogen and oxygen gases: dust (mostly dead skin cells), dirt, dust mite feces, and anything hitching a ride on the way, such as bacteria and viruses. In addition to moistening the air you breathe, mucus in a purely mechanical sense, traps microbes, thus preventing them from gaining access to the living membrane beneath (see Chapter 13). The hairs in your nose, called *vibrissae* (the same name given to the whiskers on a cat), when coated with mucus, help to trap larger particles and keep them away from the membrane. The cilia in the trachea also help to move the trapped pathogens up to the pharynx, where they may enter the gastrointestinal tract to be digested.

> ### The Big Picture
>
> Despite the acidity of vaginal fluid limiting bacterial growth, the presence of some bacterial flora also prevents yeast infections from developing, simply through the competition for space. A fascinating theory also explains the monthly menstrual flow, despite the expenditure of energy required to rebuild the uterine wall afterward. The loss of blood carries pathogens out of the vagina, thus limiting the chance of infection.

When I was little, my mother always told me that a good way to prevent infection with a small cut was to squeeze some blood out. The movement of blood out of the cut carries, if not all, at least most of the pathogens with it! Think of how many places humans release materials from the body: urethra, anus, and vagina. This outward movement of urine, feces, and vaginal fluid carries pathogens away from the body. Tear ducts release fluid in response to irritation, to flush out irritants. The stomach can also respond to pathogens by producing vigorous contractions, thus vomiting out any offensive material.

At the same time, various chemicals are produced that aid in the defense against pathogens. Gastric juices, with a highly acidic pH of 2 (give or take), kill many bacterial species, and can also break down bacterial toxins. The mild acidity of vaginal fluid helps limit bacterial growth. Bacterial and fungal growth is inhibited on the skin due to its surprisingly low pH (3 to 5) from fatty acids and lactic acid in the *sebum* released by the sebaceous (oil) glands. That's not the only chemical trick that skin has; sweat glands include *lysozyme* in their secretions, as do lacrimal glands (in tears), salivary glands (in saliva), and the nasal mucosa; this enzyme breaks down some bacterial species' cell walls.

Big Eaters

The body has built-in cleanup crews that are able to not only remove dead cells and cellular fragments, but to actually remove pathogens. First the cell must attach its membrane to that of the pathogen in a process called *adherence*, which sometimes involves long extensions of the cell membrane. Then these cells literally eat their way through the problem, which is why they are known as *macrophages*, which literally means "big eaters"! The process they use is called *phagocytosis* (see Chapter 4), in which they literally use the cell membrane to engulf their "prey," in a process called *ingestion*, forming a surrounding vacuole called a *phagocytic vesicle* or *phagosome*.

Some of these cells in the body stay put; in other words, they form part of a specific organ, as part of the normal operation of the tissues. Some examples include *alveolar macrophages* in the lungs (for the dust that gets past the mucous membrane), *microglia* in the brain, and *tissue macrophages* in bone marrow, lymph nodes, and spleen. These cells, called *fixed macrophages*, make perfect sense for those tissues that regularly make contact with pathogens or even foreign particles (see Figure 17.1).

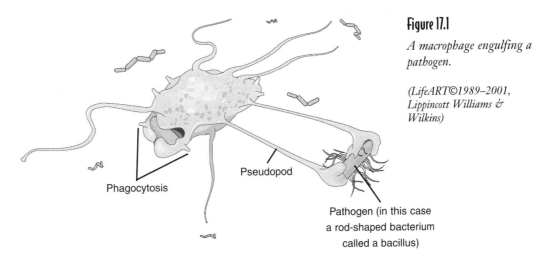

Figure 17.1

A macrophage engulfing a pathogen.

(LifeART©1989–2001, Lippincott Williams & Wilkins)

Phagocytosis

Pseudopod

Pathogen (in this case a rod-shaped bacterium called a bacillus)

On the other hand, those are not the only areas that may need macrophages from time to time. Another group, called *wandering macrophages*, have the ability to migrate to the infected site. Materials released from pathogens and from damaged tissues, as well as complement proteins (see the next section), act as chemical attractants for the macrophages in a process known as *chemotaxis*. Together the two types of macrophages are called the *reticuloendothelial system*, or the *mononuclear phagocytic system* (so called because macrophages develop from monocytes, see Chapter 10).

Flex Your Muscles

Chemotaxis makes more sense if you think about stepping on a tack. The cold metal introduces pathogens into your skin; at the same time, the tissues in the skin are damaged. The combination of the chemicals from the damaged cells and the wastes and toxins from the pathogen act as a signal beacon. *Wake up! There's an infection over there!*

Once inside the phagocytic vesicle, the vesicle fuses with a lysosome forming a *phagolysosome*. After breaking down bacterial cell walls with *lysozyme*, other hydrolytic enzymes tear the bacterium apart. The cell then absorbs the usable materials through the membrane of the phagolysosome, leaving wastes, in a *residual body*, that are removed through exocytosis (see Chapter 3).

With a name reminiscent of a slasher movie, *natural killer cells* (*NK cells*), a type of lymphocyte, attach to the cell membrane of damaged cells and pathogens (thanks to the good old MHC), and then release a protein called *perforin*, which groups together on the cell membrane, creating pores through which the cell contents escape, thus killing the cell. The NK cells themselves are protected by a protein aptly named *protectin*. The contents of the broken cells, victims of the NK cells, can be eaten by macrophages.

NK cells are responsible for *immunological surveillance*, in that they recognize the antigens presented on the surface of either pathogens or cancer cells. In that sense, it is possible that at any given time you may have a handful of cancer cells in your body, but that the NK cells will wipe them out. This is one of the reasons that certain rare forms of cancer, such as *Kaposi's sarcoma*, are found in AIDS patients, because their immune response prevents them from destroying the cells before they proliferate.

Running Interference

Sometimes the best defense is advanced warning. One way the body does this is through the release of *interferons* (*IFNs*). When a cell is infected with a virus, it is literally hijacked. Being made of only a protein coat and a nucleic acid core (DNA in viruses, RNA in retroviruses), a virus attaches to the outside of a cell and injects its core into the cell. The viral nucleic acid replicates (the RNA from a retrovirus first needs to be converted to DNA using the enzyme *reverse transcriptase*) and directs the production of more protein coats, thus turning the cell into a virus factory.

If a cell is unprepared, it is helpless to fight a viral infection, but that doesn't prevent it from practicing a form of cellular altruism! That cell can release interferons, and thus warn cells in the area of the coming viral onslaught. Since each infected cell can *lyse* (break open) and release hundreds of viruses, each capable of causing infection, it is important to stop the infection before it gets too far. By attaching to surface receptors, the interferons trigger the cell to do such things as suppress cell growth, or even call out "kill me now" to macrophages and NK cells. Slowing cell growth is important in terms of preventing some of the cancers that are triggered by viral infection, such as cervical cancer. Calling for macrophages and NK cells results in the destruction of those cells before they can release more viruses.

Inflammation

One sign of the stress of infection on tissues is inflammation. We've all experienced it, the painful redness, the swelling; you may even have felt the heat from the swelling. In some cases, such as joints, the swelling immobilizes the joint, preventing further damage. The first stage of inflammation involves dilation of blood vessels (*vasodilation*) and an increased permeability of the vessel walls. This allows for stage two, when the phagocytes migrate to the site of the injury, not to mention introduce antibodies to the injured tissue. You might remember from Chapter 10 that chemicals involved in clotting are also released here. The last stage involves the actual repair of the tissue.

Several chemicals are involved in inflammation. One of the most familiar is called *histamine*; released when tissues are injured, histamine helps with stage one of inflammation. This explains the helpfulness of antihistamines when one has swelling. *Kinins* help with stage one, as well as attracting phagocytes (chemotaxis). *Prostaglandins* also help call phagocytes, as well as strengthening the action of both histamine and kinins.

An interesting group of proteins involved in inflammation and other immune responses is called the *complement system*. In terms of inflammation, they not only stimulate the release of histamine, they also call phagocytes to the area (chemotaxis again). When some of these attach to pathogen cell membranes, they promote phagocytosis ("Eat me!") in a process called *opsonization*, or *immune adherence*. Not content to just call in the troops and say "eat that one," some complement proteins, in a decidedly uncomplimentary manner, can actually perform *cytolysis*, by using a number of proteins (the *membrane attack complex*) to break through bacterial cell walls, destroying the cell. Talk about taking matters into your own hands!

Prostaglandins and a substance called *interleuken-1* are responsible for another nonspecific defense: fever. Interleuken-1 is a *pyrogen*, or fever-causing chemical, that travels to the hypothalamus, triggering the release there of prostaglandins. The prostaglandins themselves help to turn up the thermostat. Metabolic increases (to generate heat), vasoconstriction (to prevent heat loss), and even shivering lead to higher body temperature. Fever

---

The Big Picture

When one gets a chill, the vaso-constriction in the skin will make the skin feel cold. The body will also resort to shivering in order to raise the body temperature further, which is why you can feel cold, even with a fever!

accomplishes several things, such as inhibiting pathogen growth (depending on the species), helping the work of interferons, and putting body repairs on a fast track.

The Smallest Army

Although many defenses are nonspecific, some of the glory of the immune system is the ability to mount specific defenses. In other words, the immune component of the lymphatic system can mount an attack on a specific pathogen. This is the reason why people tend to get chicken pox, or a particular cold, only once, for the body can recognize some past aggressors and stop them before they can mount a full-blown infection. The twentieth century also brought the development of vaccines, which take advantage of these immune abilities. This section explores both *cell-mediated immunity* and *antibody-mediated immunity*.

Innate Versus Acquired

Specific immunity can be divided two ways: active versus passive, and naturally acquired versus artificially acquired. Active immunity involves B and T cells (discussed in the next two sections) recognizing foreign antigens and preparing an attack response (cell-mediated and antibody-mediated) to them. Passive immunity, on the other hand, involves antibodies received from another person; in other words, the body doesn't actually actively produce the antibodies, it just receives them passively. Both types of immunity exist naturally and artificially (through human intervention).

Naturally acquired active immunity is what people usually think of in terms of immunity: You get infected, your body responds. *Artificially acquired active immunity* is the basis of vaccines. Injecting a person with a pathogen that has been altered to make it inactive, enough to introduce the antigen, but not to start an actual infection, causes the body to start to mount a response, including creating antibodies. The fact that certain cells retain a memory of the pathogen-specific antibody allows the body to respond quickly to a repeat of the infection; this not only explains why people only get chicken pox once, it also means that if you get exposed to the pathogen for polio, after being vaccinated, your body will quickly produce antibodies, destroying the pathogen before you get sick.

The development of vaccines has been made an incredible difference in the spread of disease. In the first three decades of the last century, many kids died of childhood illnesses, such as diphtheria or whooping cough (pertussis). Parents used to be afraid to have their kids go to the swimming pool in the summer out of fear of getting polio. Those fears are now gone, and graveyards have far fewer infant and childhood tombstones, all due to vaccines.

Naturally acquired passive immunity involves receiving antibodies from another person, which occurs naturally only two ways: from mother to fetus across the placenta, and from mother to baby through breast milk. The latter form is one of the reasons why breastfeeding, if medically possible, confers greater advantage to a newborn than bottle feeding. *Artificially acquired passive immunity* is similar to vaccines in that it involves injection, but differs in that the patient only gets antibodies.

Medical Records

The rise of vaccines and antibiotics led many to believe that we would conquer disease. Antibiotic resistance, which developed mainly through people misuse, explains why bacteria are, in some cases, getting the upper hand. Despite the possibility of terrorist use of the virus, smallpox was eliminated from human populations, but only because humans are the only vector for the virus. Other viruses have nonhuman vectors in the wild, making them impossible to eliminate. We might be able to keep diseases in check, but they will always be with us.

T Cells

T cells are so named because they are produced in the thymus. The decrease in size of the thymus gland as people age explains the increased incidence of cancer in the elderly, as their bodies can no longer mount a good surveillance for developing cancer. The action of T and B cells are the basis of cell-mediated immunity. One of the important things about some immune cells is the ability to present antigens on their surface, thus giving them the name *antigen presenting cells* (*APCs*).

This is important because for a cell to be foreign the DNA must be foreign, but as the DNA is within the nucleus, the body cannot go about breaking open all the cells just to scan the DNA. Luckily, since DNA is the recipe for protein, foreign DNA results in at least one foreign antigen on the cell membrane. By recognizing that antigen as foreign, the job is done. In addition, when an infected cell presents an antigen from the pathogen on its surface, it signals the body to destroy it. When macrophages eat pathogens, they also present the foreign antigens, like a trophy, on their surface ("Hey T cells, look what *I* ate!").

There are several types of T cells. My personal favorite are called *cytotoxic* (or *cytolytic*) T cells, but I like their other name: *killer T cells!* As their name implies, they *kill* through the use of perforin, like the NK cells. They also call the wandering macrophage reinforcements in, and then inhibit their migration away from the infection site. The rise of the killer Ts is stimulated by a chemical called *interleukin* 2 from T_4 helper cells, which are the ones that are targeted by HIV, the virus that causes AIDS. Ironically, the making of interleukin 2 is a result of interleukin 1 made by the macrophages after eating the pathogen.

Helper T cells also combine with B cells when a pathogen is present and stimulate the B cell to become a *plasma call*, which produces antibodies (see Figure 17.2). When the infection seems to be under control, *suppressor T cells* inhibit the immune response, even to the point of killing some of the lymphocytes. One clever little device in the body is the ability to remember foreign antigens, and be prepared for another attack. One type of such cell is a *memory T cell*, which hangs around for years.

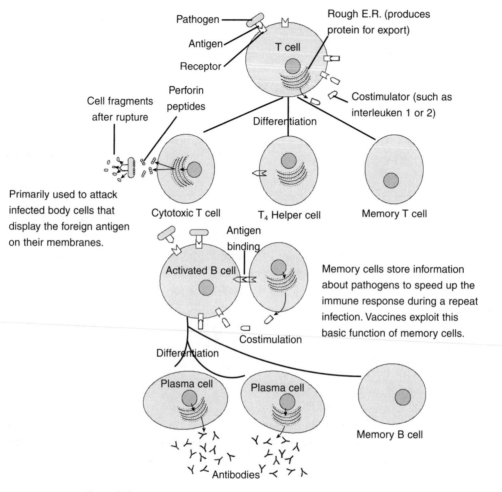

Figure 17.2

This diagram illustrates the connections between the T cells and the B cells.

(©Michael J. Vieira Lazaroff)

B Cells

B cells are so named because they are produced in the red bone marrow. This is the reason why some cancer treatments involve bone marrow transplants, or the removal of bone marrow prior to chemotherapy, with replacement later. Just like the memory T cells, there are *memory B cells*, which lie in wait for the next repeat infection. As I said earlier, the helper T cells work with the B cells to form the *plasma cell*, which is the mature B cell that makes the antibodies.

When the pathogen attaches to the B cell it becomes activated. The T_4 helper cell then joins up with the B cell, by a molecule that attaches to receptors on both cells. The helper T cell's release of the autocrines (local hormone) interleukin 2, 4, and 5 act as *costimulators* that lead to the differentiation (and subsequent proliferation) of B cells into plasma cells.

Antibodies

I have mentioned antibodies, also known as *immunoglobulins*, quite a number of times, but I provided very little information about what they are and how they work. Given their incredible variety of shapes, it makes sense that the molecule of choice for making antibodies would be proteins, the most variable of all organic molecules. Given that the mechanism for making proteins, mRNA, and ribosomes (see Chapter 3), all that the cell needs to do is to alter the instructions (DNA) for making the antibody. Each cell has a variety of genes that are used to create parts of antibodies, and combinations of the genes are used to create an antibody for a specific pathogen. The combination of genes in a person helps explain why some people may be genetically better able to fight off certain infections, for apart from individual differences in health, not everyone infected with a particular pathogen dies.

There are five classes of antibodies, but the vast majority (around 75 percent) are in a group class G immunoglobulins (IgG). They have either a roughly Y-shaped or T-shaped configuration made of two longer heavy chains, and two shorter light chains (see Figure 17.3). Most of an antibody's shape is constant, despite the pathogen it is for, but the very tip of each "Y" is variable; that variable region is known as the *antigen-binding site*. Despite being complex polypeptide chains (each heavy chain has around 450 amino acids, and each light chain around 200), three of the five classes are described as monomers because the large form is a single unit; one form, however, is a dimer (2), with antigen-binding sites on both ends, and another form is a pentamer (5) with sites on each of the five ends.

Medical Records

Have you ever gotten sick when you are under a lot of stress? Remember how mucous membranes are often the location of a pathogen's entry into the body? Well, IgA antibodies are found on mucous membranes, and in many body secretions (milk, mucus, saliva, tears, and so on). The quantity of these antibodies goes down when we are stressed, which leaves us vulnerable.

Antibodies act in more than one way. In some cases the antibodies cluster around the pathogen, which signals macrophages to chow down. In other cases the attachment of the antibody to the pathogen directly causes the destruction (lysing) of the pathogen. Antibodies sometimes break down bacterial toxins. The complement system, described earlier, is triggered by the release of some antibodies. Some antibodies are found on the surface of B cells, acting as antigen receptors; those receptors are used to activate B cells, which end up having all the fun!

Figure 17.3

The general shape of an antibody.

(LifeART©1989–2001, Lippincott Williams & Wilkins)

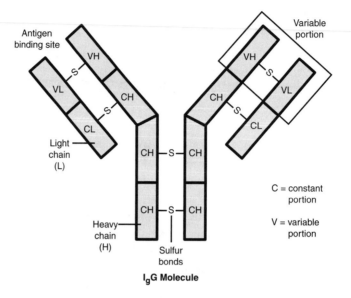

The Least You Need to Know

◆ The body has two basic ways of fighting infection: nonspecific resistance and specific resistance.

◆ Nonspecific resistance involves the use of the skin, mucous membranes, macrophages, inflammation, fever, and certain chemicals to protect the body from invasion, without the need to specifically identify the invader.

◆ Specific resistance involves the use of B cells and T cells to mount an attack on a specific pathogen, using cells and antibodies.

◆ The ability to store instructions for making specific antibodies in a cell's memory not only protects us from future infection, but also is the foundation for the use of vaccines.

Part **6**

Who's in Charge

Sounds like a simple question, doesn't it? Now think about the federal government. Who *is* in charge? A quick look at the constitution shows that this is by no means an easy question to answer. With the separation of powers, the three branches of government, executive, legislative, and judicial, are each involved in running the country.

In the same sense, our body subdivides control of itself. For the instant gratification types, there is the rapid response team of the nervous system. You have a tack that needs stepping on? Well, your nervous system will give you all the painful details! You have a pop quiz (evil little suckers, those teachers) and your heart needs to speed up. Your nervous system has got it covered.

What about all the stuff that isn't urgent? The maintenance, the regulation, the slow and easy stuff? Your blood helps out here by carrying chemical messages back and forth. Hormone highway! From announcing your presence in the womb, to driving your parents crazy as they ride your teenage mood swing roller coaster, to makin' babies, your hormones are always on the go. As we should be. So let's get on with it!

Chapter 18

Raging as You're Aging

In This Chapter

- ◆ Feedback loops
- ◆ Hormone types
- ◆ Hormone actions
- ◆ The glands and their hormones

When you are a single cell, you can easily control what you are doing, but things get awfully complicated when you are multicellular. It is very important that the body be able to control the various goings-on in the tissues. This is especially important as the organs and body systems do not function in isolation; each depends on the others for both general needs (such as oxygen) and specific needs (such as when the gall bladder releases bile after eating fat).

The body has two ways of controlling the body tissues. One is through the nervous system, with its miles and miles of neural pathways. This type of control has the advantage of being very rapid, as you know if you have ever put your hand on a hot burner! The other, slower way involves hormones, which are a collection of chemicals that alter the behavior of specific organs and tissues on the cellular level. This type of control system relies on the blood to disseminate its chemical messages. Welcome to the glory that is the endocrine system.

Feedback: Not Just for Guitars

One of the most important things to understand is that hormones do not work alone. To regulate the release of hormones, the endocrine organ that's releasing the hormone needs to be able to adjust its output by responding to chemical changes, be it altered calcium ion levels, in the case of calcitonin, or by altering levels of another hormone, as in estrogen inhibiting a hormone that regulates gamete production.

Regardless of the chemicals involved, the mechanism is pretty much the same: Hormones are regulated by feedback loops—receptor, control center, and effector (see Chapter 1). With the few exceptions, such as oxytocin and labor contractions, hormones are regulated by negative feedback loops, which is ideal in terms of maintaining equilibrium.

In terms of hormones, the effector gets another name: *target organ* or *tissue*. Every hormone is produced by an organ, some of which also belong to other systems (such as the pancreas, which you might remember is also digestive). Receptors on the surface of cells in the endocrine organ act to receive information from the environment and then signal the control center. The organ that produces the hormone is the control center; more specifically, operons in the nuclei of cells in that organ act as the control center, turning a gene on or off and thus instructing the cell to either make or stop making a hormone.

The hormone, in turn, after being picked up by veins leaving the endocrine organ, travels all over the body thanks to the cardiovascular system. Eventually some of the hormone makes it to an artery entering the target organ (or tissue), which in turn responds to the hormone. That response could be as simple as increasing the growth rate, as in the case of growth hormone, or it could mean the production of another hormone!

For example, a hormone from the *hypothalamus* (I'll call it hormone A) targets the *hypophysis* (pituitary gland), and the pituitary gland then produces its own hormone (hormone B). That pituitary hormone (hormone B), of course, targets another organ, which in turn produces its own hormone (hormone C)! As if that isn't bad enough, that last hormone (hormone C) is part of a feedback loop and targets the hypothalamus. It turns out that hormone C inhibits the production of hormone A by the hypothalamus, the one that started this whole mess off!

When the production of hormone A is inhibited, the level of hormone B drops, which in turn stops (or at least slows down) the production of hormone C. The low level of hormone C is not able to inhibit production of hormone A, which is then produced, and the whole merry-go-round starts all over again! To be fair, hormones, especially hormone C in this scenario, will affect other organs and tissues: regulating ion levels, blood pressure, you name it! It is, nonetheless, important to understand the role of hormones (such as hormone A and B) in the regulation of hormones themselves!

Types of Hormones

Hormones are classified in different ways. One way depends on the location of the target cells, which divides them into autocrines, paracrines, and endocrines (see Figure 18.1). Autocrines, simply put, are hormones that act only on the cell that produces them (*auto* = self); the T-cell costimulator interleuken-2 (see Chapter 17) is one example, which is released in conjunction with antigen attachment to trigger T cell function. Paracrines are local hormones that act on neighboring tissues (*para* = next to); histamine—which is released in response to an allergen, such as the toxin in a mosquito bite—causes swelling only in the area next to the allergen. Endocrines, the hormones of this chapter, are hormones that act on tissues and organs that may be quite distant from the organ producing the hormone, and they travel via blood vessels.

Figure 18.1

Hormones differ by the location of the target cells.

ENDOCRINES

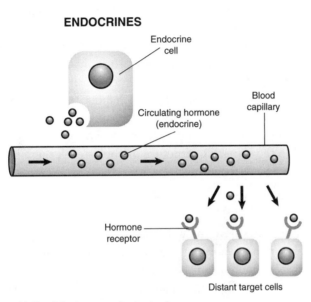

(a) Circulating hormones (endocrines)

PARACRINES

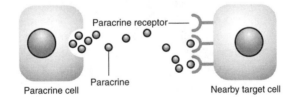

AUTOCRINES

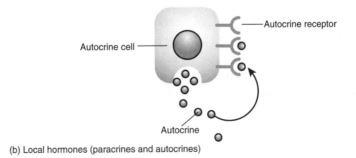

(b) Local hormones (paracrines and autocrines)

Exocrine glands, by contrast, release their products directly into ducts. Lacrimal glands (see Chapter 21), for example, release tears into tear ducts. You have also seen such glands in the mouth (salivary glands, see Chapter 14) and the skin (sudoriferous, sebaceous, and ceruminous glands, see Chapter 16). The pancreas, my favorite, is both endocrine (insulin and glucagon) and exocrine (digestive enzymes), one of many organs that fit into more than one system.

Hormones also differ in terms of their basic chemical structure. Biogenic amines are modified amino acids and are the smallest of the hormones; some examples of biogenic amines are two thyroid hormones, T3 and T4, and epinephrine (adrenaline). Eicosanoids come from arachidonic acid (one of the fatty acids) and are primarily local hormones, such as prostaglandin (see Chapter 23). Peptides (short amino acid chains) and proteins (longer amino acid chains, with more complex structure, see Chapter 2) are the largest of the hormone types; most endocrine hormones are from this group, including insulin, glucagon, and all hormones from the hypothalamus and hypophysis. Steroids are lipid-based hormones that are all related to the four-ring structure of cholesterol; these hormones include the sex hormones, estrogen and testosterone.

Before going too far, I should let you know a little bit about naming hormones. Some of the names get very long. Because of this, physiologists found a simple way to speak in a sort of shorthand. Most hormones have the word hormone, so a capital *H* always means "hormone." Since many hormones were first discovered in other animals, a handful of hormones have a lowercase *h*, which means "human." In this way you can easily distinguish human growth hormone (hGH) from its analog in cows, bovine growth hormone (bGH).

The Big Picture

bGH, also known as bovine somatotropin (bST), is given to cows in order to trigger the production of more milk. In and of itself this seems like a harmless idea, especially since none of the hormone is found in the milk that is produced. Upon further examination, however, the extra milk production leads to infections in the mammary glands called mastitis. Treatment of mastitis involves the use of antibiotics that can ultimately be found in the milk, potentially increasing the development of antibiotic resistance.

Some hormones stimulate an organ to act or to release a hormone (S = stimulating, as in FSH = follicle-stimulating hormone) or both, whereas others inhibit the release of a hormone (I = inhibiting, as in PIH = prolactin-inhibiting hormone). Stimulating hormones are a product of the pituitary, whereas hormones from the hypothalamus that stimulate the release of pituitary hormones are called releasing hormones (R = releasing, as in GnRH = gonadotropic releasing hormone). The following table provides a quick summary of shortcuts used to name hormones.

Shortcuts to Hormone Names

Shortcut	Meaning	Example
H	Hormone	GH = Growth hormone
h	Human	hGH = Human growth hormone
S	Stimulating	TSH = Thyroid stimulating hormone
I	Inhibiting	PIH = Prolactin inhibiting hormone
R	Releasing	TRH = Thyrotropin releasing hormone

How Hormones Work

The basic idea of every hormone is that a chemical from one cell can alter the behavior of another cell (except for autocrines, in which the cell alters its own behavior). To understand the basic principle of how this is done, you have to look at the structure of the cell membrane (see Chapter 3). Those ubiquitous proteins are more than just for transporting materials and for cell recognition. Some of them act as receptors for certain chemicals, such as hormones.

Remember that all cells, except gametes and red blood cells, have a complete complement of DNA, and are thus theoretically able to make all the proteins within their genome (the sum total of all their genes). During development, however, the cells differentiate; certain genes are turned on and others are permanently turned off. As some of those genes produce proteins that act as receptors for certain hormones, the presence of such a protein on a cell membrane makes that cell, and the tissues or organs that are comprised of such cells, targets for those hormones. In this way, hormones that travel throughout the bloodstream may target only a very small part of the body, as in the case of certain hypothalamic hormones acting only on the pituitary, which is no bigger than a pea!

The location of such receptors says a lot about the type of hormone involved. In the case of steroid-based hormones, for example, they are lipid soluble (being lipids themselves), and so they pass right through the cell membrane. The thyroid hormones, as biogenic amines (altered amino acids), are small enough to be able to pass freely through the cell membrane. Once the hormone enters the cell, it travels to the nucleus and binds to the receptor portion of the repressor protein attached to the operator of the operon (see Chapter 3). By doing so, the repressor becomes inactive and detaches from the operator, allowing mRNA to ultimately transcribe the gene, and a protein is ultimately made. Thus the hormone is directly responsible for the production of that protein, which might in fact be another hormone! Whatever the nature of the protein produced, it will alter the activity either of that cell, or of other cells, thus bringing about the desired response.

Protein-based hormones, on the other hand, cannot pass through the cell membrane, in part because they are not lipid soluble; their action is mainly based on their ability to trigger actions within the cell. These peptides and proteins bind to receptors on specific cells.

As a result of that binding, a number of actions are triggered inside the membrane and the cytoplasm, which, ultimately, cause the programmed effect of the hormone. In this situation the hormone acts as a *first messenger*, but since it can only act on the plasma membrane, a *second messenger* is required.

There are a number of steps, but once again, Figure 18.2 will help. After the arrival of the first messenger, a G-protein on the inside of the membrane activates an enzyme called *adenylate cyclase* to convert ATP (adenosine triphosphate) into cyclic AMP or cAMP (cyclic adenosine monophosphate). cAMP plays the role of the second messenger, which activates protein kinases. These protein kinases, in turn, help to strip a phosphate off of ATP (thus making ADP), and add it to other enzymes; the enzymes are thus phosphorylated. It is the action of these phosphorylated enzymes that results in the hormones' ultimate effect. In some cases a protein kinase may inhibit enzymes, rather than activate, depending on the hormone's action.

Give the Glands a Hand

The release of hormones is the result of several processes that make up the *endocrine reflexes*. This term makes sense, if you think about it, because these responses really need to function like reflexes; they need to be involuntary. Like reflexes, they function on a stimulus/response system, and the response is always the release of a specific hormone. The stimuli that trigger the response, however, may vary.

In some cases the stimuli may be *humoral stimuli*, which means that changes in the concentration of ions or molecules in the extracellular fluid (as in parathyroid hormone's release due to extracellular Ca^{2+} concentration) can cause a response. *Neural stimuli* refer to the arrival of a neurotransmitter at the junction of a neuron and an endocrine gland (as in the case of the release of epinephrine from the adrenal gland after stimulation from a sympathetic nerve; see Chapter 20). The last group of stimuli are called *hormonal stimuli*, in which the arrival of one hormone either stimulates or inhibits the release of another. This is the main means by which the hypothalamus and hypophysis communicate, and by which the hypophysis communicates with other glands.

Despite all these different stimuli, hormones do share a basic tendency to be regulated through negative feedback loops (see Chapter 1). In this way, hormone levels stay balanced, and the body can maintain homeostasis. Although there are some positive feedback loops, such as the oxytocin (OT) loop with the uterus, they each need to be broken by a specific event, or the release of the hormone would just keep increasing.

With the concepts of the types of hormones, and the way they work, especially the nature of the feedback loops, all that is left is to look at the glands that make up the endocrine system. The pituitary, with its bulb shape on a long stalk called the *infundibulum*, and its multiple hormones targeting multiple glands, made a great candidate for a master gland, the nickname it held for many years. The protective bone case provided by the sella turcica of the sphenoid bone (see Chapter 6) helped maintain the idea.

Further research, however, established that the release of various regulating hormones controls the hormones of the pituitary. Behind the stalk, there is another gland below the thalamus, hence the name *hypo*thalamus, that really *is* in charge. Releasing hormones and inhibiting hormones, the products of the hypothalamus, take the title of master gland away from the pituitary.

Hypothalamus: The Head Honcho

To get some perspective, you need to understand that the hypothalamus is really the main regulatory center in the body. That regulation involves both hormones and the autonomic nervous system (see Chapter 20). In fact, the hypothalamus acts as a link between the endocrine system and the nervous system. The location of the hypothalamus, as part of the most primitive portions of the brain, is related to the fact that evidence from the animal kingdom points to the ancient nature of many of our hormonal loops.

The Big Picture

To best understand the hypothalamus, you need to grasp how several concepts intersect. It's location, connecting it to both the endocrine and nervous systems, plays an important role here. The general area, surrounding the third ventricle, is related to the lack of the blood-brain barrier there (see Chapter 19), allowing full access to the contents of blood plasma. This, in turn, is necessary to the hypothalamus's role in maintaining homeostasis. By receiving info from the blood, the hypothalamus releases hormones that stimulate the pituitary, that in turn releases hormones that affect homeostasis. A nice little trick!

All endocrine organs have their hormones to enter a blood vessel (a vein from the endocrine gland) in order to ultimately reach their target organ (through an artery). The only problem with that is that the blood must travel all the way to the heart, then to the lungs, and then back to the heart, and get pumped out again before it gets to where it's going. In general, that's not much of a problem, despite seeming so inefficient, because the blood needs to go that way anyway. But what if the two organs are only a few millimeters apart, as in the hypothalamus and the hypophysis?

The answer to this dilemma is a nifty little portal system. In Chapter 12 you learned that a portal system involves a connection between two capillary beds, thus bypassing the return to the heart, as in the hepatic-portal system connecting the capillaries of the abdominal organs and those of the liver. The hypophyseal portal system is much smaller, connecting the capillaries of the hypothalamus and the hypophysis. In this sense, hormones from the hypothalamus only have to travel a short distance to get to their target (see Figure 18.2).

Figure 18.2

Note that the release of hormones from the hypothalamus affects the release of hormones from the hypophysis (pituitary).

(LifeART©1989–2001, Lippincott Williams & Wilkins)

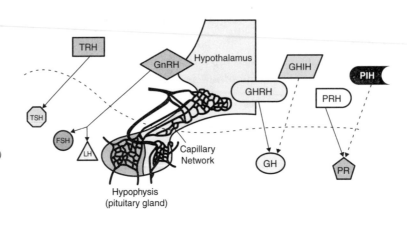

RH = Releasing Hormone, which promotes the hypophysis to produce and release a specific hormone.

IH = Inhibiting Hormone, which prevents another hormone's release. Note: The dotted arrows indicate that IH levels are low when RH levels are high (and vice versa).

The following table indicates which hypothalamic hormones *trigger* the release of pituitary (hypophyseal) hormones, and which *inhibit* their release. This brings up the point that the hypothalamus, in general, releases regulatory hormones (that is, hormones that regulate the release of other hormones). Hormones that cause the release of others are called *releasing hormones* (RH), and those that inhibit are called *inhibiting hormones* (IH). The following table lists hormones of the hypothalamus and the pituitary hormones they regulate.

Hormones of the Hypothalamus

Hypothalamic Hormone	Pituitary Hormone	Final Target
TRH: Thyrotropin RH	TSH: Thyroid SH	Thyroid gland
CRH: Corticotrophin RH	ACTH: Adrenocorticotropic hormone	Adrenal cortex
GnRH: Gonadotropic RH	FSH: Follicle SH testes cells	Follicle-ovary
	LH: Luteinizing H testes cells	Follicle-ovary
PRF: Prolactin releasing factor	PRL: Prolactin	Mammary glands
PIH: Prolactin IH	PRL: Prolactin	Mammary glands
GH-RH: Growth hormone RH	hGH: Human growth H	All body cells
GH-IH: Growth hormone IH	hGH: Human growth H	All body cells
MSH-IH: Melanocyte SH-IH	MSH: Melanocyte SH	Melanocytes

There are, however, two exceptions. Antidiuretic hormone (ADH) and oxytocin (OT) are both produced in the hypothalamus, and then stored in the posterior pituitary gland (neurohypophysis), from which they are released. For this reason they are considered pituitary hormones, although that is a bit misleading. The hormones are produced by neurons in the hypothalamus, and then carried via slow axonal transport (they don't have far to go, after all) to the basement membranes of the capillaries in the neurohypophysis, making their distribution to the blood quite easy. The name of this part of the pituitary comes from the fact that these hormones are produced by neural tissue, rather than epithelial tissue, as in the anterior pituitary (adenohypophysis).

Pituitary: The First Lieutenant

Now that you've met the "master gland's master gland," you'll meet the first lieutenant: the pituitary gland (hypophysis). As explained in the previous section, there are two parts to the pituitary: the anterior section (adenohypophysis), and the posterior section (neurohypophysis).

The anterior pituitary has many more hormones than the posterior, and they differ in that they are actually made there, making them true pituitary hormones, rather than axonal carpetbaggers (see Figure 18.3). The cells of the hypophysis are divided according to the cells that make them. If you use a little logic, I bet you can figure out which hormones are made by which cells. (Just use the little tricks to vocabulary that I have been teaching you all along!)

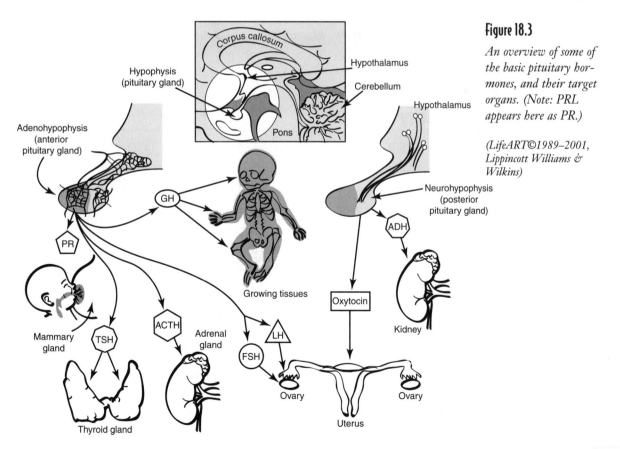

Figure 18.3

An overview of some of the basic pituitary hormones, and their target organs. (Note: PRL appears here as PR.)

(LifeART©1989–2001, Lippincott Williams & Wilkins)

Medical Records

Human growth hormone (hGH) is released throughout a person's life, with little bursts just prior to dawn (causing a rise in the fasting blood glucose of diabetics called *dawn phenomenon*). If the amount of hGH is too low in one's youth, it leads to *pituitary dwarfism*, and if it's too high in youth, it leads to *gigantism*. Once we reach adulthood, the epiphyseal plates (where bone growth occurs, see Chapter 5) have ossified everywhere except the hands, feet, and face. In adulthood, too much hGH leads to growth in those three areas, a condition called acromegaly.

Somatotrophs (*soma* = body) make the hormone that affects every cell in the body: human growth hormone (hGH). Lactotrophs (this one's easy) affect lactation (the making of breast milk), and so they make prolactin (PRL). Corticotrophs affect a cortex, in this case producing adrenocorticotropic hormone (ACTH), which targets the adrenal cortex, and melanocyte-stimulating hormone (MSH); melanocytes are found in the equivalent of the cortex of the skin—the epidermis. Thyrotrophs release thyroid-stimulating hormone (TSH), of course! The last ones, my favorites, are called gonadotrophs, which release follicle-stimulating hormone. (FSH) Thyrotrophs release and luteinizing hormone (LH), both of which target the gonads (ovaries and testes). The following table lists the effects of the multiple pituitary hormones.

The Effects of Pituitary Hormones

Hormone	Target	Effect
ACTH	Adrenal cortex	Glucocorticoids secreted
ADH	Kidneys	H_2O reabsorbed, blood volume up
FSH	Ovaries (females)	Follicle dev., estrogen is made
	Testes (males)	Sperm development and maturation
hGH	All tissues	Growth, anabolism, and catabolism
LH	Ovaries (females)	Ovulation and corpus luteum is made
	Testes (males)	Production of testosterone
MSH	Melanocytes	Increased melanin production
OT	Uterus, mammary glands (females)	Labor contractions and milk ejection
	Vas deferens and prostate (males)	Contraction of vas deferens and prostate gland
PRL	Mammary glands	Lactation (milk production)
TSH	Thyroid glands	Thyroid hormones produced

Pining for the Pineal

The *pineal gland* (*epiphysis cerebri*) is posterior to the pituitary (but higher up, as the name epiphysis implies), behind the brain stem, kind of hidden back where the two hemispheres of the cerebellum meet. This little wonder, so deep in our head, is not too far from where our visual pathways make it to our occipital lobe. Given this gland's function, it makes perfect sense!

The pineal gland controls our circadian rhythm (day-and-night cycles). As you can imagine, this depends on the amount of light that reaches our eyes. We don't think about it, because night and day are, well, out of our control. If you ever live near either pole, however, you will experience prolonged days and nights. (In the extreme, you will have one six-month day, and one six-month night a year!) As such, people respond by altering the amount of artificial light to compensate.

The hormone responsible for this is called *melatonin*, which is produced by *pinealocytes*, but only in the dark. This helps explain *seasonal affective disorder* (with the all-too-clever acronym *SAD*), in which a lack of sunlight in winter leads to the overproduction of melatonin, and ultimately to depression. The treatment, literally, is *more light*, quite simply to reduce the amount of melatonin. Although little understood, melatonin slows the release of GnRH and is thought to play a role in the onset of puberty.

The Big Picture

Many years of nature videos as a kid always left me with the question: Why do so many animals mate in springtime? It always made sense, because there was more food for the young then, but how was the body kept infertile in the winter? Yeah, I was a weird kid for asking such questions, but what can you do? The answer is elegant in its simplicity: 1) Winter is darker, 2) darker means more melatonin, 3) melatonin inhibits GnRH, 4) less GnRH makes you infertile. It's so simple it's cool!

Thyroid

Since I talked about the hypophysis and its secretion of TSH (thyroid-stimulating hormone), it seems only fitting that I talk about what TSH actually does (see Figure 18.4)! First of all, the thyroid gland is anterior and inferior to the thyroid cartilage, which itself is anterior to the larynx. The thyroid itself has *left and right lateral lobes*, connected by a thinner *isthmus*, which, in turn, sometimes has a small, vertical, pyramidal lobe extending upward. Posterior to the thyroid, next to the trachea, are four small glands called the *parathyroid* (*para* = next to).

On a microscopic scale you will see thyroid follicles. When inactive, the cells are squamous, but that good ol' TSH turns 'em into cuboidal and columnar cells that are ready to churn our hormones! In the middle of this follicle is a colloid (think egg white) filled with a protein called *thyroglobulin*, which contains many tyrosine amino acid molecules. Remember how some hormones are biogenic amines (altered amino acids)? Well, you're about to meet one.

Figure 18.4

The feedback loop between the thyroid and the pituitary.

(© Michael J. Vieira Lazaroff)

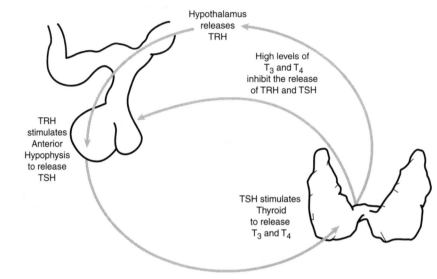

Hypothalamus
releases
TRH

High levels of
T_3 and T_4
inhibit the release
of TRH and TSH

TRH
stimulates
Anterior
Hypophysis
to release
TSH

TSH stimulates
Thyroid
to release
T_3 and T_4

Next I want you to think about how salt in the grocery store is almost always "iodized." This addition of iodine is important, because we need it for our thyroid glands to make two of its hormones: T_3 and T_4 (formerly thyroxine). A good source of dietary iodine is fish, but people in certain areas of the world rarely get fish. Without enough iodine the thyroid gland enlarges, producing a goiter in one's neck. Iodized salt was the simplest solution to this problem.

Medical Records

The release of too much thyroxine (T_4) is called *hyperthyroidism;* releasing too little is called *hypothyroidism.* Excess thyroxine greatly increases one's metabolism, and is marked by weight loss, excitability, and high blood pressure. The extremely slow metabolism from low thyroxine has the opposite effects: weight gain, lethargy, and low blood pressure.

The first step in making these hormones is to bind iodine ions to tyrosine molecules. A pair of these, when joined, becomes T_2 (a reference to the pair of iodized tyrosines). Joining a pair of this, of course, makes a T_4 (also called *tetraiodothyronine*), or a T_2 and a single iodized tyrosine makes T_3 (also called *triiodothyronine*). Most of the T_3 and T_4 releases are bound to various plasma proteins, thus keeping it in reserve for release into peripheral tissues.

T_3 and T_4 are lipid soluble, so they enter the cell directly. Much of the activity involves energy usage. These hormones attach to the mitochondria (see Chapter 3), thus increasing the amount of ATP produced. These hormones also trigger (via operons) the production of enzymes for glycolysis and ATP production. The overall effect is to increase the metabolic rate, as well as the production of body heat.

Young children, due to their small size and large surface area to volume ratio (see Chapter 3), risk losing heat in cold weather, so TSH production rises, thus raising the child's body heat! Pretty cool!

Parathyroid

There is one other thyroid hormone, *calcitonin (CT)*, which works in conjunction with *parathyroid hormone (PTH)*. In response to high blood Ca^{2+} and phosphate (PO_4^{3-} and HPO_4^{2-}) levels, CT, which is produced by the *parafollicular cells* (next to the follicles, of course) of the thyroid gland, reduces blood Ca^{2+} and phosphate levels by stimulating osteoblasts to build bone (see Chapter 5). PTH, on the other hand, stimulates osteoclasts to break bone (once again, Chapter 5) down, thus releasing it into the blood and raising blood Ca^{2+} and phosphate levels. PTH, from the parathyroid, also increases reabsorption (see Chapter 15) of Ca^{2+} and phosphate in the kidneys (thus reducing Ca^{2+} and phosphate levels in the urine). The kidneys, in addition, release *calcitrol*, which increases the absorption of Ca^{2+} in the GI tract. As such, calcitrol, PTH, and calcitonin are all regulated by blood Ca^{2+} and phosphate levels via negative feedback loops.

Thy Must? Thymus!

When we dissect the cats in my anatomy and physiology class, I always love the day we open the thoracic cavity. There, above the heart, attached to the mediastinum (see Chapters 11 and 13), is the unassuming thymus. I always get the students to compare the size of their thymuses (thymi?). This strange little organ grows, of course, as the animal does, reaching maximum size at puberty, and then this bizarre little organ actually gets smaller after puberty; so you can, in relative terms, gauge the age of an animal by its thymus.

You can learn a lot from this little tidbit, because the thymus is involved in immunity, as it is also an organ of the lymphatic system. The seven hormones of the thymus are called we *thymosins*, but they used to be considered a single hormone (thymosin). The target cells are lymphocytes, and the hormones are involved in the development and maturation of lymphocytes. When you are very young, your body is exposed to many pathogens, as well as just typical bacterial flora, for the first time. The large size is representative of the busy job the organ has at that age.

As your immune system develops, in general it has fewer new challenges as you age. In the long term, however, your immunity drops as you enter your later years, along with your shrinking thymus, making you more vulnerable to colds, which often last longer. A flu that is merely debilitating can be lethal to the elderly. Lastly, the lowered immunity makes it harder for the body to seek out and destroy errant cancer cells, thus increasing one's cancer risk.

Kidneys: Not Just for Urine!

Homeostasis is an amazing thing, especially for the sentient; to be sentient, by definition, is to be self-aware. Despite all our emphasis on our own thoughts, when we are conscious of hunger and thirst, we are merely responding to the effects of our hormones! In response to low blood volume (and thus not enough water) *antidiuretic hormone* (*ADH*) from the neurohypophysis increases our reabsorption of water (see Chapter 15) in the kidneys, which, in turn, increases blood volume.

When blood volume gets too low, the release of ADH is triggered by the release of *renin* from the *juxtaglomerular apparatus* (see Chapter 15). Renin then converts the inactive *angiotensinogen* into *angiotensin I*; then in the lungs, of all places, an enzyme (*angiotensin-converting enzyme*) converts angiotensin I into *angiotensin II*. It's a good thing it gets converted by the lungs, for angiotensin can then make its way to the thirst centers, as well as stimulating the release of ADH.

As if that weren't enough, angiotensin II stimulates the adrenal glands to release *aldosterone*, which tells the kidneys to hang on to more Na^+, which in turn helps the kidneys to reabsorb water. Don't worry though, because the kidneys aren't done yet! With so much access to RBC (red blood cells or erythrocytes, see Chapter 10), the kidneys can sense the oxygen levels in the blood (directly correlated to the number of RBCs) and can respond to low levels through the release of *erythropoietin* (*EPO*). If you assumed from the name, and from the context, that EPO stimulates erythrocyte production, good job!

The Big Picture

One thing that's ironic about the endocrine system is the fact that all the hormones, despite specific targets, travel all over the body via the blood. In some cases this is just plain silly! Think about aldosterone. Released from the adrenal glands, it must pass all the way through the body before it gets to its target, just centimeters away! Even sillier is renin, which has to travel all over the body to get converted into angiotensin II, despite taking effect just a few microns away from where renin is first made!

Adrenals: The Kidneys' Lofts

The *adrenal glands*, or *suprarenal glands*, lie on top of the kidneys, as the name implies (*ad* = attached, *renal* = kidney). These glands are about the level of the twelfth rib, although the right kidney is slightly lower, due to the space taken up by the liver. There is a fibrous capsule around the adrenal gland that helps keep it attached there. Similar to so many organs, there is an outer cortex surrounding the medulla in the middle (sounds like a TV show). I emphasize these, as the hormones are divided according to the location of their production.

The medulla is stimulated by the sympathetic branch of the autonomic nervous system (see Chapter 20). There are two medullary hormones: epinephrine (formerly called adrenalin) and norepinephrine. Both hormones function as neurotransmitters (see Chapter 19, which are released by *neuroendocrine cells* (neurons that produce hormones). These hormones target most cells in the body.

One of the main effects of these two hormones is the increase in the rate of glycolysis and cellular respiration (see Chapter 3), as well as to initiate *glycogenolysis* (the breakdown of glycogen into glucose molecules). In addition, the heart rate becomes elevated, breathing becomes faster and deeper, respiratory bronchioles dilate (hence the use of epinephrine to combat systemic allergic reactions), and lipids are released by adipose tissue (see Figure 18.5).

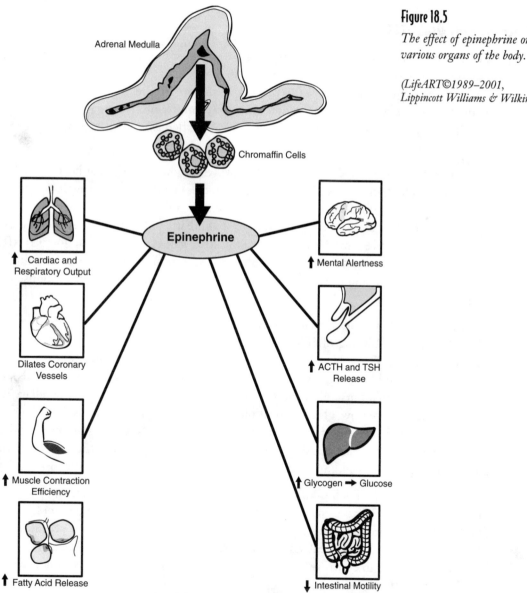

Figure 18.5

The effect of epinephrine on various organs of the body.

(LifeART©1989–2001, Lippincott Williams & Wilkins)

The adrenal cortex produces three types of hormones: mineralocorticoids, glucocorticoids, and androgens. Mineralocorticoids, such as aldosterone, target the kidneys, and help to regulate ions in the body, such as Na^+. Glucocorticoids, such as cortisol, corticosterone, and cortisone, on the other hand, are involved with glycogen production (gluconeogenesis), glucose synthesis, and the release of fatty acids from adipose tissue (thus reserving glucose for storage as glycogen). Glucocorticoids are stimulated by the release of ACTH from the adenohypophysis; without these steroids we would die within a week! Androgens are the same as those made by the testes, and in females they are converted to estrogen. Although androgen production is also stimulated here by ACTH, the amounts are very small, and are little understood.

Pancreas and Diabetes

The pancreas is one of those dual citizenship organs, acting as both an endocrine organ, and as an accessory organ of the digestive system (see Chapter 14). The digestive enzymes produced by the *exocrine pancreas* (99 percent of the organ) are, by definition, picked up by ducts near the *pancreatic acini*, which ultimately empty into the small intestine. The remaining 1 percent is taken up by clusters of cells called *pancreatic islets*, which form the *endocrine pancreas*.

These pancreatic islets have four cells, each of which produces a different hormone: alpha cells produce glucagon, beta cells produce insulin, delta cells produce growth hormone–inhibiting hormone (GH-IH) or somatostatin, and F cells release the poorly named pancreatic polypeptide (discussed in Chapter 14). Pancreatic polypeptide regulates the release of pancreatic digestive enzymes, as well as inhibiting gall bladder contraction. GH-IH, in addition to living up to its name, also inhibits the secretion of both insulin and glucagon and slows absorption and enzyme secretions from the GI tract, usually in response to meals high in protein.

Now, on to blood glucose regulation (see Figure 18.6). Blood glucose usually lives somewhere between 70–120 mg/dl. The only way to maintain such a narrow range is through negative feedback. When blood sugar gets too high (such as after a meal), it is lowered with insulin; when blood sugar gets too low (from too long a break between meals), raise it with glucagon. In a nutshell, insulin attaches to insulin receptors on the capillary walls, and is thus used to remove glucose from blood plasma; insulin also helps with the storage of extra glucose through the formation of glycogen and lipids. Glucagon, on the other hand, targets the liver cells, causing the breakdown of glycogen, and thus releasing the glucose monomers and adding them to the bloodstream (raising blood glucose levels).

Imagine drinking a half gallon of milk in a half hour, and still feeling thirsty. Imagine getting up to pee over 20 times in a night. These are the signs of untreated diabetes that I developed just before my eighteenth birthday, at the end of my freshman year in college. Although there are different causes of diabetes, the signs are the same. In each case there isn't enough insulin to control the blood glucose levels. The two most common types are type I and type II.

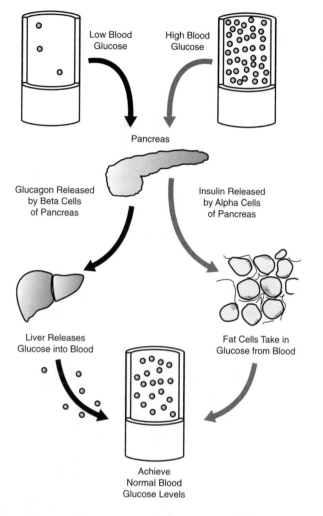

Figure 18.6

The opposing roles of insulin and glucagon in blood glucose regulation.

(LifeART©1989–2001, Lippincott Williams & Wilkins)

The Big Picture

Diabetes is a systemic illness because excess glucose can have dire consequences. If the body is starved for glucose, it will begin to break down its own tissues for food, producing toxic ketones that can lead to coma or death. On a slower scale, those tissues (the retina, the kidneys, and nerves) that do not require insulin to absorb glucose will be damaged by the high glucose levels after a number of years (possibly leading to blindness, kidney failure, and nerve damage). If you do what I do, take at least five injections of insulin a day, and measure your blood sugar six to eight times a day, you will most likely survive, as I have, after 22 years, with no complications.

In type I, juvenile onset, insulin-dependent diabetes, the beta cells of the pancreas have been destroyed in an autoimmune response (see Chapter 17); eventually, no insulin is produced and the patient must take insulin injections. Type II, adult onset diabetics, can often control their blood glucose with diet and certain pills (although insulin therapy might be necessary); these medication help the body to more efficiently use the insulin it already makes.

Testicles

In addition to churning out sperm, testicles are also little hormone factories. The *testicles*, or *testes*, produce two hormones: *testosterone* and *inhibin*. Even in the womb, testosterone production affects the development of the brain (explains a lot!); the *hypothalamic nuclei*, for example, will affect adult sexual behavior, some of which is influenced even before birth. Testosterone production, by the *interstitial cells* (*Leydig cells*), increases during puberty, ultimately affecting *secondary sex characteristics*, such as body hair, and the layout of body fat (a moment on the lips, a lifetime on your abdomen …). Testosterone has also been linked to male aggressive behavior.

FSH from the adenohypophysis stimulates the *sustentacular cells* (*Sertoli cells*) in the differentiation and maturation of spermatozoa (a process that takes about 70 days). FSH also causes these cells to release *inhibin*, which acts in a negative feedback loop, inhibiting the secretion of FSH. It is also thought that GnRH release is also inhibited this way.

Ovaries

The ovaries and the hormones of the menstrual cycle are given special treatment in Chapter 22. The female gametes are produced only one at a time, and released during ovulation. In addition to the roles of FSH and LH discussed in Chapter 22, there are other hormones called *estrogens* and *progestins*. Estrogens (*estradiol* is the main one) basically are involved in the final maturation of the secondary oocyte (see Chapter 23), and in the building of the uterine lining (*endometrium*).

The main progestin is the familiar *progesterone*. In addition to building up the endometrium, progesterone helps to speed up the movement of the secondary oocyte along the *fallopian tube* (see Chapter 23). If fertilization occurs, the division of the *zygote* to a *morula*, and then a *blastula* (a single cell, to a solid ball of cells, to a *hollow* ball of cells; see Chapter 23) is helped along by progesterone. Lastly, progesterone, with a little help from estrogen, prolactin, and even growth hormone, causes the breasts to enlarge in preparation for lactation and breast feeding.

The only hormone left would be almost funny if there weren't a powerful reason for it! This hormone, *relaxin*, which is also involved in breast development, is elevated during pregnancy. Its release is stimulated by LH in the beginning, but later it is a placental hormone called *human chorionic gonadotropin* (*hCG*) that stimulates the secretion of relaxin. Its most interesting function is to "relax" the cartilage in the pubic symphysis, making it easier for the baby to pass through the pelvic outlet (the bottom of the pelvis; see Chapter 6). As if it weren't hard enough to walk during the ninth month of pregnancy, a woman's pelvis has to feel like it's going to fall apart!

The Least You Need to Know

♦ Hormones function via feedback loops, with the vast majority controlled by negative feedback loops, usually involving the hypothalamus and the pituitary.

♦ Hormones can be steroids, altered amino acids (biogenic amines), peptides, and proteins.

♦ Lipid-soluble hormones travel directly through the cell membrane, triggering an operon in the target cell nucleus, while water-soluble hormones require a cell receptor to trigger the target cell's response.

♦ Hormonal secretion by the pituitary is dependent upon the presence or absence of regulatory hormones from the hypothalamus.

♦ The pituitary, in turn, controls the secretion of hormones by various other endocrine organs.

♦ The hormones from the other endocrine organs regulate various aspects of homeostasis.

Chapter 19

You're Getting On My Nerves!

In This Chapter

- Divisions of the nervous system
- Types of support cells
- Neuron structure and function
- Reflex arcs
- Nerve cell transmission

As the other control system, after the endocrine, the nervous system carries a great burden. It must be able to make rapid changes in the body and regulate the environment at all times. Not content to borrow the pathways of the cardiovascular system, the nervous system has its own elaborate pathways, often running roughly parallel to the blood vessels.

The response is so fast that there is no appreciable difference in reaction time, regardless of height. The great distances the messages often travel require certain tricks to speed up the nerve impulse. Time savers, such as which messages need to go all the way to the brain and which don't (such as reflexes), are part of the solution. These topics and more are discussed in this chapter.

The Great Divide

The nervous system can be divided two ways, anatomically and functionally. The basic anatomical division is between the *central nervous system* (*CNS*), which includes the brain and the spinal cord, and the *peripheral nervous system (PNS)*, which includes the cranial and spinal nerves (see Figure 19.1). You might think of the spinal cord as one large nerve with many nerve branches, but as there are nerves that extend from the brainstem, and nerves extending from the spinal cord, the spinal cord is considered part of the CNS.

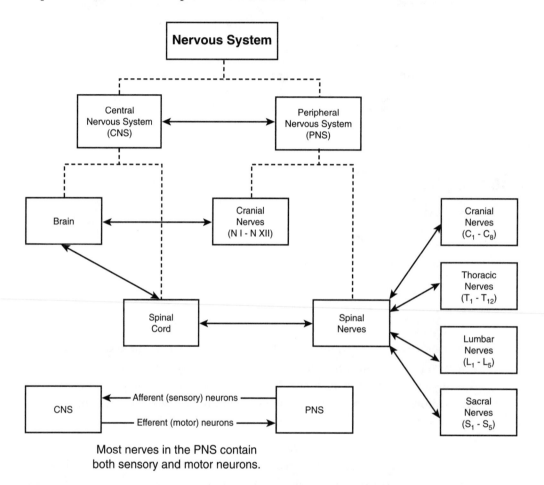

Most nerves in the PNS contain
both sensory and motor neurons.

Figure 19.1

The organization of the nervous system is based on the basic division between the central nervous system (CNS) and the peripheral nervous system (PNS).

(©Michael J. Vieira Lazaroff)

The job of the nervous system is basically to sense the environment and the body's place in it, coordinate either a voluntary or an involuntary response, and then carry out that response. As the true control center, the CNS needs to have access to information about the rest of the body, just like a kitchen manager in a restaurant needs to know about the attendance of the staff, the alcohol and food orders, the condition of the equipment, the reservations, and so on, in order to plan accordingly. The only way to get access to the information about the body is through the PNS.

Another way to divide the nervous system is in terms of function. In this organization of the nervous system each of the two branches includes parts of both the CNS and the PNS. The division is based on two factors: whether the response is voluntary, and the area of the body that carries out the instructions of the CNS. The voluntary portion is called the *somatic nervous system* (*SNS*), in which the responding organs are skeletal muscles. The other division is called the *autonomic nervous system* (*ANS*), which is involuntary. If you think about the organs that need to be working in the background, without having to give them thought, you know where the bulk of these organs are: in the thoracic and abdominopelvic cavities.

The Central Nervous System

The CNS is, of course, a little more complex than just "brain and spinal cord." In looking at the brain, numerous divisions are important. In this chapter, I plan on only introducing the cast of characters; I will cover them in more detail in Chapter 20. When people usually think about the brain, their mind quickly goes to, well, the mind, or cerebral cortex, the part of the brain where thought and perception occur, but there's more.

First of all, we divide the brain into two *hemispheres*, with the division right along the midline. True, the either/or mentality is more a sign of the hard wiring of the mammalian brain than with our hemispheres, but the left and right hemispheres, nonetheless, do divide some of the brain's activities. Neither hemisphere works completely alone, however, for messages are passed back and forth between the hemispheres in a broad band of neurons called the *corpus callosum*.

Medical Records

It has been determined through imaging technology that women actively use both hemispheres of the brain when listening to conversations; that ol' corpus callosum is just kept jumpin'. Men, on the other hand, use mainly the left hemisphere. Is this why men aren't very good listeners?

The brain is divided into left and right hemispheres, with each hemisphere divided into the same lobes: frontal (anterior, of course!), parietal, temporal, and occipital. Beneath the brain is the *diencephalons* (made of the *thalamus* and the *hypothalamus*), which is just above the *brainstem*. Arising out of the top of the spinal cord, the brainstem is made of, from top to bottom, the *midbrain*, the *pons*, and the *medulla oblongata*. Coming out of the pons is another lobe, the *cerebellum*. The spinal cord, extending through all the vertebral foramina in the spinal column, continues the CNS all the way to the end of the sacrum. Flowing around the spinal cord and brain, and through the *four ventricles*, is the nutrient-rich protective fluid known as *cerebrospinal fluid* or *CSF*. All these aspects of the CNS will be covered in Chapter 20.

The Peripheral Nervous System

The brain may be the neural control center, and the spinal cord the main thoroughfare of the nervous system, but neither of these, alone, can carry messages to and from the foot soldiers, our tissues and organs. For that we need the PNS. The neurons in the PNS are constrained by the fact that every neuron can only send messages in one direction. Within the PNS, in order to do the job properly, there need to be two sets of neurons: *afferent* and *efferent*. Afferent neurons bring messages from the body to the brain; these neurons are sensory in nature, hence also being called *sensory neurons*. Efferent neurons, on the other hand, carry messages from the brain to the body, instructing the body as to what action to take; as the majority of these actions are made by muscle tissue (skeletal, cardiac, and smooth), these are also called *motor neurons*. Most peripheral nerves carry messages both ways, hence having both afferent and efferent neurons, but certain cranial nerves are limited to either sensory function, such as the *optic nerve (N II)*, or motor function, such as the *oculomotor nerve (N III)*.

In addition to afferent and efferent neurons, the PNS can be divided according to the type of action that the efferent nerves trigger. Those voluntary actions, which involve skeletal muscles, are part of the *somatic nervous system (SNS)*. Muscles, however, are just the tip of the iceberg, when it comes to what actions the body can take. Look in either the thoracic or the abdominopelvic cavity, and, with the exception of the diaphragm, you'll see no skeletal muscles. Even the diaphragm is an interesting exception in that it is not entirely voluntary. It is the involuntary actions of the organs of these two cavities that led to the name of the area of the PNS that triggers them: the *autonomic nervous system (ANS)*, although we now know that the brain regulates these functions, and that they are not truly autonomous.

The last method of dividing the PNS is purely in terms of anatomy, according to the location of the nerves. There are 12 pairs of *cranial nerves*, indicated as N I through N XII. A number of these are completely sensory, as so many senses are located in the head (see the following table). The sections of the spine also have their own nerves, called *spinal nerves: cervical nerves (C_1–C_8), thoracic nerves (T_1–T_{12}), lumbar nerves (L_1–L_5), and sacral nerves (S_1–S_5)*.

The Functions of the 12 Cranial Nerves

Cranial Nerve	Sensory	Motor
I Olfactory	Smell	(None)
II Optic	Vision	(None)
III Oculomotor	(None)	Eyeball movement
IV Trochlear	(None)	Eyeball movement
V Trigeminal	Gums, teeth, lips, facial skin, and so on	Jaw movement
VI Abducens	(None)	Eyeball movement

Cranial Nerve	Sensory	Motor
VII Facial	Taste (anterior $2/3$ of tongue)	Facial muscles lacrimal (tear) and salivary glands
VIII Vestibulocochlear	Hearing/balance	(None)
IX Glossopharyngeal	Tongue (posterior $1/3$), pharynx, palate, 0_2, CO_2, pH, blood pressure	Pharynx, parotid salivary gland
X Vagus	Visceral sensation	Visceral response
XI Accessory	(None)	Neck, palate pharynx, larynx
XII Hypoglossal	(None)	Tongue

Support Staff

There is an old saying that you only use 10 percent of your brain. I've never understood this expression, for we use all of it, although not at the same time. The expression has something to do with the fact that we have far more (up to 50 times more!) support cells, called *neuroglia*, or *glial cells*, than we do neurons. Since those neuroglia perform essential functions for the neurons—including regulating ion concentration, insulating axons, and lining ventricles—it seems odd to consider them as unused! This section concentrates on the many ways neuroglia make themselves indispensable, allowing us to use 100 percent of our brain!

In the CNS

You might have heard the expression *gray matter*, and perhaps you equated it with the brain. But did you also know about *white matter*? The difference between the two depends on one of the forms of neuroglia. Each neuron has a cytoplasmic extension called an *axon* (see either Chapter 4, or the neuron section of this chapter) that carries the messages away from the cell body; in order to speed up the message the axon is surrounded by an insulating sheath made of myelin. The *myelin sheath* is simply a phospholipid cell membrane wrapped around and around to protect the axon, similar to the way the plastic coating on an electric wire preserves the direction of the electricity, preventing it from traveling outward.

The myelin sheath is white in color and is provided by specific neuroglia (*oligodendrocytes* in the CNS, and *neurolemmocytes* or *Schwann cells* in the PNS). The white matter in the brain and spinal cord is due to the high concentration of myelinated axons in that area, due to the white color of the myelin. The gray matter, on the other hand, contains either unmyelinated axons or mostly dendrites and cell bodies.

The central nervous system has four types of neuroglia, as opposed to only two for the peripheral nervous system. Part of the reason for this is the basic nature of the two systems, for the PNS is basically concerned with the transfer of information to and from the CNS, but the CNS is the system that must actually make the decisions (conscious or otherwise) that regulate the rest of the body. The wider range of functions in the CNS requires a wider range of glial cells.

In addition to the oligodendrocytes mentioned so far, there are three others: *ependymal cells*, *microglia*, and *astrocytes*. *Ependymal* cells are basically epithelial cells (squamous to columnar, see Chapter 4), and they are found lining the ventricles. *Microglia* are pretty cool because they play the role of macrophages (see Chapter 17), cleaning out cellular waste, attacking invaders, and so on. Although they are usually fixed, they have been known to find their way to damaged tissues.

Astrocytes are a bit more complex. These cells have numerous cellular extensions, which help to maintain a framework holding together the various axons and dendrites in the CNS. In addition, this framework apparently guides the connections of the growing neurons during development. The framework, mentioned above, is also crucial in terms of repairing damaged tissues. One of the most interesting parts of their job is to help provide what is called the *blood-brain barrier* (*BBB*).

Blood-Brain Barrier

Bizarre as it might seem, there is in fact a difference between the function of capillaries in the brain and those in the rest of the body. One of the reasons for this is, well, capillaries aren't very leaky! You might remember some of the tricks of capillaries to release plasma (see Chapter 12), including gaps between the endothelial cells that make up the thin capillary walls. Brain capillaries, on the other hand, lack those gaps. They even have tight junctions (see Chapter 3) to prevent leakage. Unlike most capillaries, those in the brain also have a continuous basement membrane on the outside of the endothelium. As if that weren't enough, those pesky astrocytes put their cytoplasmic projections on the outside of the capillaries (see Figure 19.2). Multiple astrocytes thus form a continuous outer layer to almost every capillary in the CNS.

The blood-brain barrier refers in part to the differential movement of materials. Certain substances pass through easily, such as glucose, O_2, CO_2, and H_2O. As a matter of fact, insulin is not needed for glucose to leave the capillaries around nervous tissue; the high glucose levels in the nervous tissue of diabetics leads to a form of nerve damage known as neuropathy. Lipid-soluble molecules, as you would expect, travel through quite easily; this is a good thing in the case of anesthetics, but it doesn't help in terms of some rather addictive substances that also make their way through easily (alcohol, caffeine, heroin, and nicotine). Ions such as Na+, Cl-, and K+ (all a part of neural propagation), can pass through, but they rely on carrier-mediated transport. Proteins and most antibiotics cannot pass through the blood-brain barrier; this antibiotic barrier can be troublesome in terms of cranial infections.

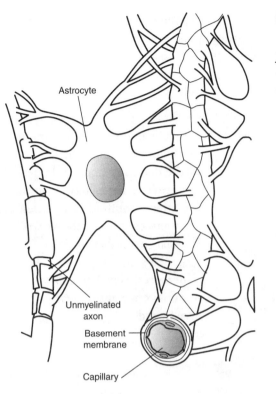

Figure 19.2

The projections of astrocytes form part of the physical barrier known as the blood-brain barrier.

Not all of the brain contains such a barrier with the blood. The regions around the third ventricle, including the hypophysis, hypothalamus, and pineal gland, have normal capillary flow. The name for these regions, circumventricular organs or CVOs, should be self-explanatory (that is, "around the ventricle"). Given that these areas are involved in the monitoring of homeostasis (fluid levels, blood pressure, thirst, and hunger), it makes sense to not have the blood-brain barrier there!

In the PNS

It might come as a bit of a surprise that there is no cell equivalent to the microglia in the PNS. A closer look, however, shows that the CNS is fairly insulated from other tissues, whereas the nerves of the PNS are embedded in other tissues, particularly the nerve endings. Those tissues are already serviced either by wandering or by fixed macrophages.

There are only two types of neuroglia in the peripheral nervous system: *satellite cells* and *neurilemmocytes*, also known as *Schwann cells*. The satellite cells are important in terms of clumps of nerve cells known as *ganglia*. Ganglia contain more than just axons and dendrites, the makeup of nerves; ganglia (singular = ganglion) also contain the more cumbersome cell bodies. These cell bodies are surrounded by the flattened satellite cells, which basically provide protection.

Remember how the axon carries messages away from the cell body? Well, the long axons of the PNS require insulation just as much, if not more so, than in the CNS. This insulation is provided by neurilemmocytes. These weird little pill-shaped cells grow around a small section of an axon, gradually producing more and more membrane, which ends up wrapping layer after layer of tightly coiled membrane.

These multiple layers of phospholipid bilayers form the *myelin sheath*. The white color of the myelin sheath is what gives myelinated nerve tissue the name *white matter*; *gray matter*, on the other hand, is made up of either unmyelinated axons, or the dendrites and cell bodies of neurons. It is important to note here that all axons in the PNS are protected from the interstitial fluid by neurilemmocytes, but that these Schwann cells don't always wrap around the axon a freakishly large number of times. When each Schwann cell wraps around a single axon multiple times, the axon is myelinated. When each Schwann cell wraps around multiple axons, but only once around each one, then the axons are not myelinated.

Flex Your Muscles

The name *neurolemmocyte* makes sense when you remember that the cell membrane of neurons is called the neurilemma (like the sarcolemma on muscle cells). These cells wrap around the neurilemma, thus insulating it!

Neurons

Okay, slow down. Axons? Dendrites? Cell bodies? Remember, neural tissue is important in its ability to communicate from one region of the body to another. Unlike the endocrine system which communicates slowly, using hormones carried by the cardiovascular system, neurons must be able to communicate *quickly!* The only way to ensure this is to make actual physical, cellular connections between different areas of the body. Neurons are designed to carry nerve impulses, basically waves of depolarization (see Chapter 8 "You've Got Potential"), along the length of the cell, which is made up of cytoplasmic extensions called *dendrites* and *axons*.

The Parts, To and From

When everyone thinks about typical human cells, they probably see a cell with a nucleus in their mind (unless they are thinking about red blood cells!). Neurons, too, have a nucleus, but it is away from where the action is! The most "normal-looking" part of the cell is the *cell body*, *perikaryon*, or *soma*, which contains the nucleus. Although the nucleus is usually a hotbed of activity, it is less so in neurons. Very little cellular activity actually happens here, which is one of the reasons the supporting work of the neuroglia is so necessary. Most neurons lack centrioles in their somas, which is the other reason why scientists believed for so long that adult neurons didn't divide.

Flex Your Muscles

The word *perikaryon* should sound a bit familiar. Remember learning about prokaryotic and eukaryotic cells? The word *karyon* means *kernel* and refers to the nucleus. Prokaryotic cells are cells that evolved before (*pro* as in prologue) a nucleus, and eukaryotic cells have a good (*eu* as in euphemism) nucleus. Well, in the same sense, the perikaryon or cell body is the area around (*peri* as in perimeter) the nucleus.

On either side of the cell body are the parts of the neuron that make it famous: the dendrite and the axon. The difference between the two is basically functional, in that it is not always possible to tell which is which under the microscope. The function of the dendrite is to carry the nerve impulse to the cell body, and the axon carries the impulse away from the cell body (see Figure 19.3). The basic nature of a neuron, unlike a phone cord, or a fiber-optic cable, involves only one-way transmission. So the transmission is always as follows: dendrites, cell body, axon.

Dendrites are usually unmyelinated (thus making dendrites part of the gray matter), whereas axons can be either myelinated (white matter) or unmyelinated. The real action takes place both along the dendrites and axon (the nerve impulse) and at the connection between the neuron and other neurons, or neurons and muscle cells, for example; these connections are called synapses (see Chapter 8). Synapses always involve the end of axons, called axon terminal bulbs, or synaptic end bulbs. Although some of the neuron to neuron synapses occur at the perikaryon, the majority of them are found at the dendrites, which account for 80 to 90 percent of the surface area of your garden variety neuron.

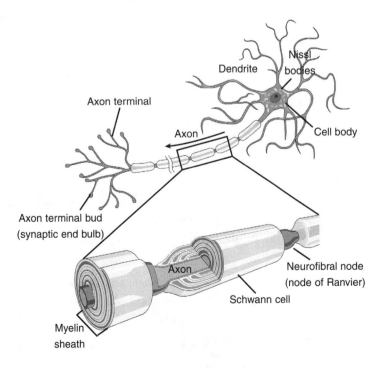

Figure 19.3

The dendrites and axons of a neuron carry neural impulses to and from the cell body.

(LifeART©1989–2001, Lippincott Williams & Wilkins)

Those long axons and dendrites don't sound very sturdy. Sure the oligodendrocytes and Schwann cells must help, as well as the astrocytes, but isn't there any inner framework? I'm glad you asked! The cytoskeleton (see Chapter 3) is alive and well here, but traveling incognito. The microtubules and microfilaments go by the names *neurotubules* and *neurofilaments* here; together they are bundled as neurofibrils, which run the length of the dendrites and axons, providing internal support.

Medical Records

Viruses and toxins have been known to take advantage of fast axonal transport by hitching a ride. The toxin from tetanus bacteria, whose effect led to the name lockjaw, is one such culprit. The herpes virus is another culprit; herpes zoster, which causes chicken pox, is extremely painful if and when it reoccurs later in life as *shingles*, because it attacks nerves, traveling through the nerve root from the dorsal root ganglion. Lastly, the rabies virus will work its way to the brain, and the speed of the transport means a bite on the arm or neck requires earlier medical intervention than a bite on the leg!

Axons have another trick up their sleeve! They have their own inner transport mechanisms, which differ from the rapid membrane-based nerve impulses. *Slow axonal transport* involves the movement of chemicals along the *axoplasm* (axonal cytoplasm); this cumbersome method only goes one way, carrying chemicals to the axon terminals at the rate of a whopping 1 to 5 mm a day (ooooh!). Fast axonal transport is 40 to 400 times faster, and has the advantage of going both ways (anterograde = *from* the soma; retrograde = *to* the soma) via the neurotubules in the axon. In addition to molecular cargo, organelles and even viruses have been known to take a ride on the fast axonal transport.

The Insulated Wire

Remember those pesky neurolemmocytes? Well, these Schwann cells, as I said earlier, insulate the axon. The insulation is very much like the insulation on an electric wire. If you have ever done projects with electricity in school, then you might have noticed that electricity running through coiled wire exerts an electromagnetic field. You can't see the field, of course, but simply watching a compass needle jump shows that it is there.

This field indicates the movement of electrons away from the wire. The insulation keeps the electrons traveling along the wire, thus eliminating "leakage." In the same sense, the movement of Na^+ and K^+ ions in and out of neurilemma is easier if there isn't too much area for them to enter and leave.

The analogy gets tricky if you take it too literally, because the transmission of electricity is along the wire, whereas the transmission along the axon takes place on the outside of the "wire." The key is the spaces between the Schwann cells. These little gaps, connected by little sections of bare axons, are called *neurofibral nodes*, or *nodes of Ranvier*. For this reason, Schwann cells, and their myelin sheath, are also called *myelinated internodes*.

Simply calling these cells *myelinated internodes* illustrates the importance of the nodes. Remember the whole process of polarization and depolarization of membranes in muscle cells (see Chapter 8)? Energy is required to power the active transport that makes a membrane polarized. Depolarization happens fast due to facilitated diffusion. The difference here is that, unlike the wave of depolarization along a myofiber (muscle cell), the depolarization jumps from node to node. This process, known as *saltatory conduction* (as opposed to *continuous conduction* in unmyelinated axons) helps to speed up the propagation of the nerve impulse.

Connecting with Synapses

Synapses have all the fun. The axons and dendrites just carry the message, but it is the ability for these cells to actually *talk* to one another that is the true glory of nervous tissue! The talking takes place at the synapses, between axons and dendrites, axons and cell bodies, or axons and effectors (such as muscles). In the case of neuron to neuron synapses, the first cell is the *presynaptic neuron* (with its *presynaptic membrane*), and the receiving cell is the *postsynaptic cell* (with its *postsynaptic membrane*).

There are actually two types of synapses: electrical and chemical. *Electrical synapses* are also called *gap junctions* and are formed by protein channels that physically connect two cells (see the image of cell junctions in Chapter 3). These protein channels, called *connexons* (because of the way they connect cells), speed up the transport of ions from cell to cell, so that a wave of depolarization can travel very quickly, for example, from one cardiac cell to another, making multiple cells act like one.

Chemical synapses, however, are the basis of the vast majority of synaptic connections in the CNS and PNS. The majority of synapses are called *cholinergic synapses*, due to their use of *acetylcholine* (ACh) as the *neurotransmitter* (the chemical that sends the message from cell to cell, initiating a depolarization of the second cell. For a detailed run-down of this process, see the discussion of the synapse between a motor neuron and a muscle cell in Chapter 8.

One aspect of acetylcholine that is not discussed in Chapter 8 is that of the recycling in the synapse. I mention the enzyme acetylcholinesterase (ACh-esterase) in the synaptic cleft. ACh-esterase breaks ACh into acetate (A) and choline (Ch). The acetate is absorbed and used by cells in the area. The choline, however, is absorbed by the synaptic end bulb. Coenzyme A (CoA) is taken up by mitochondria there, and converted into acetyl-CoA. The acetyl-CoA then combines with the choline to produce acetylcholine (ACh) and a coenzyme A (CoA), and the cycle continues.

Neurotransmitters

The field of neuroscience continues to find new neurotransmitters. Despite many of these discoveries, it turns out that the action triggered by the neurotransmitter has to do with the receptor, rather than the neurotransmitter, for the same chemical may stimulate certain neurons, but inhibit others. This leads to the idea of *excitatory neurotransmitters*, which initiate depolarization of the postsynaptic membrane, and *inhibitory neurotransmitters*, which causes *hyperpolarization*, thus inhibiting the depolarization of the postsynaptic membrane.

Medical Records _____

Physical addiction is based upon the building of extra receptors in the brain. The body, in its effort to maintain homeostasis, will build extra receptors (for example, endorphin receptors) to deal with the influx of the drug (for example, heroin). Over time, the addict will need to take more of the drug to get the same high; this phenomenon, caused by the extra receptors, is called *tolerance*. Withdrawal is hard due to the large number of now empty receptors. In the case of heroin withdrawal, since endorphins block pain, many empty receptors make quitting cold turkey very painful.

Some of these many neurotransmitters have a direct effect, as seen with ACh, but others trigger a response indirectly, using a secondary messenger not unlike that seen in the case of protein hormones (see Chapter 18). In addition to ACh, some neurotransmitters may be biogenic amines (epinephrine, norepinephrine, histamine, dopamine, and serotonin), amino acids (glutamine and aspartate), hormones (ADH, oxytocin, and so on), ATP, or even dissolved gases (carbon monoxide and nitric oxide). One interesting group of neurotransmitters is called *neuropeptides*, which may be only a few amino acids long; the most famous of these is called *endorphin* or *enkephalin*. *Endorphin* is short for *endogenous morphine*, or the body's own morphine, and it is the endorphin receptors to which opium, morphine, codeine, and heroin bind.

Types of Neurons

In a functional sense, neurons are classified according to the direction of their nerve impulse: *afferent* (or sensory) neurons travel *to* the brain, and *efferent* (or motor neurons, stimulating *effectors* such as muscle) travel *from* the brain. In addition, the majority of neurons in the body actually carry messages from one neuron to another; these middle men are called *interneurons*. Some neurons function by *stimulating* the following neuron, while others function by *inhibiting* the following neuron. Lastly, neurons also vary in terms of the layout of the dendrites and axon (see Figure 19.4).

One form is the *multipolar neuron*, in which there are multiple short dendrites, and one long axon; these neurons, which look rather like they have dreadlocks, are very common in the brain and spinal cord. *Bipolar neurons*, as their name suggests, have the cell body in the middle, with one long dendrite, and one long axon; all bipolar neurons are sensory. *Unipolar neurons* are also, and more accurately, called *pseudounipolar neurons*. These rather bizarre-looking neurons have a continuous track connecting the dendrite (which may be myelinated) to the axon, with the cell body off to the side, connected by a stalk; sensory neurons in the PNS mostly take this form. *Anaxonic neurons* really aren't without axons, but their axons and dendrites are short and similar in length and appearance; the function of these neurons is still under investigation. I discuss sensory neurons in more detail in Chapter 20.

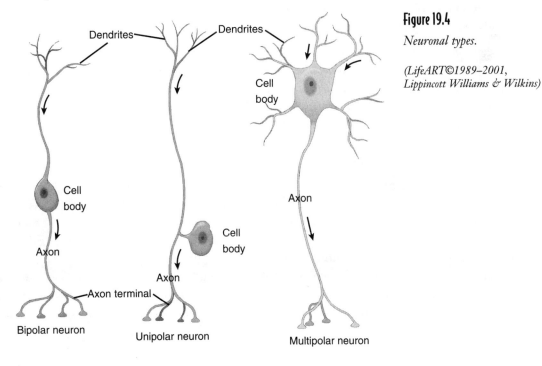

Figure 19.4

Neuronal types.

(LifeART©1989–2001, Lippincott Williams & Wilkins)

Types of Circuits

The connection between neurons leads to some interesting connections called *circuits*. The nature of the arrangement of each type of circuit says a great deal about the function of that circuit. Series circuits, the least interesting of the bunch, are merely a connection of single neurons in a line; such circuits basically just carry information from one center of the brain to another for processing.

On the other hand, circuits may either *diverge* (one neuron connecting to many) or *converge* (many neurons connecting to one). Their arrangement is an aspect of their function. A diverging circuit of sensory neurons, for example, would split in terms of the areas of the brain to which the sensory input goes. Visual images would need to respond to the amount of light to open or close the pupil, or to interpret words, or identify the emotions behind a facial expression; a diverging circuit would thus send the same signal to multiple locations in the brain.

Converging circuits, on the other hand, would most likely be motor neurons. Consider your breathing. Chances are you haven't been thinking about that as you have been reading this chapter (or if you have, you should pay more attention to what you are reading).

In that sense, isn't breathing involuntary? If you said yes, you're right, but what about holding your breath? Isn't that voluntary? If you said yes, you are *also* right! The only way that is possible is through a converging circuit, which will allow stimulus for the diaphragm by either a voluntary or involuntary neuron. In the case of holding your breath, the voluntary neuron must have an inhibitory effect on the motor neuron, or else the involuntary command will take over. Imagine suddenly breathing while under water, and you'll get the idea of the importance of this inhibition!

Parallel circuits look similar to diverging circuits, and, indeed, they must first diverge before they can become parallel. Such circuits help explain responses to sudden painful stimulus. Have you ever picked up something that was way too hot to hold? Think about what you did. Did you drop it suddenly? Did you also pull your arm back? What about your voice? Did you suddenly scream out a bunch of syllables my editor won't let me put in this book? All of those sudden responses to that stimulus were sent to different parts of the brain, and then to different effectors, through a parallel circuit.

Reverberating circuits differ from the others in that they have neurons with branches that go backward to stimulate the circuit earlier in the chain. Remember that neurons only send messages one way, so such a circuit really requires a place for an axon to backtrack to a dendrite of an earlier axon. But why would the body do such a thing. Once again, such a circuit implies a particular function. Such a circuit, for example, would be used to control a particular repetitive action, such as breathing. In higher levels of the brain, reverberating circuits control consciousness, for without a constant restimulation through the circuit, there would be no continuity to our consciousness. Such circuits require specific circumstances to break the circuit, as when we fall asleep!

The Parts of an Arc

A simple way to explore the function of neurons, both alone, and in concert, is to look at a reflex arc. To understand the parts of a reflex arc, it is helpful to think about the spinal cord, the dorsal and ventral roots to the spinal nerve, and feedback loops (see Chapter 1). Reflexes follow the basic pattern of a negative feedback loop, with a receptor, a control center, and an effector. Given that the purpose of a reflex is to reverse a stimulus, in other words negative feedback, reflexes help to maintain homeostasis.

Let's take a look at a relatively simple polysynaptic reflex, shown in Figure 19.5, for it will illustrate the concept well. First of all a stimulus, such as stepping on a tack, triggers pain receptors, thus initiating an impulse in a sensory (afferent) neuron. That impulse travels through the dorsal root ganglion to the dorsal root to the spinal cord. The posterior gray horn forms synaptic connections with two interneurons, which, as the name implies, are not the end of the line. A monosynaptic reflex, on the other hand, forms a single synapse with a motor (efferent) neuron.

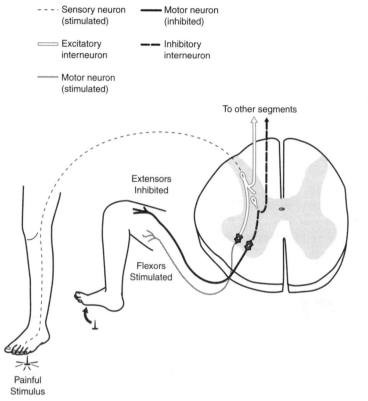

- - - Sensory neuron (stimulated)
⎯ Motor neuron (inhibited)
▭ Excitatory interneuron
⎯ ⎯ Inhibitory interneuron
⎯ Motor neuron (stimulated)

To other segments

Extensors Inhibited

Flexors Stimulated

Painful Stimulus

Figure 19.5

This polysynaptic reflex has not only a sensory neuron, but two interneurons (one to triggering a neuron that stimulates the flexors, and one that triggers a neuron that inhibits the extensors).

Although these interneurons may send messages to other segments of the spinal cord, or to the brain, here I am going to consider the synapses to motor neurons. It is important to consider here that muscles, our effectors, come in antagonistic pairs (see Chapters 8 and 9); as one muscle contracts, its antagonist relaxes, and vice versa. If both muscles contract there would be no movement! Therefore, one interneuron is called the *excitatory interneuron*, because it stimulates a motor neuron, and the other is called an *inhibitory interneuron*, because it inhibits a motor neuron.

You've Got Potential

If you remember the basics of polarized membranes from Chapter 8, this will all be familiar. An inactive neuron is not truly at rest, for it requires a membrane to have an uneven concentration of ions on both sides of the neurilemma (the nerve cell membrane); a membrane in this state is said to be polarized. Ions and molecules naturally move to a state of equilibrium through a form of passive transport called *diffusion* (see Chapter 3). To be uneven, active transport is needed, which requires energy in the form of ATP.

Nerve impulses require some form of stimulus to initiate the depolarization of the membrane. Since that depolarization is accomplished through facilitated diffusion, in which ion channels open in the membrane, causing a rapid movement of ions, this combination of active transport and facilitated diffusion helps explain the rapidity of our response.

There are many types of ion channels which will initiate the depolarization of the membrane. All of these channels are *gated channels*, which require some environmental change to open them. These are opposed to *leakage* or *nongated channels*, which are so named because they are always open. A *chemically gated ion channel* is typically found at synapses, such as the cholinergic receptors that open in the presence of ACh (in both muscle cells and neurons). Such chemical channels open directly as a result of the neurotransmitter, for the receptor becomes a channel protein when the neurotransmitter binds to it. Indirect chemically gated channels function very similar to the G-protein/secondary messenger systems involved with some hormones.

Light-gated ion channels open, as the name implies, to the presence of light, which is why they are found in the retina. *Mechanically gated ion channels*, on the other hand, open in response to pressure, or vibration, which explains their presence in the ear (for both sound and for balance), and in the skin for various forms of touch. Chapter 21 explores the various senses much further.

The last to consider is called the *voltage-gated ion channel*, which is essential to the basic conduction of nerve impulses. Its name refers to the current produced by the movement of ions, rather than the movement of electrons as in electricity. As a result of these channels, the depolarization of the membrane in one area results in the depolarization of a neighboring channel, and so on, ultimately producing an impulse. This makes cells with these channels (muscle cells and neurons) *excitable*, which means they are capable of producing an *impulse*, otherwise known as a traveling wave of depolarization and repolarization.

All for One? All or None!

Throughout this chapter you might have noticed the similarities between muscle and nerve cells. The same polarization, the same depolarization, the same neurotransmitter (ACh, at least for some neurons), and so on. Another thing they share is the all or none principle. Have you ever played with dominoes? I don't mean the game, but standing a bunch of them on end, just looking forward to the moment when you tipped the first one over, and the whole chain topples over in series.

You can tilt the first domino only so far and not have it fall over, but once you pass the point of no return, that sucker's goin' *down*, and all the others will fall down in glorious quick succession. In the same sense, muscle cells and neurons function on an all-or-none principle. Once the stimulus exceeds a certain threshold, the contraction or nerve impulse in that single cell *will* happen, and it will happen all the way. That principle is fully dependent on those gated ion channels opening, letting Na^+ ions flood into the cell.

Action Potential

Despite the differences between the various gated ion channels, they all function on the same basic principle, which involves a cell at rest—remaining polarized as a result of ATP maintaining the active transport of ions. These membranes have a basic *resting potential*, which has to do with the difference in charge on either side of the membrane. The high concentration of Na^+ and Cl^- ions on the outside, and the high concentration of K^+ ions and negatively charged amino acids on the inside of the neurilemma produce an overall voltage of –70 mV (–70 mV = –0.07 Volts). This means that the inside of the membrane is 70 mV more negative than the outside.

When the membrane potential rises to –55 mV, a threshold is reached, and this point of no return triggers the depolarization of the membrane. The membrane will reach an overall voltage of +30 mV, or 30 mV more positive on the inside of the membrane, before repolarization begins. Depolarization and repolarization form a rapid cycle that lasts about one millisecond, or one thousandth of a second, depending on the type of neuron. Inhibitory neurotransmitters make the membrane potential more than –70 mV, thus making it harder to reach the threshold of –55 mV. This state is called *hyperpolarization*. The succession of neighboring channels carrying this on creates a nerve impulse, also known as an *action potential*.

Figure 19.6 illustrates that the voltage-gated Na^+ channel has two gates, an activation gate and an inactivation gate; the K^+ channel only has one gate. In the Na^+ channel the activation gate remains closed when the membrane is maintaining its resting potential. With one gate closed, there is no need to keep the other closed. When the membrane depolarizes, both the activation gate of the Na^+ channel, and the single gate of the K^+ channel opens (although much more slowly), causing the ions to flow, thus allowing them to reach equilibrium (depolarized).

In order for the membrane to repolarize, the inactivation gate of the gated Na^+ channel closes, and the gated K^+ channel opens to help restore resting potential. When the membrane is fully restored to its polarized state, the K^+ gate closes, the Na^+ inactivation gate opens, and the Na^+ activation gate closes. During this time the Na^+/K^+ pump has been working to pump the ions to areas of higher concentration (that pesky active transport again!). Amazing that all of this happens in one millisecond! No wonder an action potential, nerve impulse, happens so fast!

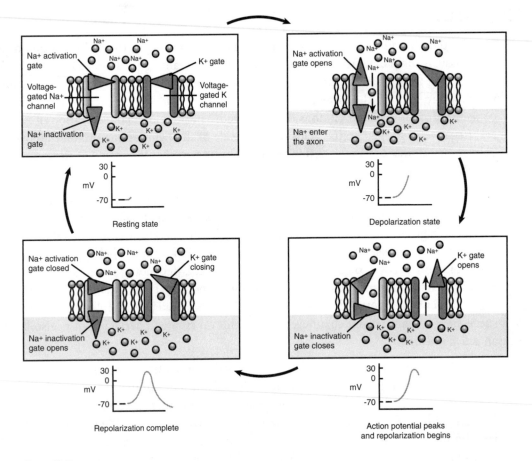

Figure 19.6

The gated ion channels involved in an action potential. The Na$^+$/K$^+$ pump, which is used to return the membrane to a polarized state, is not shown here.

Propagation

Remember those voltage-gated channels? The depolarization of one channel causes a change in the voltage of that area of the cell membrane. This causes the neighboring area of the cell membrane to depolarize. This depolarization of the ion channel is followed by an immediate repolarization of that part of the membrane. You might remember from Chapter 8 that after a cell is stimulated there is a refractory period, during which a cell cannot be stimulated again. What this means is that the depolarization of one ion channel can trigger the next one to depolarize, but the previous one must recover before it can be stimulated again.

The result of this is that depolarization, and the wave known as the nerve impulse, can travel in only one direction: forward! With the previous ion channel recovering, stimulation cannot travel backward! The ultimate result of this is the need to evolve two sets of nerve fibers, afferent and efferent, to carry messages back and forth.

Once again, propagation from one neighboring ion channel to another is called *continuous conduction*. This works fine in many circumstances, including the brain, but it is limited by its speed. In the brain the messages don't have far to travel, but in the PNS messages may have to travel 2 meters in a fraction of a second! In this case the neurolemmocytes (Schwann cells) speed up the conduction. The myelin sheath allows the impulse to jump from node to node, producing the much faster *saltatory conduction*. Just what the PNS ordered!

The Least You Need to Know

- The nervous system is divided into the central and peripheral nervous systems (CNS and PNS).

- Nervous tissue involves not only neurons, but various supporting cells called neuroglia, which vary in the CNS and PNS.

- Neurons have three basic parts: the cell body, dendrites bringing messages to it, and axons bringing messages away from it.

- Neurons communicate with one another, and with tissues such as muscle, through the use of neurotransmitters at synapses.

- Circuits created through multiple neurons differ in function by virtue of their arrangement.

- Nerve impulses are conducted through depolarization and repolarization of neuronal membranes.

Chapter 20

The Central and Peripheral Nervous Systems

In This Chapter

- ◆ The division of the central nervous system
- ◆ Meninges, gray and white matter, gyri and sulci, ventricles, and brainwaves
- ◆ The function of the brain lobes
- ◆ Sensory and motor homunculi, limbic system, thalamus, and brainstem
- ◆ Spinal cord and spinal nerves

The CNS, which includes the brain and spinal cord, has long been a source of mystery, for its functions are so complex, and the way those functions are carried out is not immediately obvious. The spinal cord is easier; using the knowledge we have learned about neurons, we can map regions and track reflexes, because its overall function is mainly to carry impulses from one place to another.

The brain, on the other hand, is far more complex and much subtler. I mean, compare it to the heart; both the function and the means of carrying it out is immediately visible, through the alternating contraction and relaxation of the atria and the ventricles. Now explain anger. How, let alone where, does the brain experience anger? We now know where (the amygdala), but we still don't know how!

A brain alone, however, is not enough. The nervous system must be able to communicate with all parts of the body. I always teach my students that three systems have physical connections with every part of the body: cardiovascular, lymphatic, and nervous. These nervous system connections allow sensory impulses to the brain (afferent), and motor impulses from the brain (efferent). Nerves of the PNS may be somatic in nature (sensing and moving the body in response), which is mainly voluntary, or those nerves may be autonomic in nature, which is involuntary.

Even if your train of thought leaves the station, at least part of your body is working in the background. This chapter discusses the architecture and workings of the two branches of the peripheral nervous system (PNS): the *somatic nervous system* (*SNS*) and the *autonomic nervous system* (*ANS*).

You've Got Some Nerve!

Sorry, I couldn't resist the bad joke at the expense of the spinal cord. One of the great fields of medical research is neurology, and the various subdisciplines, such as neuroanatomy. We can thank a Mr. Phineas Gage, whose tragic railway accident in Vermont, in which a bar was explosively propelled through his skull on September 13, 1848, forever altering his personality and temperament, for the acceleration of our knowledge of the brain. Although cases such as these, and strokes mentioned above, are very valuable, as is much of the imaging technology, there are some very real barriers to research.

We can learn a great deal by studying animal brains, but how do you ask them their feelings, or to solve a mathematical problem? We can ask that of people, but how do you measure microscopic anatomical changes without harming the patient? Certain ethical issues are insurmountable. Apart from this, how do you look at a brain in action, at the microscopic level, without altering it? It is a little like Schrödinger's famous cat and the quantum theory of superposition. Yet, despite these limitations, a great deal has been learned about the brain and the spinal cord.

Meninges

To start off, how do you protect your brain, given its consistency, somewhat akin to soft cheese? If you remember the cranial bones (see Chapter 6), that's great, but they are awfully hard on the inside, too, so you need a bit of protection from banging into the inner wall. As a matter of fact, that is what a concussion is, in which your head hits a surface, and your brain, through inertia, keeps moving until it bangs into the inside of the skull. There is no way to stop that with a major impact, but you can lessen it, and even prevent it with weaker impacts, through a series of structures called the *cranial meninges*.

The meninges, which are also found around the spinal cord, where they are called *spinal meninges*, are composed of multiple layers, which serve not only to protect the CNS by cushioning it (the way a water bed would stop you from hitting the floor), but also to help supply nutrients via the *cerebrospinal fluid* (*CSF*), which will be discussed later in this chapter (see Figure 20.1). Imagine that you are practicing an ancient medical technique called trepanation, in which you cut a hole in your skull, presumably to release the "evil spirits" within. Anyway, what would you see as you began to cut?

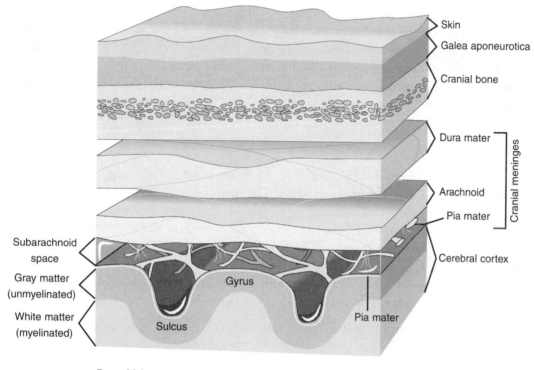

Figure 20.1

These are the many layers of tissue connecting the brain to the skull.

(*LifeART©1989–2001, Lippincott Williams & Wilkins*)

First of all, you have to cut your way through the skin, and then, depending on *where you cut*, you have to cut through either muscle or connective tissue to get to the periosteum, and then the cranial bone. This is where your saw comes in handy. My anatomy students are always a bit taken aback by how hard it is! Luckily, brain surgeons today have miniature radial saws. After cutting through the bone, things get interesting. If you expect to see the hills and valleys (gyri and sulci) right away, think again.

The first layer you'll see will be the outer *dura mater*, which forms a continuous outer layer for both the brain and the spinal cord. There are two layers for the dura mater, given these hard-to-remember names: the outer layer, and the inner layer. Around the spinal cord is an extra space above the dura mater, between it and the walls of the vertebral foramen (see Chapter 6); it has a familiar name that will suddenly make sense:

the *epidural* layer (of childbirth fame). Between the two layers there may be a dural sinus (you will meet them by name later) filled with CSF.

Just inside the inner layer of the dura mater is the *arachnoid layer*, which occasionally has villi extending into the dura mater. Under the dura, between the dura mater and the arachnoid, is the *subdural space*; it is speculated that this is just an artifact of tissue preparation, and doesn't exist in a living person. Does the name arachnoid make you think of spiders? The pattern of collagen and elastic fibers under the arachnoid layer, in the unromantically named *subarachnoid space*, look enough like a spider's web to warrant the name. CSF also flows through the subarachnoid space. At the bottom of this layer is the *pia mater*, which is immediately surrounding the *cerebral cortex* of the brain (*cortex* = the outer layer).

Shades of Gray

If you cut a brain from front to back with a deli slicer (not that I'm suggesting you do so), you will see clear layers in gray and white. The outer layer, the cerebral cortex, is gray, as are layers closer to the thalamus and hypothalamus, but much of the rest of it is white. As you learned in Chapter 19, the gray layer consists of unmyelinated nerve tissue, such as cell bodies, dendrites, and unmyelinated axons. This area is likely to have numerous synapses, and it is at these synapses that the real work of the brain is done. For example, processing of sensory input, and the coordination of muscle responses happen in this gray area (see the section on sensory and motor control later in this chapter).

The white matter, with all the oligodendrocytes (see Chapter 19) covering the axons with myelin, is basically just transmitting information (via nerve impulses), either to areas within the same hemisphere, or in areas from one hemisphere to another. Upon dissection it is actually possible to remove a great deal of the other tissue in the brain, so as to highlight the actual tracts of nerve fibers in the brain. One of the better-known areas is the *corpus callosum*, which carries the impulses between the left and right hemispheres. For this reason, the corpus callosum is made of white matter.

The spinal cord also has gray and white sections (see Figure 20.2). The gray sections, with all their synapses, make a great deal more sense in the context of the sensory and motor divisions of the spinal nerves. The white matter, however, carrying impulses up and down, is covered here.

If you look at the image showing various cross-sections of the spinal cord, you will begin to get a sense of some of the sensory (to the brain) and motor (from the brain) pathways. The concepts of sensory and motor are explored in more detail later in this chapter. In the sensory pathway, to begin with, there are a number of neurons involved. A stimulus first triggers a receptor, which may or may not be part of a neuron; in any case, that triggers a *first-order neuron* in the PNS, which will ultimately trigger a *second-order neuron* in the gray matter of the CNS.

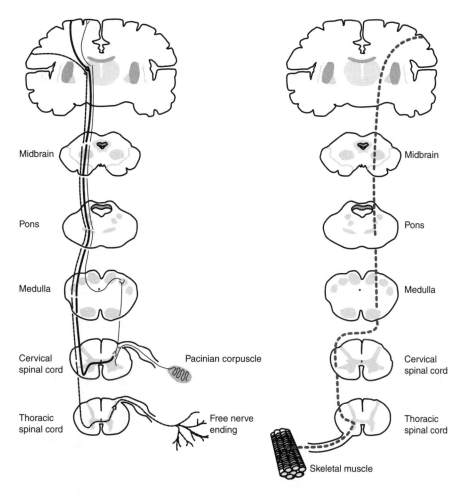

Figure 20.2

This figure shows various transverse sections of the spinal cord and brainstem, as well as the sensory and motor pathways.

(LifeART©1989–2001, Lippincott Williams & Wilkins)

Midbrain	Midbrain
Pons	Pons
Medulla	Medulla
Cervical spinal cord	Cervical spinal cord
Thoracic spinal cord	Thoracic spinal cord

Pacinian corpuscle

Free nerve ending

Skeletal muscle

The Big Picture

Writing a book like this is a bit of a false construct. Think about it. This chapter is basically about the CNS (brain and spinal cord), and the next chapter is about the nerves of the PNS. In reality they are inextricably linked. The spinal nerves also connect the CNS with every part of the body via sensory and motor nerves. If you take any one body system, or even *part* of a system, out of the picture, things start to unravel!

The axon from that neuron continues up the spinal cord (in the white matter) and crosses over to the opposite side in the *pyramids* of the *medulla oblongata*, the area immediately superior to, and continuous with, the spinal cord. The second-order neuron triggers a *third-order neuron* in the *thalamus* (the base of the brain), which in turn triggers a neuron in the primary sensory cortex, which enables you to consciously perceive the stimulus that started this relay race in the first place.

The odd butterfly shape in the center of the spinal cord is made of the anterior and posterior gray horns. Surrounding them, however, are regions of white matter. These tracts of myelinated axons form four basic areas: the *posterior column*, the *anterior column*, and the left and right *lateral columns*. Don't forget that all sensory and motor paths in the spinal column ultimately cross over from the right to the left side, and vice versa. Where this crossover occurs has a great deal to do with the origin of the impulse.

Flex Your Muscles

Once again, don't forget to break down the words. Take *anterior spinothalamic tract* as an example. A *tract* is a pathway for neural impulses (which will be white matter, since there are no synapses). Where does it go? As the name says, from the spine to the thalamus. Where is it in the spinal cord? Easy! The front! In the middle of the front, or slightly off to one side? Well, since we are bilateral, and this is only for one side of the body, it must be slightly off to the side!

The impulses from receptors for delicate sensation, such as vibration, pressure, and gentle touch, travel up the *posterior column pathways*. Crude touch and pressure (as opposed to fine touch) travel up *the anterior spinothalmic tract* in the anterior column. Sensation of pain and temperature is carried in the *lateral spinothalmic tract* in the lateral column. The fine sensations cross over in the medulla, but the latter two (crude touch and pain) cross over in the gray matter at the level of the initial spinal nerve, and then proceed straight up.

Proprioception, the perception of muscle position (see Chapter 21), travels in two separate pathways due to different destinations. The posterior column pathway takes it, along with all the other spinal sensory information, to the *primary sensory cortex* in the *parietal lobe* (see the sections on sensory and motor control, and on brain lobes, in this chapter). As you shall see in Chapter 21, the *special senses* (smell, vision, taste, hearing, and balance) all have specific cranial nerves associated with them, and do not all go to the parietal lobe.

The other pathway for proprioception goes to the *cerebellum*, which coordinates complex muscle movements (see the following section), via the *posterior spinocerebellar tract*, with the message crossing over in the gray matter at the location of the spinal nerve. What is interesting here is that sensory information that crosses over *also* goes up on the same side as it entered. This means that each side of the cerebellum receives impulses from both sides of the body. A nifty way to ensure that our movements are balanced when they need to be!

As far as muscle control goes, in addition to the cerebellum, the *primary motor cortex* in the *frontal lobe* controls muscle contraction. (Here, and elsewhere, some of the information is covered in more detail later in this chapter.) The impulses from the frontal lobe travel through large pathways called the *cerebral peduncle*, which is in the front of the *mesencephalon* and ultimately cross over mostly in the *pyramids* (a section of crossed pathways—left to right and vice versa) of the *medulla oblongata*. The neurons that cross over go along the lateral corticospinal tract (frontal cortex to the spine), but there are other neurons that don't cross over until the level of the spinal nerve. Those neurons send their impulses down the *anterior corticospinal tract*.

Gyri and Sulci

One of the things that makes a brain instantly recognizable is the shape of its surface. The long, meandering hills (*gyri*, singular = *gyrus*) and valleys (*sulci*, singular = *sulcus*) provide a greater surface area to the brain. In general, the greater the surface area of the brain, the larger the amount of the brain that is active, and the more neurons in those areas. You can see, simply by measuring the skulls of fossil hominids (our prehuman ancestors), that the cranial cavity encasing the skull has evolved to be larger. By comparing the brains of other species, you can see that the human brain has far more, and more pronounced, gyri and sulci than more primitive mammals. This means that, although the entire brain has evolved and become larger, the cerebral hemispheres, (in particular the gray matter of the cerebral cortex) have evolved faster, as mental abilities became more important to our survival than physical ones. Think of it this way: The brain's surface area is 2200 cm^2 (or 2.5 ft^2), which can only fit in our skull by folding in on itself. As the brain grows and develops, the gray matter grows much faster than the white matter, which forces it to form gyri and sulci as it folds in on itself.

Also, in terms of development, it is helpful to know about the different areas of the brain and brainstem. The developing brain only has three parts, or vesicles, by the end of the third week of life. The parts of this *primary brain* are as follows, from the spinal cord up: *rhombencephalon* ("hindbrain"), *mesencephalon* ("midbrain"), and *prosencephalon* ("forebrain"). The following table traces the development of each of those three sections.

The Development of the Areas of the Brain

Primary Brain Vesicles (3 Weeks)	Secondary Brain Vesicles (6 Weeks)	Brain Areas at Birth (40 Weeks)	Consist of These Areas
Prosencephalon	Telencephalon diencephalon	Cerebrum diencephalon	Cerebrum, thalamus, hypothalamus
Mesencephalon	Mesencephalon	Mesencephalon	Mesencephalon
Rhombencephalon	Metencephalon Myelencephalon	Cerebellum, pons, medulla oblongata	Cerebellum, pons, medulla oblongata

CSF and Ventricles

The CSF, or *cerebrospinal fluid* (removed in spinal taps), has three basic functions in the CNS: to cushion the CNS, to support the CNS (literally making the brain and spinal fluid buoyant!), and to act as an additional transport system between the CNS and the body.

As you can see from Figure 20.3, the CNS is constantly flowing around the CNS. This flow includes movement through each of the four ventricles: the two lateral ventricles (above and on either side of the thalamus), the third ventricle (running up the middle of the thalamus), and the fourth ventricle (between the pons and the cerebellum). The CSF in the fourth ventricle continues down the middle of the spinal cord in the central canal. The CSF also flows around the meninges in the following dural sinuses: *superior* and *inferior sinuses* (located above and inside the dural fold between the two hemispheres known as the *falx cerebri*), and in the *transverse sinus* (*tentorium cerebelli*, which is analogous to the bony process in the skull of the cat, between the cerebellum and the occipital bone).

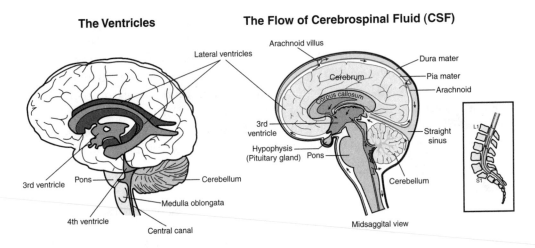

The Ventricles

The Flow of Cerebrospinal Fluid (CSF)

Figure 20.3

The flow of cerebrospinal fluid (CSF) in and around the central nervous system (CNS).

(LifeART©1989–2001, Lippincott Williams & Wilkins)

Remember those ependymal cells lining the ventricles (see Chapter 20)? Well, they are permeable to the CSF, thus making the CSF and the interstitial fluid in the brain and spinal cord chemically connected. This allows a spinal tap to measure the chemical health of the CNS because of chemicals that travel into the CSF from the brain and spinal tissue. CSF can travel into the veins at the arachnoid villi (refer to Figure 20.3). The total volume of the CSF is about 150 ml, but a choroid plexus in both the third and fourth ventricles constantly produces more, for a total of 500 ml a day. This means that the CSF is replaced approximately every eight hours, which makes a spinal tap a valuable tool to monitor changes.

EEG and Brainwaves

Remember the action potential and nerve impulses from Chapter 19? If each depolarization ranges from –70 mv to +30 mv, and each brain has billions of neurons, it makes sense that all that electrical activity will generate a bioelectric field. That field can be measured, just as the electric field is measured around the heart with an EKG. Because of measuring the brain's electric field, and with cephalic meaning "head," it makes sense that the devise is called an *electroencephalogram* (EEG).

There are four basic types of brainwaves. As you are reading this, your mind is concentrating on a task, which means your brain is producing high-frequency *beta waves* (see Figure 20.4), which it also does when you are under stress. Now, if you were to lie down and rest with your eyes closed (as is often practiced in meditation or relaxation therapy), your brain would start to produce the more relaxed *alpha waves*. If you were to be tapped by someone while resting, or if you heard a knock at your door, those alpha waves would run for the hills as the beta waves return.

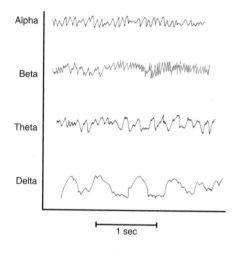

Figure 20.4

The four types of brainwaves as measured by an electroencephalogram (EEG).

Theta waves are lower-frequency, larger waves, seen in children, and briefly during sleep in adults. If you have ever been really frustrated as an adult (who hasn't?), you have produced theta waves. What kids and frustrated adults have to do with one another, other than the former causing the latter, I can't really say …. The last type of waves are the largest, and lowest frequency *delta waves*, which appear during deep sleep, regardless of your age. If either theta or delta waves are seen in adults when awake, under normal circumstances, it is seen as a sign of brain damage.

The Control Center

One of the amazing things about the brain is the idea of neuronal plasticity. Neurons have the ability to adapt and adjust to different sensory input, often within minutes. This explains how blind people have such well-developed hearing and sense of touch: Unused neurons are used to process sensory input from the other senses. This becomes important as you look at the specific areas of the brain. Many people want to know which parts of the brain are used for what functions.

Although we have a general idea, things are actually more complicated than they look on the surface. On a cellular level, apart from a fissure here and a sulcus there, few regions of the brain are clearly anatomically distinguishable. Functions can extend into neighboring areas, not to mention areas carrying out multiple functions. We might be able to specify, at least to a certain extent, the speech area, but what about consciousness, which appears to involve gray matter all over both hemispheres?

Dem Lobes, Dem Lobes, Dem Four Lobes

Now I told you those cranial bones would come in handy, because the four lobes of the cerebrum can all be found directly under the cranial bones (see Chapter 6) of the same name (see Figure 20.5). These lobes are separated by various sulci, such as the central sulcus separating the frontal and parietal lobes. Each of these lobes is, however, continuous with the others when you look at the white matter deeper within.

Figure 20.5

The four lobes of the brain, and several important areas of specialization.

(LifeART©1989–2001, Lippincott Williams & Wilkins)

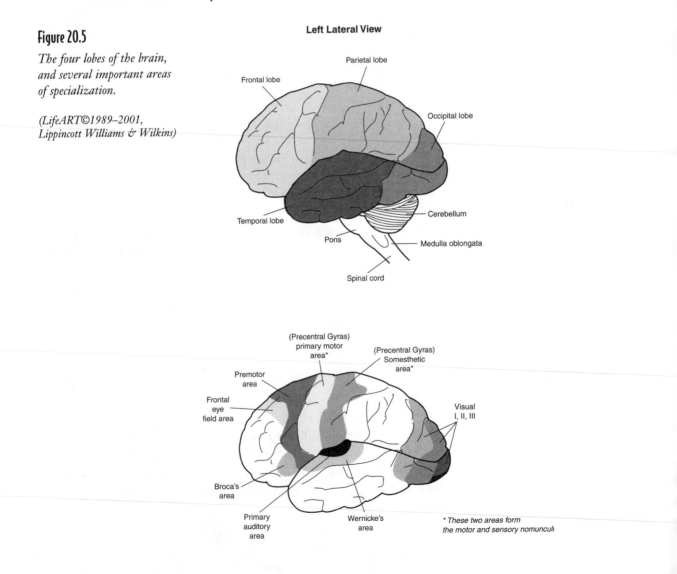

The division is really somewhat arbitrary, even, to a certain extent functionally, because the functions that appear to be on the border of one lobe may very well have active neurons in the neighboring lobe. Some functions are so obviously distant from other lobes that they can be considered unique to that lobe, and yet others, such as consciousness really involve multiple lobes. With that in mind, the following table indicates the areas of each lobe, and the functions associated with them.

The Functions of Each Lobe of the Brain

Area of the Lobe	Specific Function(s)
Frontal Lobe	
Precentral gyrus	Voluntary skeletal muscle control
Premotor cortex	Somatic motor association (interpreting complex movement)
Broca's area	Speech (muscles and breathing)
Prefrontal cortex	Coordinated information from all the association areas of the brain to predict consequences of future events (frustration/anxiety centered here)
Parietal Lobe	
Postcentral gyrus	Most senses
Somatosensory association area	Interpretation of senses to gauge the proper reactions
Wernicke's or general interpretive area (usually left side)	Connects senses with auditory or visual memory (mental connections!), thus important to personality
Temporal Lobe	
Olfactory cortex	Smell
Auditory cortex	Hearing
Auditory association area	Interpretation of hearing, such as word recognition (receptive speech)
Wernicke's area	Part of both lobes
Occipital Lobe	
Visual cortex	Vision
Visual association area	Interpretation of vision, such as written word recognition

Cerebellum

The pair of cerebellar lobes, which extend out the back of the *pons* (see the brainstem), around the fourth ventricle, is pretty cool. Similar to the cerebrum, there are two *cerebellar hemispheres* that are each divided into an *anterior lobe* and a *posterior* (*middle*) *lobe*, plus a lobe involved with balance hidden underneath, that has the coolest name: *flocculonodular lobe*.

Yeah, yeah, but what does it *do?* Well, it turns out that the primary motor cortex doesn't really do the greatest of jobs! When instructions are sent from the frontal cortex, the cerebellum coordinates the movement to make it smooth. With very complex movements, such as playing an instrument, the cerebellum is essential! Walking would not be possible if it weren't for the coordination of the muscles, and the detected balance, in the cerebellum.

Sensory and Motor Control

Our brain is wired to not only monitor the sensation of the body, but to control the movement as well. Around the middle of the cerebral cortex is a large sulcus called, well, the *central sulcus*. This marks the division between the frontal and parietal lobes. Given the central location of the sulcus, the gyri are given extremely mundane names: the precentral gyrus (of the frontal lobe), and the postcentral gyrus (of the parietal lobe). Just under each gyrus is gray matter involved in processing the sensory input, as well as motor responses.

To get a true appreciation of this, you must look at the picture of the sensory and motor homunculi (*homunculus* = little man). As you can see from Figure 20.6, certain regions of the body take up much larger regions of the brain to control. If you were to build a body with these proportions, well, let's just say that I wouldn't want to meet a person like that in a dark alley!

Let's look at each homunculus in a bit more detail. The sensory homunculus is located in the postcentral gyrus of the parietal lobe. Most of the sensory nerves, as a matter of fact, end up in the postcentral gyrus. Take a look at the size of the lips. With so much sensation located there, is it any wonder that we humans love to kiss? We see something similar in infants, who are forever exploring their world by stuffing objects in their mouths. Holding hands is also a big one, for the hands take up an enormous part of the sensory homunculus. On the other hand, you may be surprised how little of the brain involves sensation in the genitals! This makes the old expression that "men think with their penises" seem even more out of proportion!

The *motor homunculus* of the frontal lobe has some equally cool things. Notice the area that controls the movement of the face. Humans have an enormous amount of flexibility in our facial muscles. With a little practice, you can even make a living at it (think Jim Carrey). Think of how different it is reading an e-mail from someone versus seeing them say it in person. Beyond inflection of the voice, which people can mask, an enormous amount comes across in facial expressions. As such, it makes perfect sense that we also need a large part of the brain to interpret those facial expressions.

Flex Your Muscles

One thing that is important to remember here is that the division between the lobes has more to do with the location of the cranial bones than any clear physical division within the brain. Functions of the lobes that occur at the edge of a lobe often involve portions of both lobes along the border (in a form of neural NAFTA).

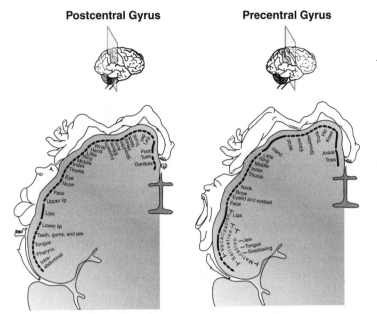

Postcentral Gyrus

Precentral Gyrus

Figure 20.6

These rather strange little figures illustrate the proportion of the cerebral cortex that is taken up by sensory and motor processing for the various parts of the body.

In terms of art, think about all of the art forms that involve the hands (drawing, painting, sculpting, playing music, and so on) and you can see that they have the brain power to back it up. On the other hand, dancers get kind of screwed in terms of brain area associated with body movement. It makes Gene Kelly's achievements even *more* amazing!

Limbic System: Seat of Emotion

The limbic system is an amazing thing, for it surrounds the thalamus and the hypothalamus, making it a relatively primitive part of the brain, and yet it controls no less than our emotions. Think about that for a minute. It seems that our emotions evolved long ago, for these structures have been in our ancestors for a long time. Remember how the gray matter is concentrated along the gyri and sulci of the cerebral hemisphere? Well, the region around the thalamus, on top of the brainstem, has a number of areas filled with a gray matter. Together they make up the *limbic system* (see Figure 20.7).

There are three basic functions of the limbic system: (1) emotional states and drives, (2) helping with memory storage and retrieval, and (3) linking the unconscious work of the brainstem with the conscious control of the cerebral cortex; the limbic system seems to be associated with pleasure and pain. A frontal section of the brain shows numerous areas of white matter carrying nerve impulses between the areas of the limbic system and the cerebral cortex. This accounts for many interesting things, such as the way a smell can trigger a long buried memory, or how we put such emotional contexts to meals; both these situations involve the more primitive senses and memory or emotion. To learn more about certain specific functions by region refer to the following table.

Regions of the Limbic System and Their Functions

Region	Function
Amygdaloid body	Connects limbic system, senses, and the cerebral cortex, anger
Fornix	Connects hippocampus to the hypothalamus
Hippocampus	Learning (storage of and access to long-term memory)
Mammillary body	Processes smell and controls chewing and swallowing reflexes
Stria terminalis	Involved in sexual behavior

Figure 20.7

The various regions of the limbic system, which is involved in emotions and memory.

(LifeART©1989–2001, Lippincott Williams & Wilkins)

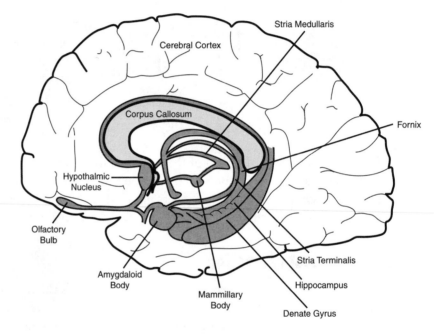

Limbic System

Thalamus: The Switching Station

You have so far learned about the spinal cord, the cerebrum, and the cerebellum, so it's time to learn about what connects these sections. Just below the cerebrum, and above the brainstem, is the diencephalon, which is made of the hypothalamus (our old friend from the endocrine system), and just above it, the *thalamus*. Just by looking at the location of the cranial nerves, it is clear that the thalamus is crucial in terms of transmitting sensory impulses to the cerebral cortex.

It turns out that a tremendous amount of sensory information is transmitted to the thalamus every second. One of the jobs of the thalamus is to filter out the majority of it and only send on what is necessary to the primary sensory cortex. Think about it. Does the skin on your back sense the clothing against your skin? I bet you hadn't really noticed before. Remember the feel of clothing against your skin after going swimming, and you get a sense of how the thalamus has to decide when the stimuli are important enough to take notice. In addition to the relay of sensory information, the thalamus also relays information about emotions from the hypothalamus to the frontal cortex.

As for the hypothalamus, it turns out that it does a little more than just crank out regulatory hormones (see Chapter 18). All of its functions are at least partially autonomic, for both the thalamus and hypothalamus function on a subconscious level. The hypothalamus regulates such things as causing a smile when you are amused (have you ever tried to keep a straight face when telling a joke?), heart rate, blood pressure (with the pons and medulla), and body temperature, to name a few. In addition the hypothalamus coordinates voluntary and involuntary functions, which is why we can feel nervous about an event that is yet to happen. The hypothalamus also produces some emotions, as well as giving you the basic drives of hunger and thirst. That's one busy little sucker!

Brainstem and Beyond

We're down to the last of it, the brainstem! Just under the diencephalon are the *mesencephalon*, the *metencephalon* (*pons* and cerebellum), and the *myelencephalon* (*medulla oblongata*). The *mesencephalon*, or *midbrain*, is mainly involved in transmitting impulses from one area to another, but it also controls certain involuntary activity of muscles, such as muscle tone and posture, as well as helping to maintain consciousness.

Standing right in front of the cerebellum and the fourth ventricle, the pons (half of the metencephalon along with the cerebellum) transmits impulses here and there, in particular with the cerebellum. The pons is also involved in aspects of the regulation of breathing.

All that's left is the oldest section of the brainstem, the *myelencephalon*, or *medulla oblongata*. More relays happen here. If you have a hard time keeping track of what cranial nerve attaches where, just look at Figure 20.8, and you can learn about the nerves by looking here for the Roman numeral, and comparing that with the table on the cranial nerves in Chapter 19. As you would imagine for such a primitive area of the brain, it controls such basics as heart rate and respiration. Much of this autonomic control is due to the connection of the N X, vagus nerve.

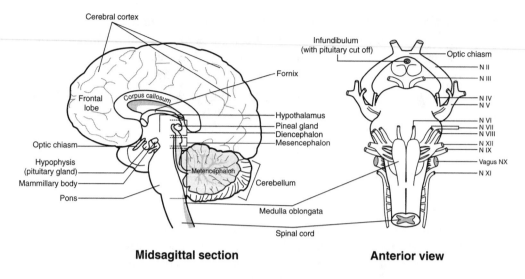

Midsagittal section   Anterior view

Figure 20.8

The regions of the brainstem, and the location where it branches into the cranial nerves.

(LifeART©1989–2001, Lippincott Williams & Wilkins)

Spinal Cord and Peripheral Nerves

A look at the spinal cord, our bridge to the PNS, shows that the sensory and motor tracts occupy areas of the cord. Remember the posterior and anterior gray horns? Well the sensory information travels through the posterior gray horn. This gray horn is further divided into *somatic sensory*, which receives information from the muscles and skin, and *visceral sensory*, which receives information from the internal viscera. This gray horn contains sensory cell bodies not found in the dorsal root ganglia. There are three pathways for sensory information, from the *ventral ramus* and *dorsal ramus* (*ramus* = branch, plural = *rami*) of the spinal nerve, or from the sympathetic nerve. As part of the posterior gray horn, it makes sense that sensory information travels through the posterior root, regardless of its source.

Flex Your Muscles

The names for the parts of a nerve are analogous to the names for parts of a muscle. If you can remember epimysium (see Chapter 8), then you will know what the *epineurium* is (the fascia around the nerve). All the nerve parts have equivalent muscle parts; they even both have *fascicles*!

Think about the sensory nerves, which have dendrites from the receptors, and axons up to the brain. The cell bodies, which are, of course, wider than the dendrites and axons, need to go somewhere! The ganglia are just the place to keep them! Sensory nerve cell bodies go in the *dorsal root ganglia*, and motor nerve cell bodies go in the *sympathetic ganglia*, which sit anterior to the ventral root, but branch off of the spinal nerve itself. (See the section on the sympathetic and parasympathetic branches of the ANS later in this chapter.) Other cell bodies can be found in the gray horns of the spinal cord. (See the sensory and motor sections later in this chapter.) In addition to each of these series of ganglia, there are some larger ganglia that form part of the sympathetic branch: the celiac ganglion, the superior mesenteric ganglion, and the inferior mesenteric ganglion.

The structure of a nerve is similar to the structure of a muscle. For example, the fascia (connective tissue; see Chapter 8) around the nerve is called *epineurium* (like the epimysium). Around each bundle of nerve fibers (neurons) called *fascicles* (so named because they are little bundles wrapped in fascia), the fascia is called *perineurium*. There is even an endoneurium around each neuron, which acts as an extra layer of protection, even in the presence of Schwann cells (see Chapter 19).

Just as there are 12 pairs of cranial nerves that attach to the brainstem, attached to the spinal cord there are four sets of peripheral nerves: 8 pairs of cervical nerves, 12 pairs of thoracic nerves, 5 pairs of lumbar nerves, and 5 pairs of sacral nerves (see Figure 20.9). These are easy to remember if you think about the divisions of the spine (see Chapter 6).

A typical college level text will explore the various functions of the spinal nerves as well as the *plexuses*, which are braids of nerves that control various functions. This is also true in terms of the muscles, in terms of the nerves that innervate them. In general, the more muscles controlled by nerves, the wider the spinal cord in that region. Since the arms are controlled by cervical nerves, and the legs are controlled by the lumbar and sacral nerves, the thoracic nerves have little to do, hence the narrowness of the thoracic spinal cord.

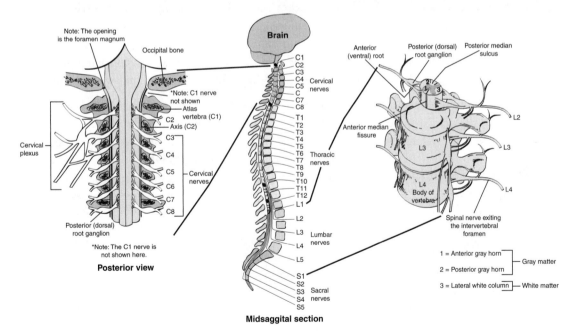

Figure 20.9

The spinal nerves, with a detailed view of the cervical nerves, and a close up of the pathways leaving the lumbar vertebrae.

(LifeART©1989–2001, Lippincott Williams & Wilkins)

Sympathetic: Yikes!

Our *autonomic nervous system (ANS)* is responsible for our unconscious control of our body, as opposed to our *somatic nervous system*, which is our thoughts, our basic conscious perception of the senses, and our movements and other reactions to those senses. The ANS is a very old section of the brain, which, despite all that it controls, does not require as much energy to function. Much of this control is centered in the medulla oblongata, which is at the base of the brain stem, just superior to the spinal cord.

Although the processing is centered in the medulla, other sections of the brain are involved. The thalamus, as a switching station for sensory input, and an emotional center (due to the limbic system), sends input to the hypothalamus. The hypothalamus is the real control center, sending instructions to the pons (higher respiratory function) and the medulla, which in turn sends instructions along the sympathetic and parasympathetic pathways. In the simplest sense, the body needs to be able to gear up in response to periods of stress or danger. Equally important is the ability to slow down when the danger is over, in order to focus on those areas ignored when under threat.

Gearing Up Organs

If you were in my classroom I would either suddenly scream out loud or surreptitiously drop a large metal lab stool. Anything to make you jump! That loud surprise is enough to get the average person's sympathetic division kicked into high gear (I could release a live tiger, but I want to keep my job!). Think about your body's reaction to such a jump, especially these organs: heart, lungs, and brain (at least that part associated with alertness).

Here's where you need to put the pieces together. I'll get you started. One of the connections in the sympathetic pathway involves stimulating the adrenal glands to release epinephrine (adrenaline). It is the epinephrine that gives you that sense of terror: elevated heart rate, deeper respiration, heightened senses (enough to make you jumpy). You've experienced the what, but *why* does it happen? To answer that, you need to think about oxygen and glucose, our quick energy source.

Flex Your Muscles

Myoglobin? Hemoglobin? It might help if you remember that hemoglobin has four subunits, each of which carries a single O_2 molecule (four in all). Each of those four subunits is similar in structure to a myoglobin molecule, which of course can carry only one O_2 molecule. Incidentally, the higher quantity of myoglobin in beef than chicken is what makes red meat red!

Since oxygen is so important for extracting energy from glucose (see Chapter 3), it makes sense that you would breathe deeply: The more deeply you breathe, the more oxygen you get. The more oxygen, the more energy. So why the fast heart rate? Easy! To get the oxygen to the muscles and to get rid of all the carbon dioxide you are releasing from your muscles. As a matter of fact, oxygen is easier to take up in the muscles due to the presence of myoglobin, which functions, like hemoglobin, to carry the oxygen.

You might remember (from Chapter 2) that the body stores some of the excess glucose as the polysaccharide (multiple sugar) glycogen. In times of stress the majority of the glucose is used by the muscles, so it makes perfect sense that glycogen is stored there, especially with all that myoglobin.

The other place where glycogen is stored is the liver. Glycogen is released as a result of the action of epinephrine from the adrenal glands, which are also stimulated by the sympathetic division. Epinephrine also works to dilate the airways in the lungs and increase the heart rate. In addition to all these subconscious actions, the brain becomes more alert, no doubt to anticipate and react to danger.

Not all organs are stimulated, however, as certain actions need to be set aside during an emergency in order to place the energy where it is most needed. The kidneys are put on hold, leaving the filtering of waste until later. It may seem paradoxical, but the digestive system spends the inning in the dugout, too. On the surface it seems that such an action is foolhardy, as you need energy in an emergency, but the energy you can absorb in a short time is very small, considering the energy needed for the rest of the digestive process. For this reason, the body opts for using energy reserves (glycogen).

Medical Records

Have you ever felt like your intestines are all knotted up when you were under stress? Those cramps are the result of peristalsis, or the movement of food along the GI tract coming to a halt. You can blame it on the sympathetic division of your ANS!

Sympathetic Pathways

The sympathetic pathways follow the general arrangement of the spinal nerves, but after leaving the spinal cord they decide to go their own way (see Figure 20.10). This makes a certain amount of sense, as the spinal nerves have branches that do need to reach the skin, stimulate the muscles, and so on. The sympathetic branches are concerned with the other organs, such as those in the thoracic and abdominopelvic cavities (see Chapter 1).

Remember those pesky anterior and posterior roots, and how the spinal nerve branched off in front, making a sympathetic ganglion (called a *paravertebral ganglion*)? All the nerve fibers before the ganglion have the rather unromantic name *preganglionic fibers*. The *postganglionic fibers* (*after* the ganglion) are also called *sympathetic nerves!* As if that's not enough, the ganglia are connected vertically, forming a pathway parallel to the spinal cord called the *sympathetic chain (or trunk)*. Along the way, a cardiac and pulmonary plexus can be found, as well as three ganglia (called *prevertebral ganglia*): the celiac (to and from the stomach, pancreas, liver, and adrenals), the superior mesenteric (to and from the small and large intestine), and the inferior mesenteric (to and from the rectum, kidneys, urinary bladder, and reproductive organs).

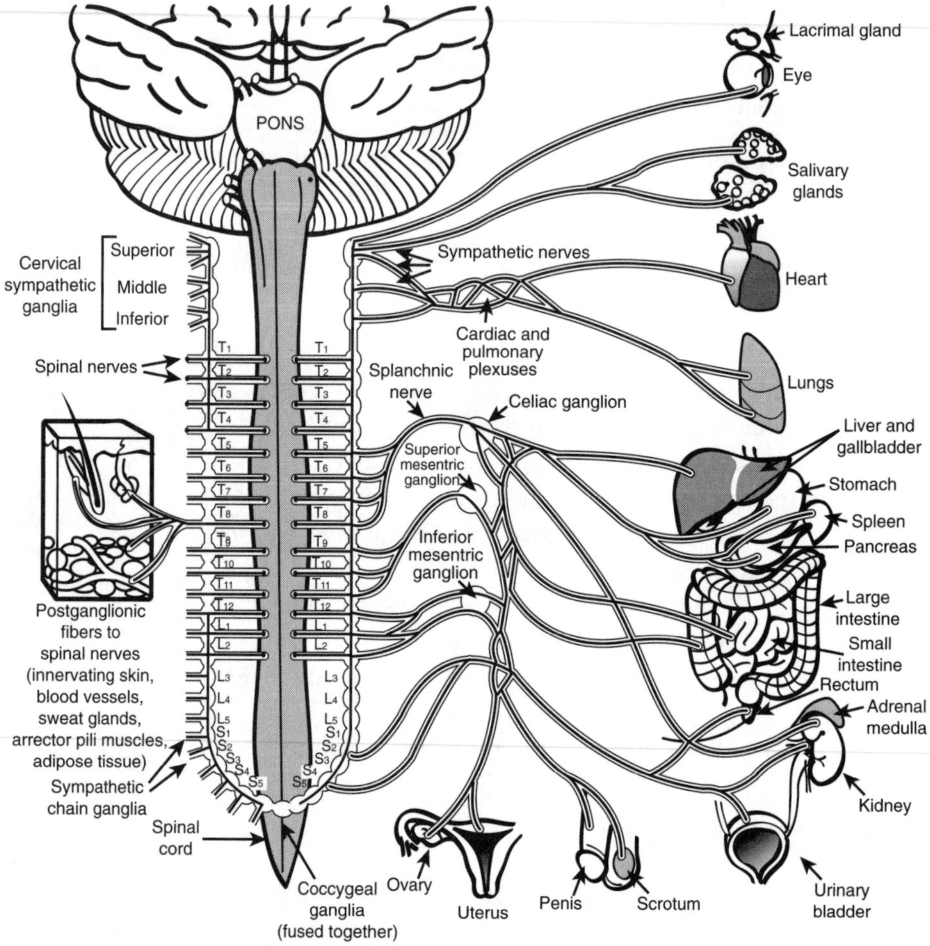

Figure 20.10

Note the parallel pathways and redundant pathways, through each ganglion and plexus, innervating the various organs.

Parasympathetic: Aaaaaaaaaaah!

Okay, okay! Calm down! No, literally. How *do* you calm down? Here comes that danged homeostasis thing again. To stay on an even keel, you need an opposite response to the fight or flight response, which is called "rest and repose."

Winding Organs Down

In addition to being used to gradually bring the body back from the fight or flight response, this branch of the ANS is more characteristic of a typical body response under normal circumstances. This may not be true of all species, or even all members within a species. Prey would most likely be using the sympathetic branch more often than predators. In addition, in social animals, those with a lower social standing in a dominance hierarchy (as in baboons) would also most likely be more sympathetic than parasympathetic. This is not to say that such animals are "wired differently," but that they are forced to live under more stressful situations, more of the time.

In any case, simply think about all of the sympathetic responses and, well, reverse them! Slow down the heart, and the breathing. There is no need to be on the alert, so all those organ systems that were set aside, such as the digestive system, can now get back to work. Despite all of this, there are actually a number of organs where there is no parasympathetic stimulation. Think about the adrenal glands; epinephrine (E) and norepinephrine (NE) are released due to sympathetic stimulation. Have you ever noticed that, once the scare is over, it takes a while for you to feel normal? There is no parasympathetic stimulation, because the adrenal glands only release the E and NE under stress, thus there is no point to sending such a message. The following table lists sympathetic and parasympathetic changes, divided according to organ.

Sympathetic and Parasympathetic Changes

Organs	Sympathetic Changes	Parasympathetic Changes
Adrenal gland	Epinephrine and norepinephrine	None (no innervation)
Digestive	Reduced activity	Increased activity
Secretion	Reduced secretions	Increased secretions
Liver	Hydrolysis of glycogen	Glycogen produced
Eye	Pupil dilation	Pupil constriction
Fat tissue	Fats broken down	None
Heart	Faster heart rate	Slower heart rate
Veins	Vasoconstriction	None
Arteries	Varies by organ	None
Kidneys	Less urine produced	More urine produced
Lacrimal glands	None (no innervation)	
Lungs	Bronchiole dilation	Bronchiole constriction
Muscles	Stronger contraction	None (no innervation)
Skin	Sweat, erect hairs	None (no innervation)

Parasympathetic Pathways

When you compare the sympathetic and parasympathetic pathways, it's as if the spinal nerves were thinking, "I've got too much to do; I just can't be bothered to slow the body down!" Seriously, the vast majority of the spinal nerves are completely irrelevant! The tenth cranial nerve (N X), also called the vagus nerve, leaves either side of the medulla oblongata and just goes straight down the body on either side of the spine (see Figure 20.11).

To be fair, other nerves are involved, such as N III (oculomotor), N VII (facial), and N IX (glossopharyngeal). Each nerve has a specific ganglion involved. As for the vagus nerve, there are a number of plexuses involved: the cardiac and pulmonary plexus (to and from the heart and lungs), the celiac plexus (to and from the liver, stomach, and pancreas), and the hypogastric plexus (to and from the small and large intestines, and all of the pelvic organs). Three of the sacral nerves (S_2 through S_4) also help out by joining the hypogastric plexus.

Figure 20.11

Note how the majority of the organs are innervated by branches of the tenth cranial nerve, the vagus nerve (N X).

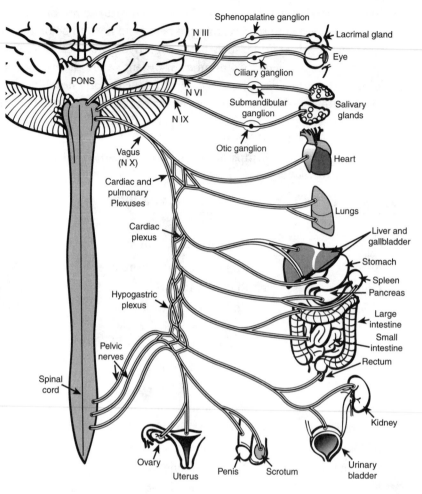

The Least You Need to Know

◆ The central nervous system (CNS) is protected by the meninges and the cerebrospinal fluid (CSF).

◆ The unmyelinated gray matter is filled with synapses, and is where the work of the brain is done; the myelinated white matter carries impulses from one region to another.

◆ The more primitive areas of the brain, thalamus, limbic system, and brainstem, control autonomic functions, and link the autonomic nervous system (ANS) to the CNS.

◆ The sympathetic division of the ANS coordinates the "fight or flight" response, and the parasympathetic division of the ANS coordinates the "rest and repose" response.

Chapter 21

Connecting to the Outside World

In This Chapter

- ◆ Vision and hearing
- ◆ Olfaction and taste
- ◆ Touch
- ◆ Balance and equilibrium
- ◆ Proprioception

Without receiving messages from the outside world, the rest of the nervous system is only so good. Our survival depends on being able to know what's happening in the outside world and about our place in that world. Students ask me now and then about extrasensory perception (ESP), of which there is no scientific support (which is why it deserves no further mention in this book), but they incorrectly label "it" the sixth sense. They were raised on the idea of five senses, when there are in fact seven. The traditional five—smell, vision, hearing, taste, and touch—are all concerned with receiving information about the outside world.

The other two senses, balance and proprioception, relate to the body's position in the world; balance is concerned primarily with the position of the head, and proprioception is concerned with the position of the muscles in relation to the body (for example, where your arms are). Despite their differences, all seven senses require some stimulus to open ion channels in neurons. So let's look around, listen for nuances, and make some sense out of the senses.

Windows to the Soul

Vision is a remarkable sense when you consider the difficulties. The lens inverts and reverses the image on the retina, and it is the job of the brain to correct it so we don't try to walk on the ceiling! We are able to perceive color (when not all our mammalian relatives can), and to see in the dark (but only in black and white). Beyond getting the message, we need to interpret it, which is a phenomenal process indeed! Many aspects of the structure of the eye deserve special mention, so let's take a look!

The Eyes Have It

First, the name eyeball really makes sense, because the shape is spherical, which allows a wide range of motion, rather akin to a ball-and-socket joint, within the orbit of the skull. Three pairs of muscle (see Figure 21.1) perform the motions themselves: the superior and inferior rectus muscles (moving the eye up and down), the medial and lateral rectus muscles (moving the eye medial and lateral), and the superior and inferior oblique (moving the eye down and lateral, and up and lateral). The majority of these are controlled by the aptly named *oculomotor nerve (N III)*.

Figure 21.1

The gross structures of the eye.

(LifeART©1989–2001, Lippincott Williams & Wilkins)

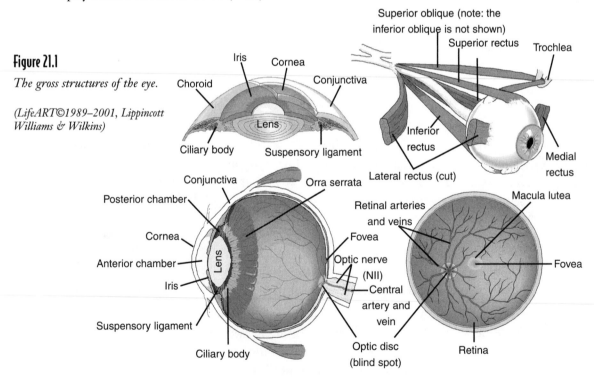

Given all the motions, the eyes require special services. The eyeball's movements all require lubrication, or you would quickly be bleeding out of your eyes as the tissues scraped against the orbit and the eyelids. The solution is the use of *lacrimal glands* (located under the later end of each eyebrow), which release the salty serous fluid (tears) through the tear ducts. Ducts alone, however, are not enough, for the fluid needs to be spread, which is the function of the eyelids. Tears drain into the nasal cavity around the lacrimal bone (see Chapter 6), which is why you get the sniffles when you cry!

The white of each eye is called the *sclera*, and its dense connective tissue provides the shape and support for the fluid-filled body (refer to Figure 21.1). The transparent *cornea*, covered by the *bulbar conjunctiva* (hence the name *conjunctivitis* for the infection), is immediately over the fluid filled *anterior cavity* (filled with *aqueous humor*). This fluid lubricates the movement of the *iris* in relation to the cornea (the *anterior cavity*) and the lens (the *posterior cavity*).

Light enters though the *pupil*, which is a circular opening directly in the front of the lens. This helps to place the image on the *macula*, which is the area of the retina responsible for the bulk of the visual perception. Attached to the lens is the *ciliary body*; the *ciliary muscles* here change the shape of your lens to actually focus the image on the macula (try looking at something *very close* to your eye and you may feel the effect of those muscles). The *central fovea* is in the middle of the macula and directly in line from the lens; the central fovea has the sharpest vision of the entire retina (refer to Figure 21.1). Between the retina and the outer sclera is the *choroid* layer, which is filled with blood vessels.

One problem is that there needs to be a way of regulating the amount of light that gets in. This is the job of the *iris*, which is where the eye gets its color (refer to Figure 21.1). The iris is a muscular structure that adjusts the aperture, or opening, of the pupil, similar to a camera. With too much light, the iris closes and thus limits the light, and when the environment is dark, it opens to allow more light. Certain drugs affect the size of the pupils, which makes the pupil a quick, noninvasive way of determining sobriety, or at least the necessity for further testing.

Although all parts of the eye contribute to our ability to see, it is the retina that is directly responsible for our vision. As I mentioned earlier, the images we see are focused on the retina, albeit upside down and backward. The back of the eye is also the most logical place to put the *optic nerve* (N II), not to mention the retinal arteries and veins. The convergence of all the neurons and blood vessels at the *optic disk* is off-center so it doesn't get in the way of the *macula*, which is the primary area for vision. Other areas are used, but with far less frequency; these areas are responsible for peripheral vision, in which the resolution is very poor, but it is nonetheless extremely useful in terms of responding to motion.

Medical Records

The shape of the eyeball varies from person to person, but this causes a problem. The spherical shape allows the lens to focus the image on the retina. A shortened eyeball places the focus behind the retina, producing *myopia*, or nearsightedness. On the other hand, an elongated eyeball places the focus in front of the retina, producing *hypermetropia*, or farsightedness.

Rods and Cones

Perception of color and of black and white involves different cells. These are the *rods* and *cones* you've heard about, and they are indeed shaped as their names suggest (see Figure 21.2).

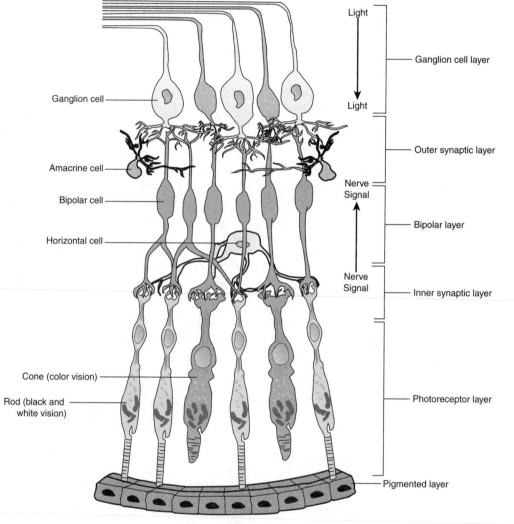

Figure 21.2

The rods and cones are toward the rear of the retina, which causes the nerve impulses to go in the opposite direction of the light.

(LifeART©1989–2001, Lippincott Williams & Wilkins)

The rods are more sensitive to light, but they are only used in perceiving the world as black and white. As such, we cannot see color by moonlight, but without light we cannot see anything, *regardless* of how long we wait for our eyes to adjust. Cones, on the other hand, perceive color, but a great deal more light is needed. The central fovea has the sharpest vision because of the high number of cones there. There are, however, no rods there; because of this, you cannot see anything in the fovea at night, but you can see objects if you look slightly off-center.

You might assume that we perceive light at the inner edge of the retina, but this is not the case. Partly to protect the retina from bright light, the rods and cones are near the back of the retina, in an area called the *photoreceptor layer*. Just behind this layer is a *pigment epithelium*, which uses melanin to prevent the reflection of light inside the eye. Light rays open ion channels that stimulate the cells of the bipolar cells, which transmit the message to the ganglion cells on the surface of the retina, and then to the optic nerve.

Can You Hear Me Now?

There is an elegant, mechanical simplicity to sound. Sound is basically based on vibrations. Vibrating objects produce vibrations, basically waves of compression, in the air that is in contact with those objects. The vibrating skin of a drum will thus produce more sound than a rigid piece of cement struck with the same stick.

The vibration of objects helps explain why you can hear a vibrating tuning fork that is touching the top of your head, because the sound travels through vibrations in your cranial bones (see Chapter 6). As a kid, did you ever make a "phone" out of two paper cups and a string? For the phone to work, the string needs to be taut, because the vibrations must travel along the string; the two paper cups function to focus the vibrations onto the string on one end, and to amplify the vibrations from the string on the other end.

This, in a nutshell, is what happens (remember to keep the string taut):

1. When you contract the muscles of your larynx (see Chapter 13), the exhaled air causes vibrations in the vocal folds.

2. Vibrations in the vocal folds cause vibrations in the air, which are amplified through the resonating chambers of our paranasal sinuses (see Chapter 6).

3. The vibrations in the air molecules cause the cup on the speaking end to vibrate.

4. Vibrations in the cup cause the string to vibrate.

5. Vibrations in the string cause the cup on the listening end to vibrate.

6. Vibrations in the cup cause the waves of compression known as sound waves at the same frequency as the original vibration!

As you will see, what happens in our ear is really just more of the same!

Not Just Corn

In many ways, the ear is little more than a slightly more sophisticated string-and-cup "phone." The shape of the outer ear, called the *pinna* (the outer edge is the *hilus*, and the bottom is the *lobule*), captures the sound waves and amplifies them. If you have never tried it, try cupping your hand behind your ear when you listen to music, or to someone talking, and you will notice how much that amplifies the sound.

Medical Records

In Chapter 16, I talked about the ceruminous glands in the ear canal that produce wax to keep objects and critters away from our eardrum. What you might not know is that an overproduction of wax, and subsequent buildup, can result in a blockage that makes it extremely difficult to hear. Due to the danger of puncturing one's eardrum, it is wise to leave the cleaning to a medical professional experienced in such matters! That's why doctors say the only thing you should clean your ear with is your elbow!

This is the principle behind the original, needless to say old-fashioned, horn-shaped hearing aids. It also explains why rabbits evolved such big ears: the better to hear predators. (Big ears, however, can also be used to radiate excess body heat.) The ear canal, or *external auditory meatus*, is also involved in the amplification and transmission of sound, serving to focus it on the next stop, the eardrum, which is analogous to the first cup in our cup-and-string "phone."

Good Vibrations

The eardrum is also called the *tympanic membrane*. As you can imagine, its role is to receive the vibrations focused by the ear canal. Amplifying the vibrations even further requires a little help. Remember the string in our "phone" from earlier? Well, there are three little bones that do the job of the string here, three bones that students often forget when building a list of all 206 bones: the auditory ossicles. These "little bones" are, in order from the eardrum inward, the malleus ("hammer"), the incus ("anvil"), and the stapes ("stirrup").

Medical Records

You've probably noticed the connection between the ear and the throat if you have ever had your ears plugged. By swallowing you lower the air pressure in your throat, thus equalizing the pressure at both ends of the Eustachian tube (pop!).

Just like any other bones with movable joints, there are ligaments involved. Apart from the articular cartilage (see Chapter 6) between the bones, there are two ligaments—a superior one and a lateral one, each of which attaches the malleus to the inside wall of inner ear (in the temporal bone)—that help to stabilize the malleus as a result of vibrations in the eardrum. The auditory ossicles further amplify the sound. The end of the stapes attaches to the oval window of the cochlea. This connection transfers the vibration from the ossicles to the cochlear fluid. Figure 21.3 shows the parts of the ear that are involved in the transmission and amplification of sound.

At this point it is helpful to know about the other lesser-known *round window*. When the vibrations make their way through the coil of the cochlea, the round window's thin membrane allows the vibrations to escape. Without it, the vibrations would bounce around the cochlea until they eventually lost energy, which would sound a bit like a fast-repeating echo. The round window is right near the opening of the auditory tube, or *Eustachian tube*, which is where the vibrations, now transferred to the air in the inner ear, finally escape to the throat.

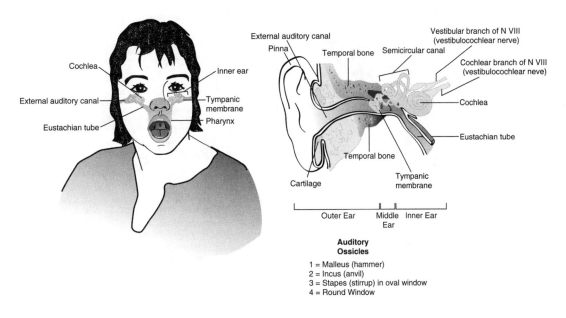

Figure 21.3

The structures of the ear are primarily involved in the transmission and amplification of sound.

(LifeART©1989–2001, Lippincott Williams & Wilkins)

Hair Cells

Now that the vibrations are in the cochlear fluid, how does that translate to nerve impulses? The cochlea is a snail-like coil (*membranous labyrinth*) filled with a fluid called *perilymph*, which is more like cerebrospinal fluid than lymph. Connected to the cochlear branch of the vestibulocochlear nerve (N VIII), this is the part of the inner ear that receives the vibrations from the auditory ossicles.

There are two regions called the *scala tympani*, that contain the perilymph, but sandwiched between the two, along the outer edge of the cochlea, is the spiral *cochlear duct* (*scala media*), which is filled with *endolymph* (similar to the fluid inside cells). Within this duct is a gelatinous glycoprotein layer called the *tectorial membrane*, which stimulates hair cells that lie beneath it (see Figure 21.4). As we shall see, these hair cells are found in the part of the ear that is responsible for our sensation of balance.

Figure 21.4

The friction of the hair cells against the tectorial membrane stimulates the neurons of the cochlear nerve.

(LifeART©1989–2001, Lippincott Williams & Wilkins)

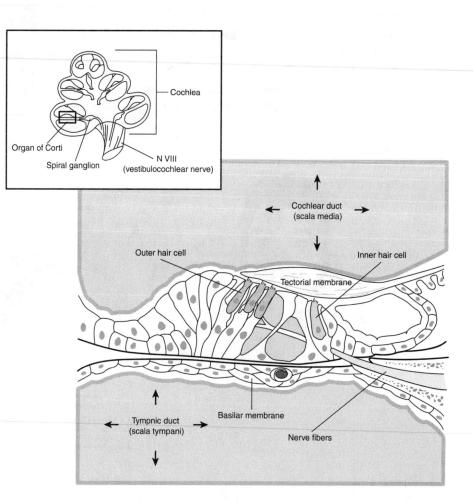

Inside the cochlear duct is a spiral *organ of Corti* with multiple rows of *hair cells* (with supporting cells) that lie just above the *basilar membrane* that separates the cochlear duct from the scala tympani. These hair cells have anywhere from 50 to 100 (depending on their location) microvilli that are surrounded by endolymph. The tips of these microvilli touch the bottom of the tectorial membrane.

When the vibrations of the stapes are transferred into pressure waves in the perilymph of the scala tympani, the *vestibular membrane* (on the upper portion of cochlear duct, opposite the basilar membrane) starts to vibrate. This vibration creates pressure waves in the endolymph, causing movement of the tectorial membrane. The movement of the tectorial membrane creates contact with the microvilli of the hair cells, which opens up ions on the basal ends, which are attached to the cochlear branch of the vestibulocochlear nerve (N VIII). *Sound!*

The Nose Knows

The sense of smell is perhaps one of the most ancient of senses. The ability to detect the presence of molecules in one's environment (for that is what smell actually is) is extremely important if you are on the search for food. This sense is highly evolved in sharks, which can sense the presence of blood (and therefore food) in the water with a concentration of one teaspoon in an Olympic-sized swimming pool! Prior to evolving as a passageway for air in higher animals, the nares, or nostrils, of a shark were (and are today) used only for smell (sharks and fish breathe through their mouths). This sense is so ancient that the *olfactory nerve (N I)* is not really a nerve at all, but an extension of an ancient olfactory lobe in fish.

A quick look at the nose reveals that the olfactory bulbs (the end of the nerves) are way up in the ceiling of the nasal cavity. This shows that the nose, with its turbinates (see Chapters 6 and 13), has long since evolved to place the warming, humidifying, and filtering of air as its top function, which is amply helped by the extra surface area of the turbinates. The sense of smell is, nonetheless, present and important.

Olfactory Bulbs

Smell is an interesting sense, for unlike vision or sound, in which interpretation is so important, smell is more straightforward. For smell to respond to molecules, the molecule needs to temporarily bind to receptors in the roof of the nasal cavity. Unlike other receptors for senses such as vision, hearing, balance, and taste, the *olfactory receptors* are actually bipolar neurons, and the chemicals that bond with the cilia, known as *olfactory hairs*, on that neuron will stimulate an action potential directly.

The olfactory nerve (N I) ends with olfactory bulbs at the base of the ethmoid bone (see Figure 21.5); axons pass through the *cribriform plate* of that bone, thus giving it a somewhat porous appearance. Beneath this plate is a connective tissue layer, with the *olfactory epithelium* beneath it, with supporting cells interspersed with the dendrites of the bipolar neurons. *Olfactory* or *Bowman's glands* in the connective tissue produce mucus that is crucial to the sense of smell.

The mucus provides a fluid that dissolves the gaseous odors. The molecules thus make contact with the olfactory hairs within this thin film of mucus. If the mucus were not continually replaced, any dissolved odor would continue to make contact with the olfactory hairs, and continue to stimulate the neuron. Talk about a smell that wouldn't go away! By replacing the mucus, the olfactory hairs are available for any new molecules that come their way!

Get a Whiff of Turbulence

The turbinates are a collection of bones (the left and right inferior nasal concha) and bone landmarks (the superior and middle nasal concha, left and right, which are extensions on the ethmoid bone, see Chapter 6) that increase the turbulence of the air inhaled. By inhaling more deeply, the molecules are forced higher up in the nasal cavity, up to the olfactory nerves. This explains why people sniff in order to better smell their world.

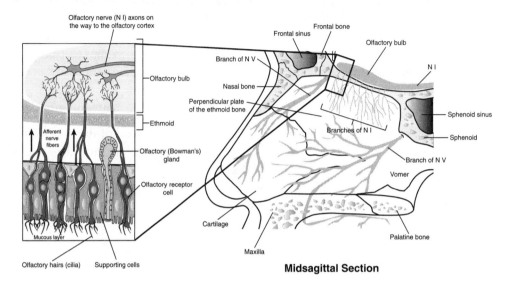

Figure 21.5

The olfactory bulbs, at the end of the olfactory nerve (N I), are found in the roof of the nasal cavity.

(LifeART©1989–2001, Lippincott Williams & Wilkins)

In Good Taste

Taste is an integral part of our lives, often to our detriment, as we tend to go toward tastes we prefer, rather than those that are beneficial to our health. There is a certain amount of irony here, for we evolved our *taste buds* to emphasize those molecules that helped us to survive. The irony melts away, however, when you examine the details. One of the greatest of our needs is energy. As such, certain molecules were so important for our evolutionary ancestors to acquire that they evolved a preference for them. (If the molecules didn't taste good, why would an animal eat them?) These high-energy molecules include simple and complex carbohydrates and fats, and it's the proportions of these compounds in our diet that is the problem. As with smell, these gustatory receptors are based on the presence of molecules binding directly with the receptors. The tongue itself is made of numerous papillae, and the tongue's 10,000 or so taste buds are found not on the upper surface of each papilla, but along the sides. The bulk of each taste bud is found in the wall of the papillae, with *gustatory hairs*, basically one microvillus on each *gustatory receptor cell*, sticking out through a hole called a *taste pore*. Each taste bud contains about 50 of these receptor cells (each of which only lasts about 10 days and so they are constantly replaced), interspersed with supporting cells, all of which are epithelial, unlike those in the nasal cavity.

Similar to smell, the molecules we taste are dissolved in a fluid, in this case saliva. Contact with the receptor cells opens ion channels that stimulate the neurons responsible for taste (see Figure 21.6). All food is a combination of four primary tastes: sour (acidic), bitter (basic), salty, and sweet. These four tastes can be perceived at any taste bud, although the taste buds in certain parts of the tongue appear to be more sensitive to specific tastes, and are thus called *taste zones*: sweet and salty in the front, bitter in the back, and sour on the sides. It may seem ironic that the *body* (the center) of the tongue has very few taste buds, but then again, the teeth we chew our food with are on the edges, so it makes sense that those edges would be rich in taste buds!

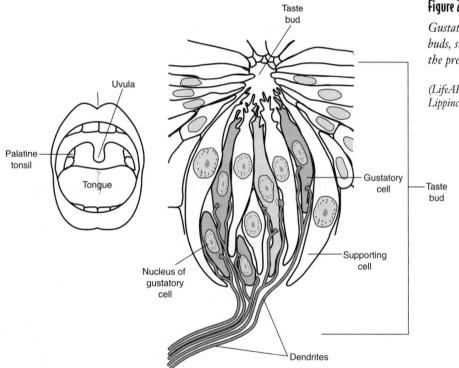

Figure 21.6

Gustatory receptors, or taste buds, stimulate neurons in the presence of food molecules.

(LifeART©1989–2001, Lippincott Williams & Wilkins)

Don't Be So Touchy!

If you are still troubled by the idea of seven senses, how about this idea: It can be argued that touch is more than one sense. In addition to touch in general, what about the perception of levels of pressure, which has its own separate receptor, so necessary when holding delicate objects? How about temperature, which requires different receptors for the sensation of hot and for the sensation of cold? What about perceiving pain, with yet another receptor?

Splitting this up into multiple senses sounds like a bit much. After all, vision has separate receptors for black and white (rods) and for color (cones). Perhaps it is best to just think of touch as a complex sense involving multiple receptor types, similar to taste. Touch was discussed in detail in Chapter 16, where you can find an image showing the location of some of the nerves and receptors in the skin.

Knowing Which Way Is Up

In the inner ear the cranial nerve that goes to the inner ear is called the *vestibulocochlear nerve* (N VIII). This name illustrates the dual nature of this sensory nerve. Unlike mixed nerves with both sensory and motor functions, this one is all sensory. The *cochlear branch* goes to the cochlea, which, as I discussed earlier, is where the sounds we hear stimulate the nerve. The *vestibular branch*, however, has nothing to do with sound. This branch sends messages to our brains about balance, from a place that should have a familiar look to it!

As I mentioned earlier, we need the three axes (X, Y, and Z) to understand the areas of the body and body position. Mathematicians probably felt pretty good about themselves when they realized the need for three axes, but nature had us beat by hundreds of millions of years. The three axes are beautifully represented by the three *semicircular canals*, which make up the *vestibular system*.

Looking like some sort of weird alien snail head, the three semicircular canals (anterior, posterior, and lateral), which are about 90 degrees from one another, stick out of the wide end of the cochlea, one for each of the three axes. Let's try a little experiment. Sit upright and close your eyes. Now point straight up and open your eyes. (Perhaps you ought to memorize these instructions first, since your eyes will be closed!) Did you point in the right direction? Close your eyes again and lean all the way left, right, forward, and back. Were you always able to point up? That's what your vestibular sense does for you!

The sense of balance or equilibrium can be divided into two parts: *static equilibrium*, which mainly has to do with the head's position in terms of gravity, and *dynamic equilibrium*, which responds to rapid movement, such as spinning, accelerating, or braking to a halt. The key to all this is the movement of either fluid or crystals (yup, crystals) as perceived by *hair cells*. As I mentioned earlier when I discussed hearing, hair cells are not really nerve cells but epithelial cells that open ion channels to stimulate neurons.

Flex Your Muscles

If you really want to get a sense of your semicircular canals, stand up and spin around 10 times very fast and then stop! Dizzying, isn't it! Didn't you ever do this as a child? Now imagine the endolymph flowing around and around with you as you spin. When you *stop*, inertia will keep the endolymph sloshing around until friction and gravity finally slows it down! Think about what you see when you are dizzy: the room doesn't look like it spins all the way around, but only part way, and then the image snaps back, as your brain tries to reconcile messages from your retina *and* your hair cells!

For static equilibrium, there are two areas where the semicircular canals join the cochlea, called the *utricle* and the *saccule*. In both miniature organs are calcium carbonate crystals ($CaCO_3$) called *otoliths*, or *otoconia*, that rest on a layer of glycoproteins called the *otolithic membrane*. Embedded in the base of that gelatinous membrane, from the surface of every hair cell, are microvilli called *steriocilia*, and a longer single cilium called a *kinocilium*. Between each of the hair cells is a supporting epithelial cell. The upper surface of the otolithic membrane is slippery, and tilting the head causes the otoliths to slide around, changing the pressure on the various hair cells, which in turn stimulate the neurons and alerts the brain to general changes in static equilibrium.

If you tilt your head slowly, the brain can still sense the change in direction (the otoliths still move), but if you are turned around very slowly (with eyes closed), or you accelerate in an airplane very slowly, you are not likely to perceive it. Rather than using otoliths, the semicircular canals use the movement of a fluid called *endolymph* to stimulate neurons as a result of a change in dynamic equilibrium. Imagine carrying a mug of hot coffee that is very full. If you don't want to spill it and scald yourself, don't you move slowly? If you move quickly, the coffee will spill. The reason for this is inertia. The coffee tends to stay at rest, but the acceleration of the mug slams into the coffee at the back of the mug, generating a wave from friction that causes the coffee to slop over the side.

In this way, the endolymph in the semicircular canals moves only to sudden changes in motion, hence the word "dynamic." The base of each canal has an arched shape, and the neurons attach to the underside of the arch. Within the fluid is a structure called a christa, which is similar to the structures used in static equilibrium, except without otoliths. There are hair cells between supporting cells, and a gelatinous layer of glycoprotein, this time called a *cupola*. Sudden changes cause the endolymph to shift the cupola, like an ocean wave will shift kelp, moving the hairs, thus stimulating the neurons. Pretty cool!

Proprioception, or Where Are My Feet?

People always seem to take this sense for granted, which I always think a shame, because it is one of my favorites. I am talking about *proprioception*, or the ability to perceive body position, or more specifically the position of your muscles, even without using your eyes. It doesn't sound like much, but think about the last time you were walking upstairs talking to someone. Did you have to look at your feet in order to reach the next step? When Mozart was a child prodigy being paraded around before European royalty, his father used to have him perform one or more parlor tricks, such as playing harpsichord with a cloth covering the keys, or even playing while blindfolded.

I called those parlor tricks because any good piano player can do that without thinking. Can you type without looking at the keys? If so, you can thank proprioception. This sense is thrown into sharp relief on those rare occasions when you are deprived of it. I remember when I was in seventh-grade I broke clean through both the radius and ulna of my right arm. When the doctor was about to set my arm it was tied up in a tourniquet, and I was injected with a "horse syringe" full of Novocain. I remember looking to the left while answering the nurse's questions, thinking my right arm was hanging loose over the edge of the table. When I looked back I was shocked to find that it was being held with my hand up in the air! That shows how much we take proprioception for granted. One last thing. Say "Happy birthday." Think you know how your tongue knows how to move without you being able to see it? That's right—proprioception!

The Least You Need to Know

- The senses all involve stimuli from the outside world generating nerve impulses by opening ion channels.

- Light is focused by the lens onto the retina, where rods and cones stimulate neurons.

- Hearing is based on vibrations in sound being physically transferred by structures in the middle and inner ear.

- Smell involves molecules in the air dissolving in nasal mucus and directly stimulating neurons.

- Taste involves molecules in our food dissolving in saliva and indirectly stimulating neurons via epithelial gustatory receptors.

- Touch is actually a complex of separate sensations involving touch, pressure, temperature, and pain receptors.

- Balance, or static and dynamic equilibrium, registers changes in position by virtue of the movement of fluid or calcium carbonate crystals against a bed of glycoprotein.

- Proprioception, the last hidden sense, is the ability to know the position of muscles without a visual reference.

Part 7

First Comes Love ...

It seems a fitting end to come full circle and focus on the start of it all. Makin' babies is a wonderful thing, as every parent knows, but how the body actually does it is truly amazing! Think about it. At any given moment, each and every one of you (of reproductive age) is in the process of chopping some of your cells in half (making gametes), chromosomes and all. Gentlemen, you are, in a most ungentlemanly fashion, doing it on overdrive.

On the chance that they meet, deep in the fallopian tube, they join and start a roller coaster ride, dividing all the way, down to the uterus, where the baby-to-be buries herself or himself, or perhaps *both*, in Mom's uterine wall. From a formless blob to a glorious expression of both Mom's and Dad's genes, the nine-month journey ends with the baby's exit into the outer world.

As it turns out, the whole process is more than just a miracle of complexity; it's also a gooey sticky mess with a fetus swimming around in its own ... well, you'll just have to read on to find out!

Chapter 22

Vive La Difference!

In This Chapter

- ♦ The making of gametes
- ♦ Female and male reproductive anatomy
- ♦ Menstrual cycle
- ♦ Components of semen

Every living thing on the planet is the result of 3.8 billion years of evolution, from the first binary fission (little more than the replication of DNA) to sexual reproduction. Asexual reproduction has the advantage of only needing one parent, but at a cost: There is a lack of genetic diversity. Greater variation, apart from making family reunions more interesting, gives a species an edge in terms of avoiding extinction, but the cost, or some would say the benefit, is the need for a mate. By contributing only half of one's DNA, the concept of sex was born. This chapter examines how it all is done.

The Gift That Keeps On Giving

By allowing two individuals to contribute DNA, the offspring retain some traits of each parent, and yet benefit from the added variation of the mixture. The only problem with this arrangement is that something new had to come along, or organisms would keep doubling their DNA every generation. The solution was an interesting one: just cut the amount of DNA in half! This brings me to the world of *gametes*, or *sex cells*, such as the female ovum, and the male sperm.

One process makes gametes possible, and it is called *meiosis*. Since meiosis produces gametes, the process is also called *gametogenesis* (creation of gametes). Although the process is for the most part the same, there are some differences between the production of ova (*oogenesis* in the female's *ovaries*) and the production of sperm (*spermatogenesis* in the male's *testes* or *testicles*).

Me Oh My Ohsis

The first concept of importance in understanding gametogenesis is the idea that the cells produced, the gametes, need to have half the DNA of the normal body or *somatic* cells. Half, yes, but not any old half! You may remember from Chapters 3 and 10 that chromosomes come in pairs—one from each parent—called *homologous pairs*, that have the same genes, but that may have different alleles, or variations of that gene. It is crucial that in dividing the DNA up that the homologous pairs are split up. For these reasons, gametogenesis (meiosis) is also called *reduction division*.

Think of our somatic cells as having not 46 chromosomes, but 23 pairs of chromosomes (diploid, or 2 n), and our gametes as having 23 individual chromosomes with one chromosome from each of the 23 types (haploid, or n). Despite all this, meiosis does not produce two gametes from one somatic cell; it produces four. To understand this, it is important to know that the process of meiosis evolved from mitosis, and DNA replication must happen before the cell can divide. The two full sets of DNA (92 chromosomes in humans) equals four half sets of 23 chromosomes.

First, let's look at the stags of meiosis. You might remember from Chapter 3 that there were four basic stages in mitosis, or normal cell division: prophase, metaphase, anaphase, and telophase. Meiosis, on the other hand, is split into two basic parts, conveniently called meiosis I (complete with prophase I, metaphase I, and so on) and meiosis II (with, of course, a prophase II, and so on; see Figure 22.1). There are, however, some interesting things that happen to the chromosomes throughout all of this.

The uncoiled chromosomes replicate during interphase, and then coil up during prophase I, forming chromatids. Since each chromosome has an identical copy, half of a homologous pair plus its identical copy together are called *sister chromatids*. In this phase the two pairs of sister chromatids for the homologous pair join up to make a group of four chromatids (called a *tetrad*) in a process called *synapsis*; this is when *crossing over* occurs (to be discussed in the next section). In males these cells are called *primary spermatocytes*. In females these cells are called *primary oocytes*, and they are present at birth, and stay in that phase until just prior to ovulation.

During metaphase I the chromatids line up in the middle of the cell (just like mitosis), but anaphase I is very different from mitosis. In mitosis each sister chromatid splits apart forming separate chromatids, but in meiosis, the tetrad splits, and one sister chromatid migrates to one end of the cell, and the other migrates to the opposite end. The two cells that are produced during telophase I may each have 46 chromosomes, but the pairs are not true homologous pairs, but rather 23 pairs of sister chromatids.

The cells produced after telophase I are called secondary spermatocytes in males, and they continue all the way through meiosis II to produce *spermatozoa*, or *sperm*; in females, however, things are a little different, as we shall see. The following prophase II and metaphase II are fairly straightforward. In anaphase II the sister chromatids separate, as in mitosis, leaving two gametes for each of the previous daughter cells, for a total of four.

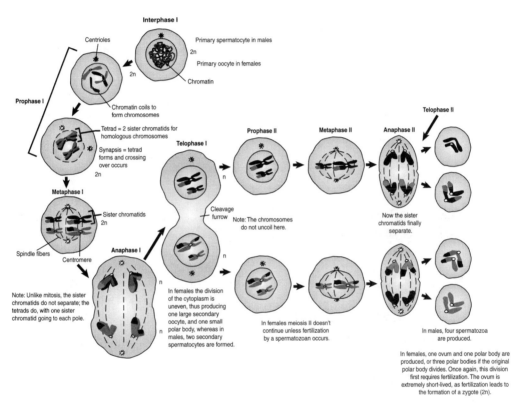

Figure 22.1

Meiosis is a multistep process, divided into meiosis I and meiosis II, which yields haploid gametes.

(LifeART©1989–2001, Lippincott Williams & Wilkins)

Flex Your Muscles

The ovum is what we call the female egg cell, but it is not, despite what you might think, what is released when a woman ovulates. Ovulation releases the secondary oocyte and one of the polar bodies. Surrounding these cells is a layer of *granulose cells*, in a layer called the *corona radiate*. Microvilli on the outside of those cells mixes with microvilli on the secondary oocyte. This area, which is filled with a clear glycoprotein, is called the *zona pellucida*.

In males all four will become spermatozoa, but in females the division of the cytoplasm is uneven in both telophase I and telophase II, leaving one large ovum, and three small *polar bodies* (which will eventually deteriorate). When ovulation occurs, the secondary oocyte, and one of the polar bodies, is released from an ovary. In females, the process of division in meiosis II is stopped during metaphase II, only to continue if it is fertilized by a sperm.

Cross on Over

Evolution is really smart, because the chromatids do something cool called crossing over that adds to the potential variety in your offspring. To understand the value of this, and everything from antibiotic resistance, to flu viruses, you should think about the parts of Charles Darwin's theory of evolution by means of natural selection, which can be summed up in the following five parts:

- There is variation in every population.

- Organisms compete for limited resources.

- Organisms have more offspring than can survive.

- Organisms pass traits on to their offspring (genetics).

- Those organisms with the most beneficial traits are more likely to survive and reproduce (natural selection).

Without variation, if all members of a species were the same, a population would be vulnerable to changes in the environment and thus to extinction. Any way to maximize variation is thus a good thing. The first way to do this is simply to have two parents each contributing half of their DNA to produce a "combo baby." With one homologous pair, each gamete has one half of that pair, producing only two possible gametes, or 2^1. With two pairs, there are four possible gametes, or 2^2. Since humans have 23 homologous pairs, 2^{23} possible gametes equals 8,388,608!

The number of potential offspring equals the number of gametes squared, so in humans, with our 23 homologous pairs, that means a grand total of 70,368,744,177,664 potential offspring! Crossing over, in turn, makes even more variation!

During synapsis, the tetrad that forms looks like two capital Xs, side by side (XX). The pinch in the middle is the centromere, where the spindle fibers, which pull the chromosomes to opposite ends of the cell, attach. Remember that each pair of sister chromatids represents chromosomes from each of the individual's parents. In crossing over, which occurs in prophase I, the arms of the chromosomes in the middle, which are from the two parents, will exchange material. What is produced, in the end, are the copies of the two original chromosomes, and two new chromosome combinations, with part of the DNA from each of the individual's parents. Variation rules!

Three's a Crowd

When everything goes perfectly, as it usually does, the sister chromatids separate during anaphase II, giving each of the gametes 23 individual chromosomes. There are two ways that this can go wrong. Sometimes, in anaphase II one chromatid may not separate, giving one gamete 24 chromosomes, and the other one only 22. When each of these combine with a gamete from the other sex, they produce a zygote with the wrong number of chromosomes.

In the first case, a normal gamete and one with 24 chromosomes produces a zygote with 47 chromosomes. Instead of all the chromosomes being in homologous pairs, one group will have three chromosomes, a condition known as *trisomy*. The other gamete will, when combined with a normal gamete, make a zygote with only 45 chromosomes. In this case, there will be one group of chromosomes that will have only one, instead of a pair, a condition known as *monosomy*.

In this way, you can see that the same mistake, in which the chromosomes fail to separate in meiosis (this failure is called *nondisjunction*), produces both monosomy and trisomy. There is one other place that nondisjunction can happen. During metaphase I, the tetrad that formed during synapsis in prophase I fails to come apart, and migrates to only one side during anaphase I. So the same point during meiosis (synapsis) that produces so much extra variety has the potential to do harm. Whether or not that harm happens is a matter of chance.

Medical Records

In humans, having an extra or missing chromosome can be disastrous for the individual's survival. In some forms of monosomy and trisomy (notably *Down syndrome*, also called *trisomy 21*) the person can live into adulthood, but in many others a baby either won't survive to term or will die within a short time after birth.

The Big Picture

Since each chromosome carries genes for many different parts of the body, each monosomy and trisomy disorder is characterized by problems in systems all over the body. In this way, some disorders can be detected using an ultrasound (see Chapter 23) to detect the development of specific organs. On a bright note, one can also tell if the baby is a girl or a boy by looking at the 23rd chromosome (XX = female, and XY = male).

Ladies' Room

During early embryonic development, you cannot tell the gender of the five- to six-week-old embryo from its physical appearance. Both females and males have internal gonads and two sets of ducts, the *mesonephric ducts*, and the *paramesonephric* or *Mueller's ducts*. In the female the mesonephric ducts deteriorate, and the paramesonephric ducts become the *uterine* or *Fallopian tubes*. In the male the opposite happens: The paramesonephric ducts deteriorate and the mesonephric ducts become the *ductus deferens* or *vas deferens*, and the developing testes descend. The tubes that develop in the female and male are involved in the transport of the gamete (female) or gametes (male) away from the organs that produced them, and so these sections of the chapter will follow that direction, from the gamete-producing organs on outward.

Medical Records

All humans are either male or female; there are no true human hermaphrodites. People with *Klinefelter's syndrome*, who are XXY, are males, and people with *Turner's syndrome*, who are XO (that is, have a single X chromosome), are female. In the strictest sense, gender is determined by the presence or absence of the Y chromosome, for all people need at least one X to survive. People who are labeled as hermaphrodites are always either female or male, but the development of their sexual organs has been affected by a hormonal imbalance, which caused the internal and external organs to develop along the lines of the opposite gender.

Unlike the male's reproductive system, which is primarily involved with the manufacture and delivery of sperm, the female's system needs also to be able to nurture a growing baby, not to mention to able to determine whether or not to flush out the system (as in menstruation) or keep the uterine lining (the *endometrium*) intact. All of this is accomplished in a fascinating system of tubes and hormonal pathways, not to mention a collect call from the developing embryo to the mother!

Her Plumbing

The two ovaries are oval-shaped organs, about the size and shape of the distal part of your thumb (see Figure 22.2). Located superior to the urinary bladder and lateral to the uterus, they are found in the pelvic cavity (see Chapter 1). The ovaries are attached to the uterus by both the *ovarian ligaments* and *broad ligaments*. The uterine tubes, ovaries, and ovarian ligaments are all encased in the broad ligaments, which are attached all the way along the sides of the uterus. The uterus is also held in place by the *uterosacral ligaments*, which, as the name suggests, attach to either side of the sacrum. Lastly, a *suspensory ligament* attaches the ovaries to the parietal surface of the pelvic cavity.

The deepest part of the ovary is called the *stroma*, and it is filled with connective tissues. The deepest part of the stroma is called the *medulla*, which is surrounded by the cortex, where the ovarian follicles are. The outer layer of the ovary is called the *germinal epithelium*. Given that ovulation involves the release of the gamete out through the surface of the germinal epithelium, it was mistakenly believed that the ova arose from those cells. Immediately beneath this layer is a connective tissue layer called the *tunica albuginea*, which connects the germinal epithelium with the follicle-rich cortex of the stroma.

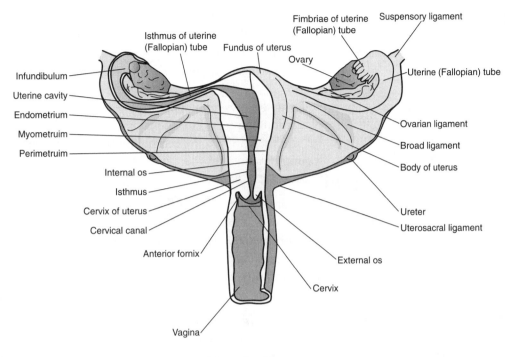

Posterior View

Figure 22.2

The layout of the female reproductive system.

(LifeART©1989–2001, Lippincott Williams & Wilkins)

Follicles go through several steps in their development. During the early phase, when the follicles remain arrested in meiosis I (primary oocytes), they are called primordial follicles. At puberty the follicles start to respond to follicle-stimulating hormone (FSH) from the adenohypophysis, and they become primary follicles. Many follicles do this each month, but most of these degenerate in a process called *atresia*. As meiosis continues, a *secondary follicle* develops, ultimately turning into a *mature follicle*, or *graafian follicle*, which is filled with fluid.

After the secondary oocyte is released during ovulation, the follicle turns into a *corpus luteum*, which becomes a mini hormone factory until either a hormonal signal tells it to continue until the placenta takes over (see the next section), or the lack of the signal causes it to degenerate, becoming the *corpus albicans*. Unlike the males, whose gametes are released directly into tubes, the secondary oocyte is released into the pelvic cavity near the blunt opening of the uterine (Fallopian) tube, with its numerous finger-like projections called *fimbriae*. The cilia in the uterine tube create a current that helps to draw the secondary oocyte inside.

Medical Records

The open-ended uterine tube can lead to a problem. Since the uterine tube is the place where fertilization occurs, it is possible for fertilization to occur outside the uterine tube. This is a form of ectopic (out of the normal location) pregnancy. A more dangerous variety occurs inside the uterine tube itself, which can cause the tube to rupture, killing both mother and baby.

The cilia on the mucosa propel the secondary oocyte, or the zygote if fertilization has occurred, along the uterine tube to the uterus. The majority of pregnancies involve implantation along, or near, the *fundus*, or top of the uterus (refer to Figure 22.2). The uterine wall is called the endometrium, and it undergoes development prior to ovulation. If implantation doesn't occur, the endometrium will slough off and exit through the cervix (the entrance to the uterus from the vagina), to the vagina, and out; this is known as the *menstrual flow*.

Fibrous connective tissue allows the vagina to expand, and muscle cells here and in the uterus help to propel the baby out during *parturition* (birth). At the deep end of the vagina is the cervix, and a recessed area around the cervix called the *fornix* (refer to Figure 22.2). At the superficial end is the vulva, where the vagina opens to the outside of the body.

Just above the *vaginal orifice* is the *external urethral orifice* that is used during urination. There are two sets of longitudinal folds of skin called *labia* that protect the area called the *vaginal vestibule*, which is between the labia and the vaginal orifice. *Greater* and *lesser vestibular glands* produce mucus that provides lubrication during sex. The smaller inner labia are called *labia minora*, and the larger outer labia are called *labia majora*, which differ from the labia minora not only in terms of having pubic hair and fat deposits, but also by having more sweat glands. Above the labia majora is the *mons pubis*, which cushions the pubic bone during sex.

At the junction of the two labia minora is the *clitoris*, which is a small organ in the shape of a cylinder involved in a woman's sexual excitement. The head of the clitoris, which is the only part exposed, is called the *glans*, which is the same name used for the head of the penis. The clitoris develops from the same undifferentiated tissue from which the male's penis develops. The two share many similarities: both contain erectile tissue (as do the nipples of both genders), both engorge with blood and enlarge in size during sexual excitement, both contain a form of foreskin. In females the foreskin is at the juncture between the labia majora, and is called the *prepuce*, which forms a kind of hood over the clitoris. The foreskin of the male is removed in many cultures in a procedure known as *circumcision*.

The breast, which provides nutrition for the baby, is discussed in Chapter 23.

Menstrual Cycle

The menstrual cycle is a marvelous collection of hormonal pathways that not only determine a woman's peak fertility, but also controls the lining of the uterus based on the presence or absence of an implanted embryo (see Figure 22.3). The pathways follow typical negative feedback loops (see Chapters 1 and 18). There are a number of characters in this little drama, which runs about every 28 days (although the actual number may vary), but if you take the time to learn them, you will see that together they make a glorious mosaic that is well worth the effort.

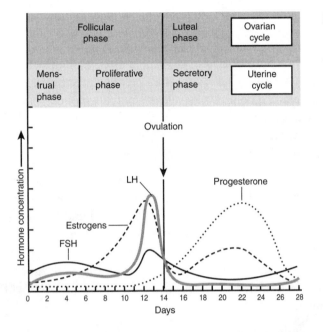

Figure 22.3

The various hormonal levels in the menstrual cycle.

The menstrual cycle affects predominantly only two parts of the body, the ovaries and the uterus. For that reason we refer to both a uterine and an ovarian cycle. Both cycles start with the onset of the menstrual flow and are divided in the middle by ovulation, which happens on approximately day 14 of the 28-day cycle. The first half of the ovarian cycle is called the *follicular phase,* for the hormones act to stimulate meiosis and to prepare the follicle for ovulation. After ovulation, since the follicle becomes a corpus luteum the cycle enters the luteal phase.

Medical Records _____

The ovaries typically alternate months in terms of ovulation, first one side then the other. Some women experience a pain from ovulation given the nifty term *mittelschmerz,* which means pain in the middle, on one side one month, and then the other the next month. When a woman ovulates on both sides, or releases two secondary oocytes from one ovary, that produces *fraternal twins,* or *dizygotic twins. Identical,* or *monozygotic twins,* are the result of one zygote that splits during embryonic development. A partial split results in *conjoined twins* (formerly called *Siamese twins*).

The uterine cycle starts with the *menstruation*, or *menses*, which lasts for about the first five days of the cycle. After that the hormones work to build the uterine wall in preparation for implantation. For this reason, this phase is called the *proliferative phase*, and as it ends with ovulation, it is also called the *preovulatory phase*. The first half of the cycle can vary a great deal, but the second half, called the *postovulatory phase* of course, is very consistent, lasting 14 days (until the onset of menstruation). Because of the increased amount of secretion by the glands of the endometrium, this phase is also called the secretory phase.

Starting it all we have a hormone from the true master gland, the hypothalamus, called *gonadotropic releasing hormone* (*GnRH*). This target organ for GnRH is the adenohypophysis, or anterior pituitary. The pituitary releases two hormones, *follicle-stimulating hormone* (*FSH*), which stimulates the development of the follicle prior to ovulation, and *luteinizing hormone* (*LH*), which stimulates the follicle to become the corpus luteum. Over the course of the follicular phase there is a rise in estrogen levels.

The estrogen level creates a temporary positive feedback loop that causes a rise in both FSH and LH. The surge of these hormones on day 13 of a 28-day cycle results in ovulation the following day. After ovulation the levels of FSH and LH drop to the lowest level in the cycle. The developing corpus luteum becomes a mini hormone factory, producing larger amounts of estrogen and progesterone. Around day 22 one of two things can happen, either the estrogen and progesterone levels will drop, and the endometrium will decrease and start to slough off after day 28, or the hormone levels will stay high, maintaining the endometrium.

The difference depends on whether implantation of the developing embryo has occurred. It takes approximately a week for the zygote to make its way along the Fallopian tube to the endometrium. During that time it will divide numerous times, becoming a solid ball of cell called a *morula* (after four days), and then a hollow ball of cells called a *blastocyst* (after five days). When it implants in the endometrium, it will start to release digestive enzymes to eat the uterine wall until the placenta develops. Yup, you started out life as a cannibal!

The mechanism by which the body knows it's pregnant is very simple. The developing placenta is called the *chorion*, and it releases a hormone called *human chorionic gonadotropin* (*hCG*). The role of the hCG is to keep the corpus luteum alive until the placenta can develop to the point where it can take over the production of estrogen and progesterone.

Flex Your Muscles

The knowledge of these hormones has led to the development of over-the-counter diagnostic tests. LH and hCG are released into the urine, so tests have been developed to detect them in urine. A surge in LH, around mid-cycle, indicates that ovulation will soon occur; this is especially useful if you are trying to get pregnant, as you are the most fertile about 12 to 24 hours after ovulation. A rise in hCG indicates that pregnancy has occurred, and I hope congratulations are in order!

So what's the big deal about estrogen and progesterone? Well, they both are necessary for the body to maintain the pregnancy. In addition, the progesterone helps to enlarge the breast by developing the breast tissue prior to producing milk, a process known as *lactation*. The estrogen is also important in terms of the whole cycle. High estrogen levels are part of a negative feedback loop with GnRH. When the level is high, as in pregnancy, the production of GnRH in the hypothalamus is inhibited.

Without GnRH there will be no rise in FSH and LH, and ovulation cannot occur. Ovulation, of course, must occur before a new pregnancy can start. The irony of this is that one of the best ways not to get pregnant while being sexually active is, well, to be pregnant in the first place! Your body cannot start a new pregnancy while you are pregnant!

Men's Room

The male's role in reproduction is a fleeting one. From the standpoint of pregnancy, the males need not only to produce sperm, but also to produce *seminal fluids*, the liquid components of *semen*, which are necessary to nourish the sperm and to help them to survive in the vagina, which is not that hospitable to sperm. The typical ejaculation contains anywhere from 100 to 200 million sperm on average; those numbers might seem excessive, but out of that number only about 50, not 50 million, but *50* ever reach the egg.

Some of the sperm die in the vagina. Some of the sperm die in the uterus. If they make it as far as the Fallopian tubes, half of them go up the wrong one! Given that enzymes in the head of the sperm are necessary to break down the zona pellucida and allow penetration, a single sperm is unlikely to be able to do the job. The high numbers are clearly important, for a low sperm count makes it harder to achieve pregnancy.

His Plumbing

The testes, or testicles, are multi-lobed organs that produce sperm (see Figure 22.4). There is an outer serous membrane called the *tunica vaginalis*, which covers a fibrous capsule called the *tunica albuginea*. Within each of the 200 to 300 lobules are densely coiled seminiferous tubules, where the sperm are produced. The actual process of spermatogenesis takes about 74 days. The sperm formation starts toward the outer edge of each tubule, with the sperm themselves being released, tail-first, into the lumen.

Between each of the tubules are the *interstitial endocrinocytes*, or *interstitial cells of Leydig*; these are the cells that produce testosterone. Between the developing sperm cells are Sertoli, or sustentacular cells, which extend from the lumen to the basement membrane. At the basement membrane the cells form tight junctions that prevent exposure to the antigens on the surface of the sperm; this is called the blood-testis barrier, and it is important because an immune response could destroy the sperm.

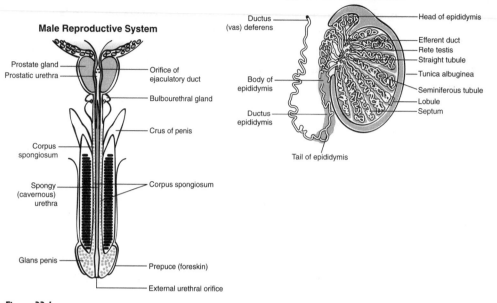

Figure 22.4

The structure of a testicle, and the glands that contribute to seminal fluid

(LifeART©1989–2001, Lippincott Williams & Wilkins)

The FSH and LH that we saw in the menstrual cycle are here, too, for the FSH stimulates spermatogenesis, and the LH stimulates the production of testosterone. The sperm, once they are fully formed, then travel out of the coiled seminiferous tubules to the short *straight tubules*, then into a tube network called the *rete testis*, and finally into the tightly coiled tube called the *ductus epididymis*. This duct, whose name always reminds me of a Roman centurion, is where the sperm come to maturity.

From the epididymis the sperm travel along the *ductus*, or *vas deferens*, to the urethra (refer to Figure 22.4). These tubes extend from the epididymis, at the posterior edge of the testicle, around the back of urinary bladder, to the urethra. The first part of the vas deferens is the part that is cut during a *vasectomy* (see Chapter 23). Just before the urethra is a widening called the *ampulla of the ductus deferens*, where the sperm are stored.

The end of the ampulla connects with the seminal vesicle duct, which contributes seminal fluid, and becomes the ejaculatory duct. This duct empties into the urethra at the point of the prostate gland, which also contributes to the seminal fluid. The prostate gland wraps around the urethra as it leaves the urinary bladder, and before it enters the penis (refer to Figure 22.4).

The penis itself is composed of erectile tissue, similar to that in the clitoris. It has a dual purpose, both for urination and for depositing sperm into the vagina during intercourse (sexual preferences aside, this was why it evolved); the urethra, which continues all the way down the penis to the *external urethral orifice*, functions also for the delivery of semen. In order to keep these functions separate, a smooth muscle sphincter at the junction of the bladder and urethra closes during sexual excitement prior to ejaculation; this not only prevents urination during ejaculation, but prevents the entry of semen into the bladder.

The penis itself has three parts: the proximal *root*, the *body*, and the distal *glans* (refer to Figure 22.4). The body contains three roughly cylindrical structures, the *corpus spongiosum penis* on the underside (ventral and medial), which contains the *spongy (cavernous) urethra*, and two *corpus cavernosa penis* on the topside (dorsal and lateral); all of these contain erectile tissue. During sexual excitement, an increase in the amount of blood entering the penis through the pairs of *dorsal arteries* and *deep arteries* causes numerous sinuses to fill with blood. The expansion of these sinuses exerts pressure on the two *superficial veins* and the single *deep dorsal vein*; this increase in pressure, allowing more blood in than out, causes the penis to become erect.

Medical Records

In some cultures the foreskin of the penis is removed either at birth or after entering puberty as a rite of passage. This operation is called *circumcision*, which was routinely performed in the United States, until a reevaluation deemed the practice unnecessary, although many parents still opt for the procedure.

The base of the penis contains the *bulb of the penis*, which is an expansion of the corpus spongiosum penis, as well as the two *crura* (singular = *crus*) of the penis which are the laterally directed proximal ends of the corpus cavernosa penis (refer to Figure 22.4). The distal end contains the acorn-shaped glans of the penis, the outer edge of which is called the *corona*. At the most distal portion is the medial, vertical slit called the *external urethral orifice*. Males are born with a *prepuce*, or *foreskin*, similar to that on the female's clitoris.

Do You See Men at Sea?

Spermatozoa have three parts: head (with the precious DNA cargo and an *acrosome* with enzymes used to break through the surface of the egg), midpiece (which contains the mitochondria), and the tail (which is a flagellum powered by the mitochondria). Semen, however, contains a lot more than just sperm. In addition to the sperm are secretions from the seminal vesicles, prostate, and bulbourethral (Cowper's) glands. For one thing, fluids from these glands make the pH of the semen slightly alkaline (pH 7.2 to 7.6). This is necessary to help the sperm survive in the mildly acidic environment of the vagina (pH 3.8 to 4.2); this environment is necessary to hold the vaginal flora in check. The fluid also contains fructose, which provides much of the necessary energy for the sperm to propel themselves all the way up the Fallopian tubes. Because of the bacterial flora mentioned previously, an antibiotic named seminalplasmin exists to help the sperm. Fibrinogen from the seminal vesicles causes coagulation in the semen (to aid in delivery) when combined with clotting enzymes from the prostate, which is in turn broken down later (to ensure sperm movement) due to the action of fibrinolysin, also from the prostate.

One of the first things measured when a couple suspects infertility is the composition of the man's semen. Not only is such a diagnostic test a much simpler one than similar tests on the female, but it also helps to illustrate the fact that fertility issues are by no means only those of the woman. A low sperm count, poor motility, abnormal morphology (shape can affect motility), or chemical imbalance can all affect fertility.

The Least You Need to Know

- Meiosis, which occurs in a woman's ovaries and a man's testes, divides the number of chromosomes in half by splitting up the homologous pairs.

- Events in meiosis increase genetic variation in gametes (crossing over) and cause monosomy and trisomy (nondisjunction).

- The female's gametes are produced in the ovaries, carried through the uterine tube, to the lining of the uterus, where implantation occurs if the ovum is fertilized.

- The menstrual cycle is an elaborate system of negative hormonal feedback loops involving the hypothalamus, adenohypophysis, ovaries, and placenta (during pregnancy).

- The male's gametes are produced in the seminiferous tubules of the testes, transported through the epididymis into the vas deferens, which passes by the seminal vesicle, through the prostate, and past the Cowper's gland, before it is released through the urethra during ejaculation.

- Semen is a combination of spermatozoa and the secretions of the seminal vesicles, prostate, and Cowper's glands.

Chapter 23

Birds Do It! Bees Do It!

In This Chapter

- ◆ Conception, pregnancy, labor, and delivery
- ◆ Prenatal testing
- ◆ Birth control methods

The drive to reproduce is a powerful one, consuming many a person's thoughts during the teen years (when one is at the height of one's fertility). An understanding of what is really going on in terms of conception, pregnancy, and delivery, far from removing the sense of wonder, only adds to it. Pregnancy is not a condition to be entered into lightly, but one that should be viewed in terms of the awesome responsibility it entails. We are living in a time when modern medicine allows us many options in terms of determining the health of an unborn child, as well as many options in terms of avoiding pregnancy.

This chapter also deals with issues of sexuality. In a biological sense, I will focus on sexuality as it relates to reproduction, knowing full well that an estimated 10 percent of the population, one of my sisters included, is gay. Please understand that there is no hidden agenda in my concentration on heterosexual behavior. So far, this book has explored the ins and outs of the 11 body systems, in terms of structure and of function. In this our final chapter, I will explore the ins and outs of how each of us got here.

When You Are Ready

The heading of this section is a little misleading, as the start of a pregnancy obviously does not require the parents to be ready. In some cases, all it requires is carelessness, or in the worst cases, a violent act against the woman's will. The body will react the same way. With that understood, this section explores conception, pregnancy, and birth from the perspective of a parent or parents who are joyfully expecting a child. Call me a romantic .... Anyway, given the great advances in modern medicine, this is the safest time to have a baby, for both the mother and the child, in the history of medicine. So if you *are* ready, take heart!

Chance Encounter

Conception seems like such a simple concept, and yet many things can get in the way. In Chapter 22, I mentioned that a number of obstacles are in the sperm's way: the mildly acidic environment of the vagina, the cervical mucus, getting lost in the uterus, going up the wrong Fallopian tube! All in all, though, placing conception in the Fallopian tube is a pretty smart idea, for the narrow pathway makes it easier for the gametes to meet.

Certain chemicals help the sperm get where they are going. *Acrosin*, from the acrosome of the sperm, makes the sperm, well, better swimmers by stimulating their flagella. In addition, prostaglandins, which are present in semen, stimulate uterine contractions, which not only propel the sperm upward, but also shorten the distance the sperm need to travel. Given where the sperm are going, it's a good thing those chemicals are there.

The Fallopian tubes are lined with cilia that are beating in the direction of the uterus, slowly moving the secondary oocyte or zygote. They also force the sperm to swim uphill, thus helping weed out the poorly formed, and possibly genetically flawed, sperm. The acrosome at the tip of each spermatozoon undergo changes in the female reproductive tract, called *capacitation*, gradually breaking down the membrane and allowing the secretion of enzymes—hyaluronidase, neuraminidase, and acrosin—which are necessary for fertilization to occur. Sperm use the enzymes on their acrosomes to gradually break their way through the cells of the corona radiata and the proteins of the zona pellucida (see Chapter 22). If too few sperm reach the egg, it is not possible to break down enough of the outer layers for fertilization, or *syngamy*, to occur (see Figure 23.1).

Once fertilization has occurred, meiosis continues in the secondary oocyte, forming the large ovum and a very small polar body that is ultimately broken down. At this point mitosis begins. From that one fertilized ovum, now called a zygote, with its precious diploid DNA cargo, are the instructions for the construction of a new human being. During its journey down the Fallopian tube, the zygote, and the various stages prior to implantation, still preserves the zona pellucida of the secondary oocyte as an outer layer.

Performing replication as it starts to move, after 36 hours the zygote cleaves into two identical diploid cells, and then four after two days, until a ball of identical cells called a *morula* is formed after four days (see Figure 23.2). Up to this point the cells are undifferentiated. By the end of day five a hollow ball of cells called a blastocyst forms. The outer layer of this ball is called the *trophoblast*, which becomes the fetal placenta, and the fluid-filled cavity inside is called the blastocoel. During this period all the nutrition comes from the cytoplasm provided by the secondary oocyte.

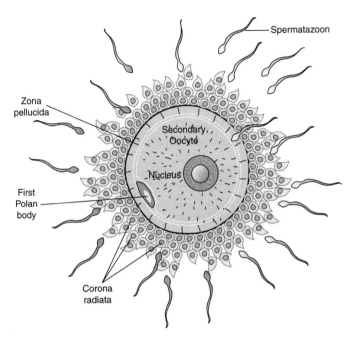

Figure 23.1

Multiple sperm are required to break through the zona pellucida, before a single sperm can fertilize the secondary oocyte.

(LifeART©1989–2001, Lippincott Williams & Wilkins)

By day 6 the outer zona pellucida is shed by a process known as hatching. Our attention, however, is called to a rather unromantic cluster of cells inside the trophoblast called the inner cell mass, which ultimately becomes, well, you! At this point the glory that is cell differentiation begins, as operons (see Chapter 3) turn on and off certain genes, sealing the direction their development takes.

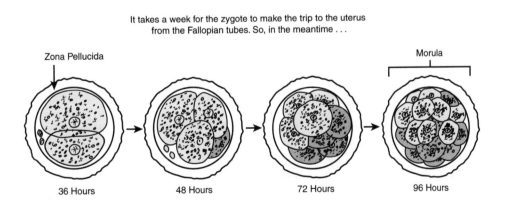

Figure 23.2

The earliest stages of development, prior to implantation.

(©2003 www.clipart.com)

Getting Your Foot in the Door

It's a bit ironic that we start our life independently, only to become fully dependent on Mom; it's as if our teenage years find us coming full circle! *Implantation* really is more than just attachment to the endometrium; it actually involves, in effect, burrowing (albeit using enzymes produced by the trophoblast) beneath the surface of uterine wall (see Figure 23.3). The portion of the trophoblast above the inner cell mass is the part to first come in contact with the endometrium, often near the fundus (or top) of the uterus.

Figure 23.3

After the blastocyst implants in the endometrium, the chorion begins to form.

(LifeART©1989–2001, Lippincott Williams & Wilkins)

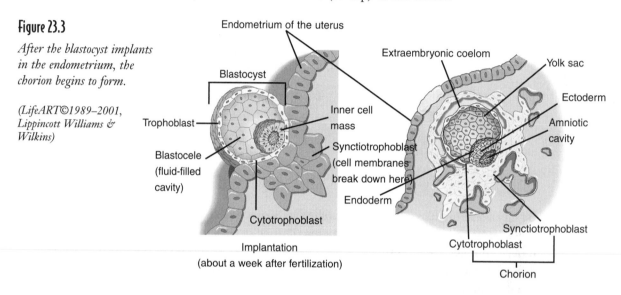

Endometrium of the uterus

Extraembryonic coelom

Yolk sac

Blastocyst

Ectoderm

Trophoblast

Inner cell mass

Amniotic cavity

Blastocele (fluid-filled cavity)

Synctiotrophoblast (cell membranes break down here)

Cytotrophoblast

Endoderm

Synctiotrophoblast

Cytotrophoblast

Chorion

Implantation
(about a week after fertilization)

The enzymes help to produce two layers from the trophoblast: the *cytotrophoblast*, which has clear cell divisions, and the *syncytiotrophoblast*, which doesn't have clear divisions, and is in direct contact with the endometrium. The products of the enzymes, basically from chewing up Mom's uterine wall, fed you in the period right after implantation.

Hormone Factories

The maintenance of pregnancy, as I mentioned in Chapter 22, depends on high levels of estrogen and progesterone, which in turn prevent ovulation during pregnancy. Initially these hormones are made by the corpus luteum, which is what is left of the ovarian follicle after ovulation. After implantation the developing embryo starts producing human chorionic gonadotropin (hCG), which is released into the interstitial spaces in the endometrium, absorbed by the lymphatic system, delivered to the blood, and then delivered to the corpus luteum, keeping it from breaking down. After about three months the developing placenta, in addition to providing nutrition for the fetus, takes over the job of producing hormones.

The endometrium changes after implantation, becoming the *decidua*, which has three parts: the *decidua basalis* beneath the embryo, the *decidua capsularis* around the amniotic sac, and the *decidua parietalis*, which is the remainder of the uterine wall. Remember the cytotrophoblast and the syncytiotrophoblast? Together they form the *chorion*, which will eventually develop into the placenta; the chorion is attached to the decidua basalis. Multiple fingerlike projections develop outward from the cytotrophoblast; these are called *chorionic villi*, and they increase the surface area for the absorption of nutrients from the endometrium.

Ultimately these villi will be filled with fetal blood vessels running to and from the umbilical cord, providing exchange between the mother and the fetus. The blood of the mother and baby do not mix during pregnancy, for the maternal blood empties into sinuses that surround the chorionic villi, forming *intervillious spaces*. It is here that all maternal/fetal exchange happens.

Pregnancy

An incredible amount happens during pregnancy (or *gestation*), more than I could possibly hope to cover in a book this size, so I'm going to concentrate on the highlights. Soon after implantation, as the chorionic villi are just starting to grow, the cells of the inner cell mess really start differentiating. A *body stalk* forms that will ultimately become our first lifeline, the umbilical cord.

At the end of the stalk, there are three divisions of what was the inner cell mass: the *amnion*, the *embryonic disc*, and the *yolk sac*. Yup, you heard right—the yolk sac. This is somewhat akin to a vestigial organ, as it is a holdover from ancestral species that received most of their nourishment from the yolk sac. Blood is formed there early on, particularly in a part called the *allantois*, which contributes its blood vessels to the developing umbilical cord. In a moment of poetic simplicity some cells from the yolk sac, a symbol that implies reproduction and fertility, migrate to the developing gonads, ultimately becoming primitive germs cells, oogonia in the female, and spermatogonia in the male.

The amnion is a thin membrane that surrounds a fluid-filled cavity called the *amniotic cavity*; the amnion and the cavity together form the *amniotic sac*, which will ultimately completely surround the embryo and later the fetus. The inner lining of the amniotic sac, close to the body stalk, is called the *ectoderm*, and the part of the yolk sac opposite the ectoderm is called the *endoderm*; between the two is the *mesoderm*. These three layers, called *germ layers*, together form the embryonic disc. It is from these three simple, unassuming layers that all of our tissues, organs, and systems develop (see the following table).

Structures of the Three Primary Germ Layers

Ectoderm	Mesoderm	Endoderm
Epidermis of skin, nervous tissue, epithelium of nasal and oral cavities, and the anal canal	Muscles, heart, skeleton, dermis of skin, epithelium of kidneys, ureters, and gonads	Epithelium of most of the GI tract and glands, vagina, urethra, bladder, and respiratory tract

Pregnancy itself is divided into three trimesters of three months each (see Figure 23.4). The first two months of the first trimester are called the *embryonic period*, during which all 11 body systems begin to form. During the first two months especially, the developing child is extremely vulnerable to dietary and medicinal influences, such as *neural tube defects* (an incomplete folding in of the ectoderm to create the tube from which the nervous system develops) as a result of low dietary folic acid levels. Alcohol also affects neurological development, as well as causing low birth weight, in a condition known as *fetal alcohol syndrome* (*FAS*).

By the beginning of the second trimester the fetus has started to move and will soon be felt by the mother. The development of the body systems accelerates. Bones become ossified, and the head, which was disproportionately large, by newborn standards, slows its growth so that the body may catch up. The third trimester adjusts the head to body proportion still further, while the fetus gains a significant amount of weight (from about 1½ lbs to 7½ lbs, or 800 g to 3400 g, from the end of the sixth month to birth). During the third trimester the fetus starts producing surfactant, which is so important to breathing (see Chapter 13), and in general develops its systems so that it becomes better able to survive outside the womb.

Figure 23.4

The changes in the size of the fetus and the uterus are very dramatic over the course of the three trimesters.

(LifeART©1989–2001, Lippincott Williams & Wilkins)

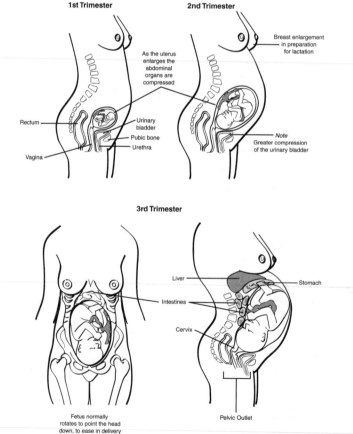

Delivery: More Than a Stork

Labor and delivery (also called *parturition*) are a long process, although the difficulty varies a great deal from person to person, and even from pregnancy to pregnancy. In addition to the effects of oxytocin (OT) covered in Chapter 1 (see Figure 1.8), other hormonal factors must be taken into account, such as prostaglandin, which stimulates uterine contractions (see Chapter 22). Progesterone, for one, inhibits contractions in the uterus, so that hormone must decrease, in part due to the release of fetal cortisol, before labor can begin. Some women may experience *false labor*, in which the contractions appear at irregular intervals. In *true labor*, not only are the intervals of peristaltic contraction (from the fundus down) regular and steadily decreasing in the intervals, but they also dilate and *efface* (flatten) the cervix. False labor does not progress, and the cervix does not dilate.

There are three stages to labor and delivery: dilation stage, expulsion stage, and placental stage. The dilation stage involves the dilation (to 10 cm) and effacement of the cervix and the rupture of the amniotic sac. Expulsion is the stage from full dilation to delivery of the baby. The expulsion phase is when the fontanels (see Chapter 6) come into play, allowing the fetal skull to adjust its shape to the birth canal. Newborn babies are still tethered to the placenta by the umbilical cord, which continues into the uterus through the vaginal canal. The third and final stage, the placental stage, involves the delivery of the placenta, or *afterbirth*, and strong uterine contractions, which are necessary to prevent blood loss after delivery.

Most babies are born within a four-week window, two weeks before and after the due date. Beyond the tenth month, the placenta can no longer provide enough oxygen to the fetus, who becomes more at risk for brain damage or death. For that reason, most pregnancies that last two weeks past the due date involve inducing labor. After the baby is born, it takes about six weeks for the mother's reproductive organs to return to their normal state, in a process called *puerperium*. This stage starts with the release of *lochia*, a mixture of blood and serous fluid (the latter ultimately replacing the former) for the first two to four weeks. The most dramatic change is in the uterus, which greatly decreases in size as well as increasing in firmness, a process known as *involution*.

Food on the Go!

The survival of one's offspring is greatly increased by being able to provide food, especially without having to go out and look for it! Breastfeeding was the solution. The mammary glands and mammary papillae (nipples to you) are responsible for breastfeeding the child. The hormone responsible for milk production, or *lactation*, is called *prolactin* (see Chapter 18). The levels of prolactin increase during pregnancy, which results in breast enlargement as the glands ready themselves for lactation (see Figure 23.5). Estrogen and progesterone levels are sufficiently high during pregnancy to limit prolactin's effectiveness. After parturition, or birth, the drop in estrogen and progesterone levels as a result of the loss of the placenta allows the prolactin to take full effect.

When the baby suckles on the mother's nipple, neurons connected to the hypothalamus stimulate that organ to decrease *prolactin-inhibiting hormone* (*PIH*) and to increase the production of *prolactin-releasing hormone* (*PRH*). The altered levels of those two hormones stimulate the mammary gland to increase its production of milk. Suckling also stimulates the release of oxytocin (OT) from the posterior pituitary. OT stimulates the contraction of myoepithelial cells, and thus *milk ejection*, or the *let-down reflex*. One of the coolest things about the hormonal pathway is that as the baby grows, and the suckling grows stronger, the production of milk increases. In the same way, when the baby is weaned, and the nipple is no longer stimulated, the milk production eventually ceases.

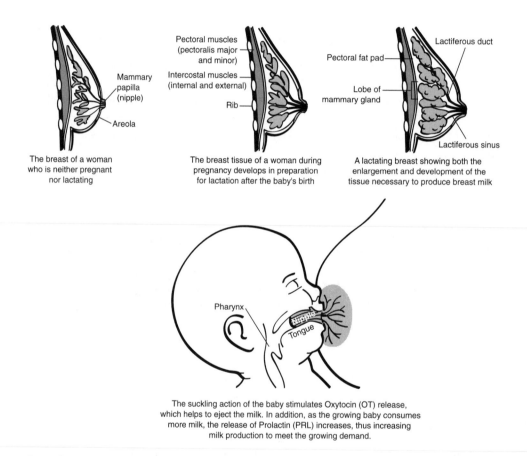

Figure 23.5

The breast increases in size during pregnancy as the tissues prepare for lactation.

(LifeART©1989–2001, Lippincott Williams & Wilkins)

As explained earlier, lactation does not begin immediately after birth. For the first three days the mammary glands secrete *colostrum*, which is lower in fat and lactose but still rich in antibodies that protect the newborn from infection. Antibodies are also released into the milk that appears after the fourth day, and it is these antibodies that are a significant part of the argument in favor of breastfeeding (although this is occasionally not possible).

When You Can't Wait

Modern *prenatal testing* (that is, before birth) has changed much of the uncertainty of pregnancy. Many basic questions, in terms of gender or of chromosomal abnormalities, can be answered now. Others are harder to diagnose, although technology and greater knowledge of the human genome will eventually change all of that.

Prenatal testing takes three basic forms: visual and auditory information (ultrasound and heart monitoring), genetic sampling, and maternal blood markers. Each form has its advantages, and its limitations (based on the nature of what is being explored). In addition, some forms of testing carry greater risks. Knowledge of the advantages, limitations, and risks will help people to make informed decisions.

Ultrasound

Ultrasound is basically a medical version of sonar, in which sound waves are sent through the mother's abdomen, and the differential reflection of the sound waves is used to create real-time images of the growing fetus. Although there are limitations to the procedure, a remarkable amount can be learned. The development of the four-chambered heart can be monitored for defects, as can the brain. Certain features that are linked to chromosomal disorders, such as the thickness of the neck and altered umbilical cord attachment in Down syndrome (trisomy 21; see Chapter 22), can be diagnosed early.

Fetal heart monitoring, apart from its use in determining the presence of twins or triplets, is important in determining whether a fetus is still alive or has miscarried. This is also an important part of modern labor and delivery, for it is an indicator of fetal stress. If the heart rate is too rapid or too slow during labor, a caesarian section may be indicated.

Amniocentesis

A remarkable technique, known as *karyotyping*, is used to determine the chromosomes in an individual. Any deviation from the standard 46 has implications, anywhere from sterility to early death, depending on the chromosomes affected. With about 33,000 genes in the human genome, and only 46 chromosomes, each chromosome must contain many genes. An extra chromosome (trisomy), or a missing chromosome (monosomy), carries with it extra or missing copies of those genes, which ultimately affects how the body functions.

If such genetic defects can be determined by looking at cells, it stands to reason that it would be useful to be able to do this before birth, if for no other reason than to emotionally prepare parents for the early death of a child, as is the case with some chromosomal defects. To perform such a test it is important that living cells be harvested from the fetus, so that the chromosomes may be observed during mitosis.

To put it bluntly, taking a chunk out of the fetus is not a good idea. Luckily, the amniotic fluid, which surrounds the fetus, not only may contain some living skin cells, but also some cells from the inside of the bladder. You realize, don't you, that this means we started out our lives swimming in our own urine!

Harvesting the cells thus means no more than harvesting the amniotic fluid with a needle through the abdomen (see Figure 23.6). The actual procedure is done with an ultrasound, as they don't want to accidentally pierce the baby! This test, apart from determining defects, can also determine the gender of the child. There is, however, a small risk of miscarriage, about 1 percent.

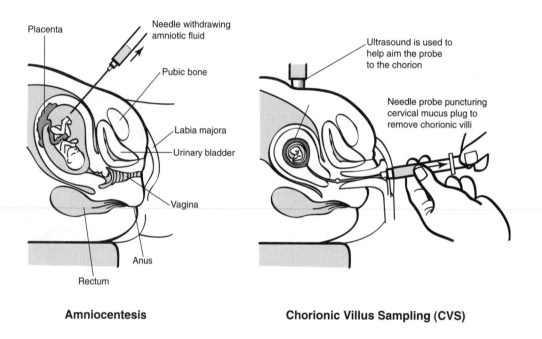

Amniocentesis   **Chorionic Villus Sampling (CVS)**

Figure 23.6

Amniocentesis and chorionic villus sampling allow fetal tissues to be examined for genetic abnormalities.

(LifeART©1989–2001, Lippincott Williams & Wilkins)

Chorionic Villus Sampling

Since the fetal placenta is genetically identical to the fetus, a more recently invented procedure involves removing the villi from the chorion, which is called chorionic cillus sampling, or CVS (refer to Figure 23.6). One advantage to this procedure is that it can be done much earlier in the pregnancy, at about 8 to 10 weeks, as compared to 14 to 16 weeks for amniocentesis, although it has a higher risk for miscarriage (2 percent). This procedure is done either through the abdomen, or through the vagina. The vaginal method requires that the thick, protective, cervical mucus plug be pierced. As a result of this, the mother must abstain from intercourse, as well as avoid baths or swimming, until the mucus plug can reform (about two weeks).

But If You're Not Ready ...

Unlike most animal species, humans do not have a limited mating season; males are fertile all year round, while females are fertile for a brief period in every 28-or-so-day menstrual cycle, after ovulation. Given that over the majority of human history, the successful raising of a child was much easier with two parents, sexual activity evolved in humans as an act of bonding. For this reason, even monogamous humans throughout history have looked for ways to be sexually active without the risk of pregnancy. Although some form of *condom*, from either animal or plant tissues, has been around since Egypt in the thirteenth century B.C.E., from its more modern (1564) "invention" by Gabriel Fallopius (you got it, the discoverer of the Fallopian tube), the condom has also been concerned with preventing sexually transmitted diseases (STD). Before going too far, I think I should tell you about some relatively ineffective methods, each of which involves not going too far. One is called *coitus interruptus*, which literally means that the male removes his penis just before ejaculation. Apart from the fact that a miscalculation can result in ejaculation into the vagina (defeating the purpose, wouldn't you say?), the lubrication that is released through the external urethral orifice may contain sperm. All in all, there is a one in five chance of getting pregnant. Not the greatest odds.

The other method is called the *rhythm method*, and it is based on knowledge of the menstrual cycle. Given that a woman is most fertile around the middle of her cycle, the rhythm method involves avoiding intercourse for about a week in the middle of the cycle. Apart from the fact that menstrual cycles can vary, making it hard to judge the period of greatest fertility, even those with regular "clockwork" cycles still have a one in five chance of getting pregnant. The odds are not really improved by noting the symptoms of ovulation, such as elevated body temperature, an increase in cervical mucus, or mittelschmerz (see Chapter 22), despite the more scientific basis of the timing. This method is known as the *sympto-thermal method*.

In general, apart from either abstinence or noncoital sexual behaviors (some of which still carry a risk of STD), no method is 100 percent effective (although some come close). To be fair, the effectiveness of any birth control method has a lot to do with how well it is used, but even so there is still a risk of pregnancy. As a result, many heterosexual couples use more than one form of birth control, or at the very least double up when ovulation is near. This is a good way to hedge your bets, using the "better safe than sorry" scenario.

Barrier: None Shall Pass

There are many types of birth control, but the simplest involves what is known as the *barrier method*. In a nutshell, pun unintended, the idea is to prevent the gametes from meeting, in a "none shall pass" maneuver. Although such a method could involve the secondary oocyte, the placement of the uterine tube makes this impractical. Given the sperm's entry into the vagina, it is far more practical to keep them out of the uterus. The condom, as I mentioned before, is the oldest such method. Keeping the ejaculate inside the condom prevents the sperm from even entering the vagina; this is also what makes this method so useful in preventing STD.

A recent invention, called the vaginal pouch, is worn by the female; in this method the male ejaculates into the pouch, which still prevents entry into the vagina. The problems with both these methods are many. Couples might delay before using it, in effect a variation of coitus interruptus. The device may be put on or in improperly, increasing the likelihood of tears or ruptures. Or, given that the penis becomes flaccid after ejaculation, the condom may slip off when the male removes his penis. All in all, there is a 10 to 20 percent chance of pregnancy.

The other two barrier methods do not work alone but in conjunction with a chemical spermicide. A *diaphragm* is dome-shaped, with a flexible rim to make it easier to insert. Its size allows it to fully cover the cervix by having the rim around the fornix, which is the depression around the edge of the cervix. The spermicidal jelly is placed inside the dome; in this way, any sperm that make their way around the rim will be killed by the spermicide. For the best protection it should be left in place for at least 6 hours, and up to 24.

A *cervical cap* is similar in basic function, but not in design. Although the thimble-shaped cap also fits over the cervix, and has spermicide inside, cervical caps use suction to stay in place over the convex cervix. Cervical caps can stay in place up to two days. Both diaphragms and cervical caps need to be fitted by a health professional, with periodic refits, especially after pregnancy. Improper placement, not to mention being dislodged during sex, can lead these devices to fail. Each device has anywhere from a one in five to a two in five chance of leading to pregnancy.

Chemical spermicides, although a part of the use of both the diaphragm and the cervical cap, can also be used in another way. The *contraceptive sponge* is used to place the spermicide in the vagina and over the cervix without the need for a medically fitted device. Although the spermicide may be released for 24 hours, the best protection lasts for about an hour. The effectiveness is similar to that of the diaphragm and cervical cap.

Chemical: Stop the Cycle

Given that hormones regulate ovulation, one of the most effective means of birth control, the *hormonal method*, involves the use of hormones to alter the menstrual cycle (see Chapter 22). By maintaining a higher level of estrogen and progesterone, the release of GnRH is prevented, thus preventing FSH and LH increases that lead to ovulation. These *oral contraceptives* ("the pill") have only a $1/50$ chance of pregnancy. Another variety involves slow-release cylinders of *progestin*, which last for five years and have only a $1/100$ failure rate.

The downside to both these methods is the systemic nature of the medication. The other methods so far mentioned are local in their action, but these medications are carried by the blood to the entire body. As hormones may have multiple effects on the body, any increase can lead to side effects, such as nausea, weight gain, breast tenderness, and so on. It is important that every patient be aware of the risks before taking them.

Post-Fertilization

All the previous methods have involved somehow preventing the gametes from meeting, either through a physical or chemical barrier, or through prevention of ovulation altogether. Other means involve actions after the fact, or *post-fertilization*. One form is called the *intrauterine device*, or *IUD*. This device is made of stainless steel, copper, or plastic, and may be kept in the uterus for a number of years. An IUD works by preventing implantation of the blastocyst into the uterine wall. Certain forms have been taken off the market due to the excessive bleeding they caused. Some bleeding and discomfort, even expulsion by the body, can still be a problem with the remaining models. They are only about 5 percent ineffective.

Abortion, either by chemical means or by surgical means, causes the removal of an implanted blastocyst, embryo, or fetus, depending on the time during the pregnancy that the surgical procedure is done. Chemical means are limited to very early in the pregnancy. One of the more common surgical methods is called *dilation and curettage*, or a D & C. This method involves dilation of the cervix, and a scraping of the uterine wall. This form of birth control, though rather late in the game, is very controversial, despite being occasionally medically necessary to save the mother's life; educating young people about options has been shown to lead to more birth control that involves preventing fertilization, and fewer abortions.

The Last Method You'll Ever Need

Sometimes a person either has no desire for children, or after having children, wants to permanently prevent future pregnancies. In this case, there is a long term solution, the last one you'll ever need, that is 99 percent effective: *surgical sterilization*. Despite gender differences, the methods are remarkable similar, for they both involve cutting the tubes that transport the gametes. In the female it involves cutting and tying off the uterine, or Fallopian tubes, a procedure called a *tubal ligation*. This procedure involves general anesthesia.

The male version, called a *vasectomy*, is much simpler, due to the location of the vas deferens. Because sperm production requires a temperature slightly lower than body temperature, the descended testicles make this an easy procedure. A small cut is made on either side of the scrotal sac, near the junction of the epididymis and the vas deferens. Similar to a tubal ligation, each vas deferens is cut and tied off, preventing the transfer of sperm to the urethra. I always advise my male students to consider this before asking a woman to have her tubes tied.

The Least You Need to Know

- ◆ Fertilization and implantation must be successful for pregnancy to occur.

- ◆ Development is based on cell differentiation, which starts with the three germ layers: ectoderm, mesoderm, and endoderm.

- ◆ Labor and delivery involve changes in the cervix and contraction of the uterus to push the baby through the vaginal canal.

- ◆ Prenatal testing allows parents to learn a tremendous amount about the unborn child.

- ◆ Birth control methods fall into three categories: barrier method, hormonal methods, and post-fertilization.

Appendix A

Further Readings

Anderson, Kenneth N., Louis E. Anderson, and Walter D. Glanze. *Mosby's Medical, Nursing, and Allied Health Dictionary*. Fifth Edition. St. Louis: Mosby-Year Book, Inc. 1998.

Campbell, Neil A., Lawrence G. Mitchell, and Jane B. Reece. *Biology Concepts and Connections.* First Edition. Redwood City, California: Benjamin/Cummings, 1994.

Crichton, Michael. *Five Patients*. New York: Ballantine Books, 1970; reissued 1989.

Gray, Henry. *Anatomy, Descriptive and Surgical*. This book is in the public domain, so numerous editions are available. My personal edition, for example, is a reprint of the "New American Edition" (from 1901), which was revised from the fifteenth English edition: New York: Crown Publishers, 1977.

Iazzetti, Giovanni, and Enrico Regutti. *Atlas of Anatomy.* Surrey, England: TAJ Books, 2002.

Karlen, Arno. *Man and Microbes: Disease and Plagues in History and Modern Times*. New York: G.P. Putnam's Sons, 1995.

Marieb, Elaine M. *Human Anatomy and Physiology*. Fifth Edition. San Francisco: Benjamin Cummings, 2001.

Martini, Frederic H. *Fundamentals of Anatomy and Physiology*. Fourth Edition (including the Applications Manual). Upper Saddle River, New Jersey: Prentice Hall, 1998.

Miller, Sue Ann, William Perotti, Dee U. Silverthorn, Arthur F. Dalley, and Kyle E. Rarey. "From College to Clinic: Reasoning Over Memorization is Key for Understanding Anatomy." *The Anatomical Record.* Volume 269:2, 2002. Pages: 69-80. (Special Issue: Meeting the Challenge: Modern Anatomy Education.)

Seeley, Rod R., Trent D. Stephens, and Phillip Tate. *Anatomy and Physiology.* Fourth Edition. Boston: McGraw Hill, 1998.

Starr, Cecie, and Beverly McMillan. *Human Biology.* Belmont, California: Wadsworth Publishing Company, 1997.

Tortora, Gerald J., and Sandra R. Grabowski. *Principles of Anatomy and Physiology.* Seventh Edition. New York: HarperCollins, 1993.

Van De Graaff, Kent M. *Human Anatomy.* Second Edition. Dubuque, Iowa: Wm. C. Brown Publishers, 1984.

Notes

A few sources above deserve special mention, for they are not traditional college texts, but are nonetheless of great value. Michael Crichton's *Five Patients*, although dated (it was written in 1969), is a fascinating description of the evolution of modern medicine and the modern hospital, using five individual patients as the starting point. Part of the interest is not just the development prior to 1969, but the development since the book was written. This book will help you gain a healthy historical perspective.

Arno Karlen's *Man and Microbes* gives an excellent historical perspective on the pathogens and their role in history, medical treatments throughout history, with special emphasis on our co-evolution with these microbes. Two splendid examples are the development of antibiotic resistance by bacteria, and emerging viruses (HIV, Ebola, SARS).

The journal article, "From College to Clinic: Reasoning Over Memorization is Key for Understanding Anatomy" by Miller et. al., is a wonderful description of my basic philosophical and pedagogical approach to anatomy. It seems that I am in good company! You can find this article online at www3.interscience.wiley.com/cgi-bin/fulltext/93517373/FILE?TPL=ftx_start&mode=html.

Lastly, you cannot escape the value of *Gray's Anatomy*. Although by no means light reading, it is still used today as a reference, especially when you need that specific detail, such as the blood vessels that attach to the esophagus. It is an essential and inexpensive part of your library. The 1918 edition is also available, complete with images, online (see Appendix B).

Appendix B

Cool Anatomy and Physiology Websites

If you have spent any time at all on the World Wide Web, you know that websites can proliferate like fungi. Any website you see here may or may not exist by the time you read this. If the link fails, you can always do a search using some of the information in the descriptions below. There is an incredible amount of information online, much of it highly technical, but much of it with wonderful images and animations.

As a general rule of thumb, be specific in your searches, for a search for *skeleton* will yield *a lot* of junk, but a search for *radiology* or *epiphysis* will be far more fruitful. Use this list as a brief springboard, starting off with the reference sites, including both glossary and dictionary sites. Enjoy!

Anatomical and Medical References

Gray's Anatomy
www.bartleby.com/107/
This website contains the complete text and digital versions of all 1,247 engravings from the 1918 edition of this classic anatomy text. As I said in Appendix A, this book is both inexpensive, and essential, but what could be easier than the whole text online!

Histology Images (from the University of Delaware's Histology Department)
www.udel.edu/Biology/Wags/histopage/histopage.htm
This is an excellent collection of histology images.

Medical Dictionary Online

www.online-dictionary.net/medical/

Rather than a specific dictionary, this is rather comprehensive listing of useful sites.

Medical Glossary (The University of Maryland's Medical school)

www.umm.edu/glossary/

This glossary also contains links that put the information into context, which is a must in terms of gaining understanding.

General Information

Access Excellence

www.accessexcellence.org/

This website was designed as a resource for biology teachers, but there is plenty of helpful information for the anatomy student.

The Visible Human Project

www.nlm.nih.gov/research/visible/visible_human.html

A remarkable breakthrough in anatomical study came with the Visible Human Project, in which a man on death row willed his body to science, and upon his death his body was frozen, encased in wax, and sliced into over 1,800 one-millimeter *transverse* slices. Those slices were then digitally photographed, and the data was then used to construct animations, three-dimensional views of organs, and blood vessels without the organs, not to mention every conceivable type of section. All of the data is now available online. You can also check out a guided online tour of the project with animations at www.madsci.org/~lynn/VH/.

WebAnatomy

www.gen.umn.edu/faculty_staff/jensen/1135/webanatomy/

This is a great site with interactive quizzes and links, maintained by professor Murray Jensen from the University of Minnesota.

Michael Lazaroff's Websites

There is quite a bit of information on my website, but it is nothing if not a work in progress.

Anatomy and Physiology

shs.westport.k12.ct.us/mjvl/anatomy/a&p-home.htm

Anatomy and Physiology Links

shs.westport.k12.ct.us/mjvl/lazaroff/cool_web.htm#a&pweb

Biology Links

shs.westport.k12.ct.us/mjvl/science/bioweb.htm

How to Use a Microscope Properly

shs.westport.k12.ct.us/mjvl/biology/microscope/microscope.htm

Index

C

X–Y–Z